TECHNIQUES AND INSTRUMENTATION IN ANALYTICAL CHEMISTRY — VOLUME 4

EVALUATION OF ANALYTICAL METHODS IN BIOLOGICAL SYSTEMS

PART A
ANALYSIS OF BIOGENIC AMINES

TECHNIQUES AND INSTRUMENTATION IN ANALYTICAL CHEMISTRY

TECHNIQUES AND INSTRUMENTATION IN ANALYTICAL CHEMISTRY — VOLUME 4

EVALUATION
OF ANALYTICAL METHODS
IN BIOLOGICAL SYSTEMS
general editor: R.A. de Zeeuw

PART A

ANALYSIS
OF BIOGENIC AMINES

edited by

Glen B. Baker and **Ronald T. Coutts**

Neurochemical Research Unit, Department of Psychiatry and Faculty of Pharmacy and Pharmaceutical Sciences, University of Alberta, Edmonton, Alberta, Canada

ELSEVIER SCIENTIFIC PUBLISHING COMPANY
Amsterdam — Oxford — New York   1982

ELSEVIER SCIENTIFIC PUBLISHING COMPANY
Molenwerf 1
P.O. Box 211, 1000 AE Amsterdam, The Netherlands

Distributors for the United States and Canada:

ELSEVIER SCIENCE PUBLISHING COMPANY INC.
52, Vanderbilt Avenue
New York, NY 10017

Library of Congress Cataloging in Publication Data
Main entry under title:

Evaluation of analytical methods in biological systems.

   (Techniques and instrumentation in analytical chemistry
chemistry ; v. 4)
   Bibliography: p.
   Includes index.
   Contents: pt. A. Analysis of biogenic amines.
   1. Amines--Analysis. 2. Amines in the body.
3. Biological chemistry--Technique. I. Baker, Glen B.,
1947-   . II. Coutts, Ronald Thomson. III. Series.
QP801.A48E86 1982   574.19'285   82-11553
ISBN 0-444-42110-6 (v. 1)

ISBN 0-444-42110-6 (Vol. 4)
ISBN 0-444-41744-3 (Series)

Printed in The Netherlands

CONTENTS

VIII

Contributors

Baker, Glen B.
Neurochemical Research Unit
Department of Psychiatry
University of Alberta
Edmonton, Alberta T6G 2G3, Canada

Baker, Judith M.
Neurochemical Research Unit
Department of Psychiatry
University of Alberta
Edmonton, Alberta T6G 2G3, Canada

Candy, John M.
MRC Neuroendocrinology Unit
Newcastle General Hospital
Newcastle-upon-Tyne NE4 6BE, U.K.

Chiu, Andrew S.
Clarke Institute of Psychiatry
University of Toronto
Toronto, Ontario M5T 1R8, Canada

Coutts, Ronald T.
Neurochemical Research Unit
Faculty of Pharmacy and
    Pharmaceutical Sciences
University of Alberta
Edmonton, Alberta T6G 2N8, Canada

Davis Bruce A.
Psychiatric Research Division
University Hospital
Saskatoon, Saskatchewan S7N 0X0, Canada

Dayton, Mark A.
Medical Sciences Program
Myers Hall
Indiana University
Bloomington, Indiana 47405, USA

Dewhurst, William G.
Neurochemical Research Unit
Department of Psychiatry
University of Alberta
Edmonton, Alberta T6G 2G3, Canada

Dryden, William F.
Department of Pharmacology
University of Alberta
Edmonton, Alberta T6G 2E1, Canada

Durden, David A.
Psychiatric Research Division
University Hospital
Saskatoon, Saskatchewan S7N 0X0, Canada

Enna, S.J.
Departments of Pharmacology and of
    Neurobiology and Anatomy
University of Texas Medical School
Houston, Texas 77025, USA

Ferkany, John W.
Departments of Neuroscience,
    Pharmacology and Psychiatry
Johns Hopkins University School
    of Medicine
Baltimore, Maryland 21205, USA

Gelpí, Emilio
Analytical Neurochemistry Unit
Instituto de Quimica Bio-Organica
Consejo Superior de Investigaciones
    Cientificas
Jorge Girona Salgado S/N
Barcelona-34, Spain

Godse, Damodar D.
Clarke Institute of Psychiatry
University of Toronto
Toronto, Ontario M5T 1R8, Canada

Hubbard, John W.
Faculty of Pharmacy
University of Manitoba
Winnipeg, Manitoba R3T 2N2, Canada

LeGatt, Donald F.
Neurochemical Research Unit
Department of Psychiatry and
    Faculty of Pharmacy and
    Pharmaceutical Sciences
University of Alberta
Edmonton, Alberta T6G 2G3, Canada

Locock, R. Anthony
Faculty of Pharmacy and
    Pharmaceutical Sciences
University of Alberta
Edmonton, Alberta T6G 2N8, Canada

Martin, Ian L.
MRC Neurochemical Pharmacology Unit
Medical Research Council Centre
Medical School
Cambridge CB2 2QD, U.K.

Midha, Kamal K.
College of Pharmacy
University of Saskatchewan
Saskatoon, Saskatchewan S7N 0W0, Canada

Warsh, Jerry J.
Clarke Institute of Psychiatry
University of Toronto
Toronto, Ontario M5T 1R8, Canada

Wightman, R. Mark
Department of Chemistry
Indiana University
Bloomington, Indiana 47405, USA

PREFACE

Analytical chemistry has grown spectacularly over the last two decades. As a result, it has become very difficult, if not impossible for many scientists to keep abreast of the many developments, innovations and new horizons. This does not only hold for specialists involved in the basic aspects, but perhaps even more so for the non-specialists who wish to apply analytical chemistry as a tool to provide answers to the problems that they face in the course of their daily work, particularly in the area of biomedicine and other life sciences.

Although the present analytical literature contains various excellent monographs and review series, appliers of analytical chemistry to biological specimens often find that they focus too much on individual analytical techniques and on their basic aspects, with little or no comparison of the potentials and limitations of a particular technique as compared to others. This leaves a need for a series which will evaluate the various analytical techniques and approaches that can be used in a particular bioanalytical field, taking into account the type of the problem, the type of the answer needed and the impact of the biological matrix in which the measurements are to be performed.

Thus, Evaluation of Analytical Systems in Biological Systems will be problem-orientated, rather than technique-orientated with the emphasis on evaluation in that the volumes in the Series will focus on a given bioanalytical problem and then provide a critical and comprehensive discussion of the particular merits and pitfalls of the various techniques and approaches available for solving that problem. This requires a team of expert authors having not only an in-depth knowledge regarding the techniques as such but also a broad and balanced overview over the entire domain. It is hoped that in this way the reader will be provided the necessary information to select the most suitable methodology for his particular application and also to properly interpret the analytical answers produced by that methodology.

It is not surprising that the first volume deals with the analysis of biogenic amines. There exists a heavy demand for reliable, sensitive, selective and accurate analyses of these compounds in biological specimens. On the other hand, major analytical innovations and applications have been reported recently in this area, so that a thorough evaluation seems very timely. We feel fortunate that Drs. Baker and Coutts have been willing to accept the challenge to edit this volume and that they have been able to find a team of outstanding authors to assist in this endeavour. The advice and suggestions of Dr. Karel Macek, Prague, Czechoslovakia during the early stages of this work are gratefully acknowledged.

Obviously, analysis in biological systems represents a rather broad field, from which topics will be selected on the basis of their importance, timeliness and feasibility. Books on the analysis of metals in human toxicology and on benzodiazepine analysis are under preparation, whereas volumes on alcohol analysis and on carbon monoxide are being set up. Additional topics will not only be selected from the biomedical sciences, but also from other areas such as environmental analysis, food and agricultural chemistry, etc. Suggestions to this end as well as comments and suggestions for improvements will be greatly appreciated.

Rokus A. de Zeeuw

Chapter 1

AMINES OF BIOLOGICAL INTEREST AND THEIR ANALYSIS

GLEN B. BAKER AND RONALD T. COUTTS

Neurochemical Research Unit, Department of Psychiatry and Faculty of Pharmacy and Pharmaceutical Sciences, University of Alberta, Edmonton, Alberta T6G 2G3 (Canada)

1.1   INTRODUCTION

Biogenic or naturally-occurring amines have been the subject of a great deal of research, particularly in the neurosciences. Much of this research has centered around the catecholamines dopamine and noradrenaline and the indolalkylamine 5-hydroxytryptamine (5-HT; serotonin). There is now a reasonable body of evidence supporting the role of these amines as neurotransmitters, and the term "biogenic amine" has become synonymous with the catecholamines and 5-HT. However, in recent years, there has been increased interest in histamine and the 'trace' amines, and in this book these are also included in the term biogenic amines. 'Trace' amines are defined as a number of naturally-occurring amines which are present in the central nervous system in much lower absolute quantities than the catecholamines and 5-HT, and include β-phenylethylamine, m- and p-tyramine, phenylethanolamine, m- and p-octopamine and tryptamine. The role of these substances in neurotransmission is unclear, but it is known that their concentrations in brain are affected markedly by a number of psychotropic drugs. They have also been implicated in a variety of psychiatric and neurologic disorders, and this will be discussed in further detail below.

There is now a voluminous literature on biogenic amines, and space does not allow for a comprehensive review of the research which has been done on these substances in biological systems. Rather, it is the purpose of this introductory chapter to highlight important aspects of the biogenic amines and to give a brief description of the format of the remainder of the book.

1.2   CATECHOLAMINES AND THEIR O-METHYLATED AMINE METABOLITES

The catecholamines (Fig. 1.1) have been the most extensively studied of the biogenic amines because of interest in their role as neurohumors. A number of excellent books and reviews are available describing the synthesis, metabolism and localization of the catecholamines in nervous tissue (e.g. 1-14). It has been proposed that one or more of the catecholamines are involved in the etiology and/or symptomatology of a number of neurological and psychiatric disorders, including migraine, schizophrenia, anxiety, depression, aggression, pheochromocytoma and parkinsonism (1-3,7,9,14,15-22). In addition, a wide variety of psychotropic

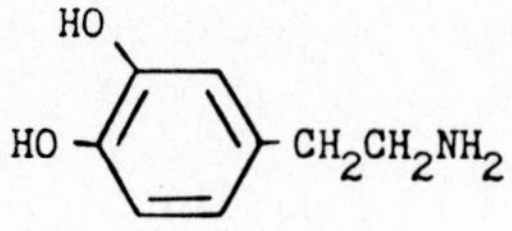

a

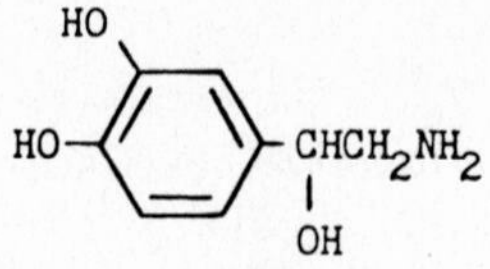

b

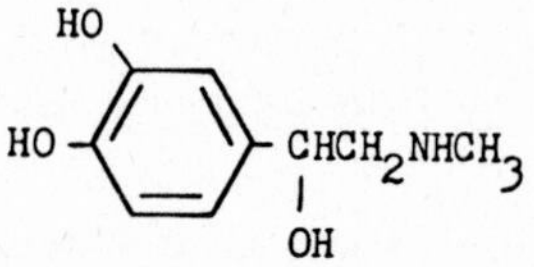

c

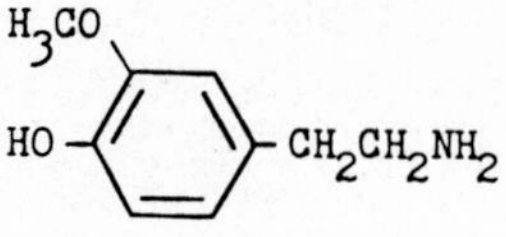

d

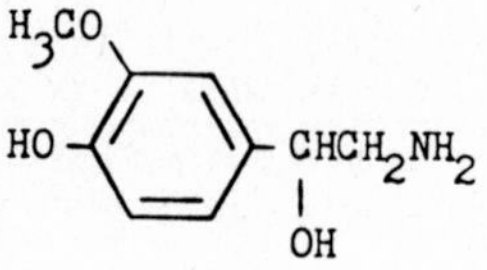

e

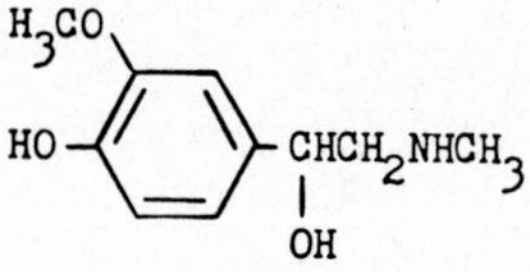

f

Fig. 1.1. Structures of the catecholamines and their O-methylated amine derivatives: a) dopamine, b) noradrenaline, c) adrenaline, d) 3-methoxytyramine, e) normetanephrine and f) metanephrine.

and cardiac drugs are thought to interact with the synthesis, storage, metabolism and/or receptor activity of the catecholamines (1-4,7,11,13,15,19-30). The ability to form fluorescent derivatives of catecholamines with relative ease and specificity has meant that much is now known about the localization of nerve tracts containing these substances (see Chapter 5 of this volume). Electrolytic or chemical lesions of these tracts as well as the administration of catecholamine-depleting drugs have been used in animal models to study the involvement of the catecholamines in a number of behavioural and physiological functions such as loco-motion, exploratory activity, hunger, thirst and sexual activity (1-3,7-9,15,18, 21,26,31).

The distribution of the catecholamines dopamine (DA), noradrenaline (NA) and adrenaline (A) and the ratios of each one to the others vary widely between the peripheral and central nervous system, within the central nervous system itself and amongst various species (1,2,6,7).

3-Methoxytyramine (3-MTA), normetanephrine (NMN) and metanephrine (MN) (Fig. 1.1) are produced by the enzymatic action of catechol O-methyl transferase (COMT) on DA, NA and A respectively. These O-methylated amines have been identified in various tissues and body fluids (32-39). Concentrations of these compounds in these loca-tions can be altered dramatically in certain disease states and after administration of a number of psychotropic drugs.

1.3   5-HYDROXYTRYPTAMINE

5-Hydroxytryptamine (Fig. 1.2), a substance which possesses strong vasocon-strictive properties on smooth muscle, is found in many parts of the body, and is located in nonneuronal sites (platelets, mast cells, enterochromaffin cells) as well as in neurones (7,40,41). Although only 1-2% of this indolealkylamine in the body is located in the brain (7), 5-HT has been of interest to neuroscientists for some time because of its putative role as a neurotransmitter and its apparent

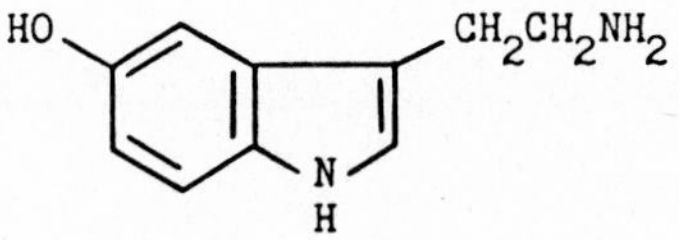

Fig. 1.2.   Structure of 5-hydroxytryptamine.

involvement in a number of psychological and physiological functions, including emotion, sleep, hunger, thirst and sexual activity (5,7,8,15,26,31,40-45). As with the catecholamines, a wide variety of psychotropic drugs affect the synthesis, metabolism, storage and receptor activity of 5-HT (7,11-13,15,21-25,30,40-45). The localization of 5-HT-containing nerve tracts has also been investigated extensively (4,7,8,15,41-44). The involvement of 5-HT in a number of nervous disorders, including migraine, depression and schizophrenia has been suggested (15, 16,21,22,41-49).

1.4 TRACE AMINES

The trace amines (Fig. 1.3) have not been investigated as thoroughly as the catecholamines or 5-HT. The primary reason for this is the very low concentration of these substances in the nervous system, which has necessitated the development of highly sensitive and specific analytical techniques. The amount of literature on these amines has increased enormously in recent years as these methods have been applied to brain and body fluids.

Although they are present in the CNS in minute quantities, the turnover rate of most of the trace amines is very rapid (50-53) and β-phenylethylamine and tryptamine can cross the blood-brain-barrier with ease (54); these two factors may be very important from a physiological standpoint. Boulton (55) has pointed out that the trace amines, like the catecholamines and 5-HT, are distributed heterogenously amongst brain areas.

Most of the trace amines exert rather strong effects on uptake and release of the catecholamines and 5-HT (56-59) and metabolic conversions between some of the trace amines and the catecholamines have been reported (60-63). Recent microiontophoretic studies in brain have revealed that application of trace amines at currents insufficient to affect baseline firing rates of neurones can alter dramatically the response of cells to DA, NA or 5-HT which are applied subsequently (64,65). Although much yet remains to be known about the role of trace amines in the CNS, the above findings and structural similarities to the putative neurotransmitter amines suggest that their function may be associated intimately with DA, NA and 5-HT.

1.4.1 β-Phenylethylamine

The neurochemistry, metabolism and pharmacology of β-phenylethylamine (PEA) and its effects on animal behaviour have been reviewed extensively (66,67). A number of monoamine oxidase inhibitors are known to cause dramatic increases in brain concentrations of PEA (68-72). One of the biochemical abnormalities resulting from phenylketonuria is an increased production of PEA and hence a greatly elevated urine level of PEA and its major metabolite, phenylacetic acid (PAA) (73,74).

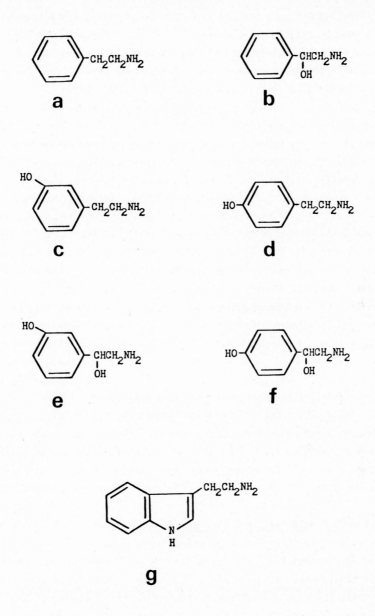

Fig. 1.3. Structures of some 'trace' amines: a) β-phenylethylamine, b) phenyl-ethanolamine, c) m-tyramine, d) p-tyramine, e) m-octopamine, f) p-octopamine and g) tryptamine.

Urinary levels of PEA are reported to be decreased in depression and increased in
mania and certain types of schizophrenia (75-77). The involvement of PEA in the
etiology of migraine, depression and certain types of schizophrenia has been suggested (78-82). It has also been reported that plasma levels of free and conjugated
PAA are elevated in aggressive psychopaths (83).

1.4.2 Meta- and para-tyramine

The ability of para-tyramine (p-TA) to be taken up into noradrenergic terminals
appears to explain the hypertensive crises experienced by some patients who are
receiving monoamine oxidase inhibitors and who have ingested foods which contain
p-TA (84). In about 30% of patients with classical migraine, headaches can be
induced by p-TA and p-TA-containing foods. These TA-sensitive patients appear to
suffer from a deficiency in the enzyme responsible for the formation of the sulfate conjugate of p-TA (85,86). A correlation between abnormalities in the EEG
record and higher than normal urinary excretion of p-TA has been reported in
schizophrenia and Parkinsonism (87-89). Abnormal urinary excretion of p-TA and/or
its principal acid metabolite, p-hydroxyphenylacetic acid (p-HPAA), has been observed in hypertyrosinaemia (90), coeliac disease, cystic fibrosis (91), pheochromocytoma (92) and phenylketonuria (93). The mean 24 h urinary excretion of
p-HPAA by adult male and female patients with primary depression has been reported
to be significantly lower than in controls (94); the authors suggested that this
indicated a deficient production of p-TA in depression.

Juorio (95,96) has reported that acute administration of clinically active
neuroleptics to mice reduces concentrations of p-TA in striatum while having no
effect on meta-tyramine (m-TA) and DA levels. In cases where structural isomers
exist (e.g. α- and β-flupenthixol), the isomer with neuroleptic activity was found
to reduce striatal p-TA levels while the inactive isomer had no effect.

1.4.3 Phenylethanolamine, meta- and para-octopamine

The primary route of formation of phenylethanolamine (PEOH), meta-octopamine
(m-OA) and para-octopamine (p-OA) appears to be by β-hydroxylation of PEA, m-TA
and p-TA respectively. As with most of the other amines, these compounds are distributed heterogenously amongst brain areas (97-100). Saavedra et al. (101) have
reported higher PEOH/NA and p-OA/NA ratios in fetal rats than in adult rats. There
is now a reasonable body of evidence in support of a role for p-OA in invertebrate
nervous systems (102-105), but little is known about the function of the OAs and
PEOH in mammalian nervous systems. Lesioning experiments have indicated that at
least part of the OAs and PEOH present in certain tissues is associated with nerve
endings (106-109).

Plasma levels of PEOH are reported to be significantly increased in patients
with hepatic encephalopathy and in the plasma and CSF of dogs in hepatic coma

(110). David (111) found that p-OA levels in the hypothalamus and in brain stem were 2-3 times greater,and m-OA levels were 5-6 times greater, in hypertensive rats than in normal controls. Elevated levels of p-OA or its metabolites have been found in body fluids from patients with hepatic coma (112-116) and decreased levels in sufferers of primary depression (94). Higher than normal urinary concentrations of p-OA in hyperthyroid children have been reported (117), and elevated levels of p-OA have been found in the CSF of epileptics (118).

1.4.4 Tryptamine

Abnormally low amounts of tryptamine (T) have been reported to be excreted by depressed patients (119,120) and Coppen et al. (120) have found that urinary T increases in patients upon recovery from depression. Dewhurst (78) reported a marked increase in urinary T in depressed patients following administration of phenelzine, the increases being much greater than those observed for 5-hydroxytryptamine (5-HT) or for O-methylated catecholamines. Several workers have demonstrated that administration of monoamine oxidase (MAO) inhibitors results in dramatic increases in brain concentrations of T in rodents (70,71,121-124). In a study of individual schizophrenics, Brune and Himwich (125) found that T tended to increase when behaviour worsened and to decrease when behaviour improved. Herkert and Keup (126) suggested that high T levels accompanied by low 5-HT levels may be a factor in the development of psychosis. Elevated urinary T has been reported in phenylketonuria (127) and in cases of carcinoid tumor (128) and thyrotoxicosis (129).

1.5 HISTAMINE

Histamine (HA) (Fig. 1.4) is a ubiquitous compound which is involved in a number of bodily functions. In recent years, the possible role of HA as a neurotransmitter in the CNS has been an active area of research (130-135). It fulfils a number of the criteria for a neurotransmitter substance, and an ascending histaminergic pathway in rat brain has been described which passes through the forebrain bundle and diffuses over the entire telencephalon (131).

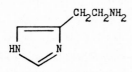

Fig. 1.4. Structure of histamine.

Several workers (136-138) have now demonstrated that a number of antidepressants and neuroleptics of diverse structure have in common the ability to block central HA receptors. Snyder (137) has suggested that the CNS HA receptor blocking action of antidepressant drugs may result, by an unknown mechanism, in an increase of synaptic levels of NA and 5-HT.

Histamine may also be involved in other mental diseases. It has been reported that schizophrenics have a relatively low incidence of allergies and the onset of schizophrenia (and manic depressive psychosis) is sometimes accompanied by a remission of asthma (139). It has also been observed that schizophrenics display a marked tolerance to intradermally administered HA, and schizophreniform psychoses have been reported (140) following antihistamine overdose.

The studies mentioned above have dealt primarily with the presence of biogenic amines in nervous tissue and body fluids and this is certainly a most active area of research. However, many of these amines may also be present in a number of agricultural products and analysis of such compounds in foodstuffs is an expanding area of interest. As mentioned above, it has been known for some time that foods rich in p-TA can cause hypertensive crises in individuals taking MAO inhibitors. It is conceivable that other structurally similar amines present in foods can also contribute to such effects. Histamine and related compounds are thought to be responsible for the unpleasant symptoms arising in scombroid fish poisoning (140, 141). Amines in foods may also be responsible for misleading results in urinary studies on the excretion of biogenic amines or their metabolites. In such studies diet should be carefully controlled.

1.6 PURPOSE OF THE BOOK

Our understanding of the localization and functioning of biogenic amines has increased rapidly in recent years with the development of new assay techniques and the refinement of old ones. Many analytical techniques are presently available, and the needs of a researcher studying the biogenic amines will depend on the amine(s) of interest, the finances, manpower and instrumentation facilities available, and the particular problem to be solved. The purpose of this book is to gather in one place a number of review articles by recognized experts. As well as discussing the historical development of the techniques in question, contributors have also dealt to some degree with possible future applications of these methods and have discussed the merits and limitations of the techniques relative to other procedures available. It is hoped that such discussions will be useful to both workers experienced in the analysis of biogenic amines who are contemplating changing or expanding techniques, and to researchers planning to enter the field of biogenic amine research and requiring an overview of the methodologies available. It should be emphasized that although this book deals primarily with analysis of biogenic amines, many of the techniques described can be modified for measurement of metabolites of these amines, other physiologically important substances, and a number of drugs. Some of the techniques, such as bioassays, radioimmunoassays and radioreceptor assays have been used much more extensively in quantitating substances other than biogenic amines and some of these substances have been included in the descriptions of the techniques because of their importance in the development of these procedures. It is hoped that the discussions of the "state of the art" of the techniques included in this volume will give the reader an overview of the principal procedures now in use for analysis of biogenic amines and an idea of likely future developments.

REFERENCES

1 H. Blaschko and E. Muscholl (Eds.), Catecholamines, Springer-Verlag, Berlin, 1972, 1054 pp.
2 L.L. Iversen (Ed.), Brit. Med. Bull., 29, Medical Dept., The British Council, London, 1973, 184 pp.
3 E. Usdin and S.H. Snyder (Eds.), Frontiers in Catecholamine Research, Pergamon Press, New York, 1973, 1219 pp.
4 E. Usdin (Ed.), Advances in Biochemical Psychopharmacology, Vol. 12, Neuropsychopharmacology of monoamines and their regulatory enzymes, 1974, 462 pp.
5 S.D. Iversen and L.L. Iversen, Behavioral Pharmacology, Oxford University Press, New York, 1975, 310 pp.
6 L.B. Geffen and B. Jarrott in J.M. Brookhart, V.B. Mountcastle and E.R. Kandel (Eds.), Handbook of Physiology, Section 1: The Nervous System, Vol. 1 Cellular Biology of Neurons, Part 1, Waverly Press, Baltimore, 1977, pp. 521-571.
7 J.R. Cooper, F.E. Bloom and R.H. Roth, The Biochemical Basis of Neuropharmacology, 3rd edn., Oxford University Press, New York, 1978, 327 pp.
8 L.L. Iversen, S.D. Iversen and S.H. Snyder (Eds.), Handbook of Psychopharmacology, Vol. 9, Chemical Pathways in the Brain, Plenum Press, New York, 1978, 410 pp.
9 P.J. Roberts, G.M. Woodruff and L.L. Iversen (Eds.), Advances in Biochemical Psychopharmacology, Vol. 19, Dopamine, Raven Press, New York, 1978, 422 pp.
10 K. Fuxe and T. Hokfelt, in W.G. Clark and J. del Giudice (Eds.), Principles of Psychopharmacology, Second Edition, Academic Press, New York, 1978, pp. 71-95.
11 N. Weiner, in A.G. Gilman, L.S. Goodman and A. Gilman (Eds.), Goodman and Gilman's The Pharmacological Basis of Therapeutics, 6th edn., Macmillan, New York, 1980, 1843 pp.
12 E.S. Vizi (Ed.), Advances in Pharmacological Research and Practice, Vol. II: Modulation of neurochemical transmission, Pergamon Press (Oxford) and Akademiai Kiado (Budapest), 1980, 479 pp.
13 C.J. Pycock and P.V. Taberner (Eds.), Central Neurotransmitter Turnover, Croom Helm, London, 1981, 197 pp.
14 P. Riederer and E. Usdin (Eds.), Transmitter Biochemistry of Human Brain Tissue, Macmillan, London, 1981, 332 pp.
15 M.A. Lipton, A. Di Mascio and K.K. Killam (Eds.), Psychopharmacology: A Generation of Progress, Raven Press, New York, 1978, 1731 pp.
16 N.J. Legg (Ed.), Neurotransmitter Systems and their Clinical Disorders, Academic Press, London, 1978, 240 pp.
17 B. Eichelman, in M. Sandler (Ed.), Psychopharmacology of Aggression, Raven Press, New York, 1979, pp. 149-158.
18 A.J. Friedhoff (Ed.), Catecholamines and Behavior. 1. Basic Neurobiology, Plenum Press, New York, 235 pp.
19 D.C. Remy and G.E. Martin, in J. McDermod, Annual Reports in Medicinal Chemistry, Vol. 16, Academic Press, New York, 1981, pp. 11-20.

20 C. Kaiser and W.E. Bondinell, in J. McDermod (Ed.), Annual Reports in Medicinal Chemistry, Vol. 16, Academic Press, New York, 1981, pp. 1-10.
21 S.J. Enna, J.B. Malick and E. Richelson (Eds.), Antidepressants: neurochemical, behavioral and clinical perspectives, Raven Press, New York, 1981, 261 pp.
22 T.A. Ban, Psychopharmacology of Depression, A Guide for Drug Treatment, S. Karger, Basel, 1981, 127 pp.
23 E. Costa and S. Garattini (Eds.), International Symposium on Amphetamine and Related Compounds, Raven Press, New York, 1970, 962 pp.
24 M.J. Antonaccio and R.B. Robson, in S. Fielding and H. Lal (Eds.), Industrial Pharmacology, Vol. 2, Antidepressants, Futura Publishing Co., 1975, pp. 18 - 207.
25 L.L. Iversen, S.D. Iversen and S.H. Snyder (Eds.), Handbook of Psychopharmacology, Vol. 6, Biogenic Amine Receptors, Plenum Press, New York, 1975, 307 pp.
26 J.D. Barchas, P.A. Berger, R.D. Ciaranello and G.R. Elliott (Eds.), Psychopharmacology: From Theory to Practice, Oxford University Press, New York, 1977, 577 pp.
27 H.M. van Praag, Psychotropic Drugs: A Guide for the Practitioner, Van Gorcum, Assen, 1978, 466 pp.
28 S. Garrattini and R. Samanin (Eds.), Central Mechanisms of Anorectic Drugs, Raven Press, New York, 1978, 487 pp.
29 D.M. Paton (Ed.), The Release of Catecholamines from Adrenergic Neurons, Pergamon Press, Oxford, 1979, 393 pp.
30 H.I. Yamamura and S.J. Enna (Eds.), Neurotransmitter Receptors, Part 2, Biogenic Amines, Chapman and Hall, London, 1981, 273 pp.
31 L.L. Iversen, S.D. Iversen and S.H. Snyder (Eds.), Handbook of Psychopharmacology, Vol. 8, Drugs, Neurotransmitters and Behavior, Plenum Press, New York, 1977, 590 pp.
32 M. Greer, T.J. Sprinkle and C.M. Williams, Clin. Chim. Acta, 21 (1968) 247-253.
33 W. Kehr, Naunyn-Schmiedeberg's Arch. Pharmacol., 284 (1974), 149-158.
34 M.G. Bigdeli and M.A. Collins, Biochem. Med., 12 (1975) 55-65.
35 B.H.C. Westerink and J. Korf, J. Neurochem., 29 (1977) 697-706.
36 L.M. Nelson, F.A. Bubb, P.M. Lax, M.W. Weg and M. Sandler, Clin. Chim. Acta, 92 (1979) 235-240.
37 N.D. Vlachakis, N. Alexander and R.F. Maronde, Life Sci., 26 (1980) 97-102.
38 D.F. LeGatt, G.B. Baker and R.T. Coutts, Res. Commun. Chem. Path. Pharmacol., 33 (1981) 61-68.
39 A. Chiu, D.D. Godse and J.J. Warsh, Progr. Neuro-Psychopharmacol., 5 (1981) 559-563.
40 M.D. Gershon, in J.M. Brookhart, V.B. Mountcastle and E.R. Kandel (Eds.), Handbook of Physiology, Section 1: The Nervous System; Vol. 1. Cellular Biology of Neurons, Part 1, Waverly Press, Baltimore, 1977, pp. 573-623.
41 W.B. Essman (Ed.), Serotonin in Health and Disease, Vols. I-V. Spectrum Publications, Inc., New York, 1978-79.
42 J. Barchas and E. Usdin (Eds.), Serotonin and Behavior, Academic Press, New York, 1973, 642 pp.
43 L. Valzelli, Psychopharmacology, An Introduction to Experimental and Clinical Principles, Spectrum Publications, Inc., Flushing, New York, 1973, pp. 39-57.
44 E. Costa, G.L. Gessa and M. Sandler (Eds.), Advances in Biochemical Psychopharmacology, Vol. 10, Serotonin: New Vistas. Histochemistry and Pharmacology, Raven Press, New York, 1974, 329 pp.
45 E. Costa, G.L. Gessa and M. Sandler (Eds.), Advances in Biochemical Psychopharmacology, Vol. 11, Serotonin - New Vistas. Biochemistry and Behavioral and Clinical Studies, Raven Press, New York, 1974, 428 pp.
46 Z.M. Kawka, Am. J. Pharmacol., 139 (1967) 136-154.
47 G. Curzon, Adv. Pharmacol., 6 (1968) 191-200.
48 G.W. Bruyn, The Biochemical Basis of Migraine. A Critique, in H.L. Klawans (Ed.), Clinical Neuropharmacology, Vol. 1. Raven Press, New York, 1976, pp. 185-213.
49 L.E. DeLisi, L.M. Neckers, D.R. Weinberger and R.J. Wyatt, Arch. Gen. Psychiatry, 38 (1981) 647-659.
50 J.L. Meek, A.R. Kroll and M.A. Lipton, J. Neurochem., 17 (1970) 1627-1635.

51 L. Lemberger, J. Axelrod and I.J. Kopin, J. Pharmac. Exp. Ther., 177 (1971) 169-176.
52 P.H. Wu and A.A. Boulton, Can. J. Biochem., 52 (1974) 374-381.
53 D.A. Durden and S.R. Philips, J. Neurochem., 34 (1980) 1725-1732.
54 W.H. Oldendorf, Am. J. Physiol., 221 (1971) 1629-1639.
55 A.A. Boulton, Lancet ii (1974) 7871.
56 A.S.V. Burgen and L.L. Iversen, Br. J. Pharmac. Chemother., 25 (1965) 34-49.
57 A.S. Horn, Br. J. Pharmac., 47 (1973) 332-338.
58 G.B. Baker, I.L. Martin and P.R. Mitchell, Br. J. Pharmac., 61 (1977) 151P.
59 M. Raiteri, R. del Carmine, A. Bertollini and G. Levi, Eur. J. Pharmacol., 41 (1977) 133-143.
60 A.A. Boulton and L.E. Quan, Can. J. Biochem., 48 (1970) 1287-1291.
61 A.A. Boulton and P.H. Wu, Can. J. Biochem., 50 (1972) 261-267.
62 K. Brandau and J. Axelrod, in E. Usdin and S.H. Snyder (Eds.), Frontiers in Catecholamine Research, Pergamon Press, New York, 1973, pp. 129-131.
63 A.A. Boulton, in E. Usdin and M. Sandler (Eds.), Trace Amines and the Brain, Marcel Dekker, New York, 1976, pp. 21-40.
64 R.S.G. Jones, Br. J. Pharmac., 73 (1981) 485-493.
65 R.S.G. Jones, J. Neurosci. Res., 6 (1981) 49-61.
66 E. Usdin and M. Sandler (Eds.), Trace Amines and the Brain, Marcel Dekker, New York, 1976, 301 pp.
67 A.D. Mosnaim and M.E. Wolf (Eds.), Noncatecholic phenylethylamines, Part 1. Phenylethylamine: Biological Mechanisms and Clinical Aspects, Marcel Dekker, Inc., New York, 1978, 536 pp.
68 J.M. Saavedra, J. Neurochem., 22 (1974) 211-216.
69 S.R. Philips, in A.D. Mosnaim and M.E. Wolf (Eds.), Noncatecholic Phenylethyl-amines, Part 1. Phenylethylamine: Biological Mechanisms and Clinical Aspects, Marcel Dekker, Inc., New York, 1978, pp. 113-138.
70 S.R. Philips and A.A. Boulton, J. Neurochem., 33 (1979) 159-167.
71 S.R. Philips, G.B. Baker and H.R. McKim, Experientia, 36 (1980) 241-242.
72 H.R. McKim, D.G. Calverley, S.R. Philips, G.B. Baker and W.G. Dewhurst, in P. Grof and B. Saxena (Eds.), Progress in Canadian Neuropsychopharmacology, S. Karger, New York, 1980, 7-13.
73 J.B. Jepson, W. Lovenberg, P. Zaltzman, A. Sjoerdsma and S. Udenfriend, Biochem. J., 74 (1960) 5P.
74 J.A. Oates, P.Z. Nirenberg, J.B. Jepson, A. Sjoerdsma and S. Udenfriend, Proc. Soc. Exp. Biol. Med., 112 (1963) 1078-1081.
75 A.A. Boulton and L. Milward, J. Chromatogr., 57 (1971) 287-296.
76 E. Fischer, A.B. Spatz, J.M. Saavedra, H. Regianni, A.H. Miro and B. Heller, Biol. Psychiat., 5 (1972) 139-147.
77 S.G. Potkin, F. Karoum, L.-W. Chuang, H.E. Cannon-Spoor, I. Philips and R.J. Wyatt, Science, 206 (1979) 470-471.
78 W.G. Dewhurst, Nature (Lond.), 218 (1968) 1130-1133.
79 M.B.H. Youdim, S. Bonham-Carter and M. Sandler, Nature, 230 (1971) 127-128.
80 E. Fischer and B. Heller, Behav. Neuropsychiat., 4 (1972) 8-10.
81 H.C. Sabelli and A.D. Mosnaim, Am. J. Psychiat., 131 (1974) 695-699.
82 M. Sandler and G.P. Reynolds, Lancet, i (1976) 70.
83 M. Sandler, C.R.J. Ruthven, B.L. Goodwin, H. Field and R. Matthews, in M. Sandler (Ed.), Psychopharmacology of Aggression, Raven Press, New York, 1979, pp. 149-158.
84 E. Marley, in D.G. Grahame-Smith (Ed.), Drug Interactions, University Park Press, Baltimore, 1977, pp. 171-194.
85 I. Smith, A.H. Kellow, P.E. Mullen and E. Hanington, Nature, 230 (1971) 246-248.
86 M. Sandler, M.B.H. Youdim and E. Hanington, Nature, 250 (1974) 335-337.
87 A.A. Boulton, G.L. Marjerrison and J.R. Majer, J. Acad. Med. Sci., U.S.S.R., 5 (1971) 68-70.
88 A.A. Boulton and G.L. Marjerrison, Nature, 236 (1972) 76-78.
89 G.L. Marjerrison, A.A. Boulton and A. Rajput, Dis. Nerv. Sys., 33 (1972) 164-169.
90 H.J. Bremer, U. Jaenicke and D. Leupold, Clin. Chim. Acta, 23 (1969) 244-246.

91 U. Jaenicke and H.J. Bremer, Z. Kinderheilk, 97 (1966) 281-286.
92 R. Robinson and P. Smith, Clin. Chim. Acta, 7 (1962) 29-33.
93 T.L. Perry, Science, 136 (1962) 879-880.
94 M. Sandler, C.R.J. Ruthven, B.L. Goodwin, G.P. Reynolds, V.A.R. Rao and
    A. Coppen, Nature, 278 (1979) 357-358.
95 A.V. Juorio, Life Sci., 20 (1977) 1663-1668.
96 A.V. Juorio, Br. J. Pharmac., 66 (1979) 377-384.
97 J.M. Saavedra and J. Axelrod, Proc. Natl. Acad. Sci. U.S.A., 70 (1973) 769-
    772.
98 J.M. Saavedra, Anal. Biochem., 59 (1974) 628-633.
99 J. Willner, H.G. LeFevre and E. Costa, J. Neurochem., 23 (1974) 857-859.
100 T.J. Danielson, A.A. Boulton and H.A. Robertson, J. Neurochem., 29 (1977)
    1131-1135.
101 J.M. Saavedra, J.T. Coyle and J. Axelrod, J. Neurochem., 23 (1974) 511-515.
102 H.A. Robertson and A.V. Juorio, Int. Rev. Neurobiol., 19 (1976) 173-224.
103 T.P. Hicks, Can. J. Physiol. Pharmacol., 55 (1977) 137-152.
104 P.D. Evans and M. O'Shea, Nature, 270 (1977) 257-259.
105 A.J. Harmar, in A.D. Mosnaim and M.E. Wolf (Eds.), Noncatecholic Phenylethyl-
    amines, Part 2. Phenylethanolamine, Tyramines and Octopamine, Marcel Dekker,
    New York, 1980, pp. 97-149.
106 P.B. Molinoff and J. Axelrod, Science, 164 (1969) 428-429.
107 P.B. Molinoff and J. Axelrod, J. Neurochem., 19 (1972) 157-163.
108 H.A. Robertson, J.-C. David and T.J. Danielson, J. Neurochem., 29 (1977)
    1137-1139.
109 H.A. Robertson, in M.B.H. Youdim, W. Lovenberg, D.F. Sharman and J.R. Lagnado
    (Eds.), Essays in Neurochemistry and Neuropharmacology, Vol. 5, John Wiley &
    Sons, New York, 1981, pp. 47-73.
110 F. Rossi-Fanelli, J. Escourrou, A.R. Smith and J.E. Fischer, in A.D. Mosnaim
    and M.E. Wolf (Eds.), Noncatecholic Phenylethylamines, Part 2. Phenylethanol-
    amine, Tyramines and Octopamine, Marcel Dekker, New York, 1980, pp. 231-244.
111 J.-C. David, Experientia, 35 (1979) 1483-1484.
112 J.E. Fisher and R.J. Baldessarini, Lancet, ii (1971) 75-79.
113 K.C. Lam, A.R. Tall, G.B. Goldstein and S.P. Mistilis, Scand. J. Gastroenterol.
    8 (1973) 465-472.
114 J.E. Fischer, in F. Plum (Ed.), Brain Dysfunction in Metabolic Disorders, Res.
    Publ. Assoc. Nerv. Ment. Dis. No. 53, 1974, pp. 53-73.
115 K. Manghani, M.R. Lunzer, B.H. Billing and S. Sherlock, Lancet, ii (1975)
    943-946.
116 L. Capocaccia, C. Cangiano, A.F. Attili, M. Angelico, A. Cascino and F. Rossi-
    Fanelli, Clin. Chim. Acta, 75 (1977) 99-105.
117 A.J. Prange, F.S. French, R.L. McCurdy, J.J. Van Wyk and M.A. Lipton, Clin.
    Pharmacol. Therap., 9 (1967) 195-203.
118 H. Kishikawa, J. Okayama Med. Soc., 87 (1975) 463-480.
119 R. Rodnight, Int. Rev. Neurobiol., 5 (1961) 251-292.
120 A. Coppen, D.M. Shaw, A. Malleson, E. Eccleston, G. Gundy, Br. J. Psychiat.,
    111 (1965) 993-998
121 C.A. Marsden and G. Curzon, J. Neurochem., 23 (1974) 1171-1176.
122 J.W. Sloan, W.R. Martin, T.H. Clements, W.F. Buchwald and S.R. Bridges, J.
    Neurochem., 24 (1975) 523-532.
123 J.J. Warsh, D.D. Godse, H.C. Stancer, P.W. Chan and D.V. Coscina, Biochem.
    Med., 18 (1977) 10-20.
124 B. Tabakoff, F. Moses, S.R. Philips and A.A. Boulton, Experientia, 33 (1977)
    380-381.
125 G.G. Brune and H.E. Himwich, Arch. Gen. Psychiat., 6 (1962) 324-328.
126 E.E. Herkert and W. Keup, Psychopharmacologia (Berl.), 15 (1969) 48-59.
127 T.L. Perry, Science, 136 (1962) 879-880.
128 D. Eccleston, T.B.B. Crawford and G.W. Ashcroft, Nature, 197 (1963) 502-503.
129 R.J. Levine, J.A. Oates, A. Vendsalu and A. Sjoerdsma, J. Clin. Endocrinol.
    Metab., 22 (1962) 1242-1250.
130 K.M. Taylor, in L.L. Iversen, S.D. Iversen and S.H. Snyder (Eds.), Handbook
    of Psychopharmacology, Vol. 3, Plenum Press, New York, 1975, pp. 327-329.

131 J.C. Schwartz, Life Sci., 25 (1979) 895-912.
132 J.P. Green, C.L. Johnson and H. Weinstein, in M.A. Lipton, A. Di Mascio and
    K.F. Killam, Psychopharmacology: A Generation of Progress. Raven Press, New
    York, 1978, pp. 319-332.
133 M. Garbarg, G. Barbin, S. Bischoff, H. Pollard and J.C. Schwartz, Brain Res.,
    106 (1976) 333-348.
134 J.C. Schwartz, G. Barbin, M. Baudry, M. Garbarg, M.P. Martres, H. Pollard and
    M. Verdiere, in W.B. Essman and L. Valzelli, Current Developments in Psycho-
    pharmacology, Vol. 5, SP Medical & Scientific Books, New York, 1979, pp. 173-
    261.
135 J.C. Schwartz, H. Pollard and T.T. Quach, J. Neurochem., 35 (1980) 26-33.
136 J.P. Green and S. Maayani, Nature, 269 (1977) 163-165.
137 P.D. Kanof and P. Greengard, Nature, 272 (1978) 329-333.
138 S.H. Snyder, J. Contin. Educ. Psychiat., 39 (1978) 21-30.
139 F.P. Rauscher, H.A. Nasrallah and R.J. Wyatt, in E. Usdin, D.A. Hamburg and
    J.D. Barchas (Eds.), Neuroregulators and Psychiatric Disorders, Oxford University
    Press, New York, 1977, pp. 416-424.
140 D. Roman, Br. J. Psychiatr., 121 (1972) 619-620.
141 Canada Diseases Weekly Report, Vol. 7-16, Health and Welfare Canada, Ottawa,
    1981, pp. 77-78.
142 Canada Diseases Weekly Report, Vol. 7-24, Health and Welfare Canada, Ottawa,
    1981, pp. 117-118.

Chapter 2

BIOLOGICAL ASSAY METHODOLOGY

WILLIAM F. DRYDEN

Department of Pharmacology, University of Alberta, Edmonton, Alberta   T6G 2E1
(Canada)

2.1   GENERAL PROBLEMS IN BIOASSAY

The use of living tissue in the investigation of biogenic amines predates the
identification of the amines themselves. It was the ability of extracts or tissue
eluates to influence the activity of frog hearts, isolated pieces of intestine or
animal blood pressure that convinced the early pioneers of the chemical nature of
nervous transmission, and spurred them into efforts to identify the substances
responsible. The passage of time has seen the resolution of the earlier questions
concerning most, but by no means all, of the peripheral amines in the body, only
to see them replaced by questions of greater subtlety concerning the amines in the
central nervous system.

The biogenic amines are generally neurotransmitters or intermediates in the
metabolism of neurotransmitters. It is a remarkable feature of the animal kingdom
that, despite the diversity of form and physiology encountered across the phylla,
there is enormous conservatism in the choice of compounds to act as neurotransmitters. The functions served by these compounds may, however, differ in different
groups. Acetylcholine, for example, is the excitatory transmitter in the skeletal
muscles of chordates, while glutamate performs the same role in arthropods. Even
within the same organism, the various transmitters may perform different roles
depending on the nature of the receptor present. Adrenaline will cause vasocon-
striction when acting on a α-adrenoceptors, and yet produce vasodilation in blood
vessels with β-adrenoceptors.

Receptors for the amines are widespread in the body, and muscular tissue in
particular provides a means of observing and quantifying responses to the amines
without recourse to sophisticated equipment or experimental protocols requiring
the passage of many days. Accordingly the methods used to assay the biogenic
amines have frequently measured the contractile response of a muscle when bathed
in or exposed to a solution containing the substance of interest. Such an approach
has the virtue of simplicity, but this is counterweighed by the necessity to make
a number of assumptions and compromises.

Firstly there is the supposition that concentration of drug is reflected in the
number of drug molecules interacting with receptors in the tissue and that this
in turn is proportional to the extent of the contractile response. Such a view

is not applicable in all cases, and possibly not in any case. Drug molecules in solution in the external bathing fluid (including blood plasma and extracellular fluid) must penetrate tissue spaces and often be exposed to a variety of inactivation mechanisms before finally coming in contact with their receptors. Thus the concentration in the fluid surrounding the receptor need not always be that of the fluid bathing the tissue as a whole. However if times of exposure to drugs are kept constant, then variations in concentration attributable to diffusion and inactivation are usually reduced to a minimum.

This caveat itself, however, contains an assumption in that the response of a tissue to a concentration of drug is presumed to be constant on every occasion. This is only so under narrowly defined conditions since otherwise tissues have an annoying habit of changing their sensitivity, especially when large concentrations of drugs are used. These changes in sensitivity are the result of receptor changes (desensitization), metabolic changes (sensitization, tachyphylaxis) or physiological compensation involving many organs in the whole animal (tolerance). The time scale of these changes ranges from a few minutes to several days or even weeks in the case of tolerance. Thus a rule of thumb is to use low concentrations of drug for brief periods of exposure that allow a measurement of the response to be made. Yet another assumption is that the response of the tissue is directly related to the number of receptors activated. An examination of physiological systems reveals that this is a treacherous supposition since the measured effect, usually muscle contraction, is removed from the primary event, drug-receptor interaction, by a sequence of steps that may involve the synthesis of secondary messengers such as cyclic nucleotides, phosphorylation of proteins, calcium fluxes, etc. Given such a multiplicity of potentially variable stages, it is obvious that a single concentration of a drug rarely produces a constant quantitative response in different tissues even when tissue weight is taken into consideration. It is always necessary to provide a set of reference responses in each tissue, and responses to unknown drug concentrations can be compared with them. Thus drug assay in tissues is a relative or comparative exercise lacking the accuracy demanded by the analytical chemist, but providing, by way of compensation, the reassurance that the compounds under investigation do indeed have profound and rapid effects on living tissues.

The use of bioassay methods in evaluating bio-active compounds was classically preferred where the identity of the active compounds was obscure. This remains true, although current interest has moved to include not only amines, but also peptides, fatty acids and their derivatives, the prostaglandins, thromboxanes and leukotrienes. Insofar as the amines are concerned, the content of a single amine in complex mixtures is still more amenable to simple bioassays in the absence of expensive physical chemical apparatus and hence it is appropriate to incorporate a chapter on bioassay in a book otherwise devoted principally to chemical methods.

This is not a respectful genuflexion towards tradition, but a reminder that analytical sophistry remains ultimately dependent on confirmation that identified compounds possess the ability to interact with tissues and produce rapid changes in activity or metabolism.

2.2 DESIGN OF EXPERIMENTS

In designing experiments to assay biogenic amines it is important to remember that the unknown amine in all likelihood has been obtained in an extract or eluate from biological material, and its guarantee of purity is in no way assured. Bioassay, although strictly defined as a quantitative procedure, nonetheless in practice encompasses a qualitative step, in which the identity of the active substance present is confirmed. While the methods differ in detail for each substance, and will be dealt with separately in the individual sections on these substances, there are a few general principles which can be enunciated here.

In general, the identification of an unknown substance rests on the nature and specificity of the response observed and on sensitivity to specific antagonist drugs. The principal stratagem is to eliminate the impossible candidates until only one possibility remains. Rarely is there a single biological response which is unique to a particular amine and which can provide positive identification of a compound. Our approach is greatly aided by the present availability of antagonist compounds or blocking drugs that show great specificity for one particular class of receptor. Thus while most biogenic amines are capable of interacting with more than one receptor often occurring together in the same tissue, the preference for each receptor type may be determined by selective blockade of the others. Hence noradrenaline can be distinguished from adrenaline or even dopamine, acetylcholine from histamine and so on.

The general procedure is to add a solution of the unknown substance to a tissue suspended in an organ bath and observe the nature of the response. Since muscle is the most frequently used tissue, the response is contraction, or relaxation if there is a degree of pre-existing contraction or tone. From a single observation such as this it is possible to eliminate whole groups of substances. It is then possible, by using appropriate blocking agents, to determine the class to which a substance belongs, and ultimately to its identity. However, throughout such procedures the assumption remains that the tissue will respond normally to known agents, and this assumption must be tested, not only at the beginning of an experiment, but also at frequent intervals throughout the identification process. This is particularly important when blocking agents are in use since many of them will act as local anaesthetics, suppressing membrane electrical activity and thus preventing any mechanical response from muscle. Thus the over-zealous application of propranolol or atropine to a tissue can depress any activity for considerable periods of time, leaving the investigator in doubt as to whether he is blocking a response specifically at the level of receptors, or non-specifically by depressing the ability of the tissue to respond to any stimulus. A useful guideline, therefore, is to use concentrations of antagonist which reduce, but do not completely inhibit,the responses evoked by agonist substances.

The choice of tissue rests largely upon the nature of the substance(s) under investigation. The comments of the later sections of this chapter therefore apply. Where the nature of the amine is completely unknown, a more general screen must first be used, and to this end tissues from the intestinal tract offer some advantages. They are usually receptive to a wide range of biogenic amines, although the sensitivity is in general less than in some other tissues. The duodenum of the rabbit is a conventional simple preparation that offers the advantages of regular autonomous contractile activity, the pendular movements, so that agents which promote relaxation can be detected by depression or cessation of this activity. A typical protocol for the investigation of an unknown substance is shown in Fig. 2.1.

The assay procedures themselves have been the subject of considerable standardization. Because many biological extracts required assay before therapeutic use, the recommended procedures are now incorporated in pharmacopoeias. In particular the British Pharmacopoeia (3) offers detailed advice on how tests should be conducted and lays particular stress on the statistical aspects of measurement. This necessity stems from the variability inherent in biological tissue. It is a matter of simple observation that animals differ from each other in most parameters. The differences are small and when a large number of animals are examined, the differences resolve themselves into a continuous bell-shaped spectrum about a modal point. If the sample size is large enough, the modal point is also the mean point of the population, and we term the distribution "normal". It is thus amenable to analysis by conventional means to produce values of variance, standard deviation and standard error of the mean. Comparisons of similar parameters from other populations are possible, and can be analyzed by Student's t-test, analysis of variance or the χ^2 test. These tests are standard tools of simple statistics and need little further explanation here, other than to emphasize that they apply when the parameter being measured is infinitely variable. Where an event such as animal death, as in LD_{50} determinations, is of interest, different statistical considerations apply. Death is not infinitely variable; it is an all-or-none phenomenon, and populations conform to binomial rather than normal or Gaussian distributions. The statistical treatment is called non-parametric and encompasses ranking experiments where animals are ranked according to performance, etc. Many examples can be found even in the published literature where the improper use of t-tests and χ^2-tests has accompanied ranking experiments. Should the reader wish a more detailed consideration of statistics as applied to biological sciences, the book by Colquhoun (4) is one of many recommended.

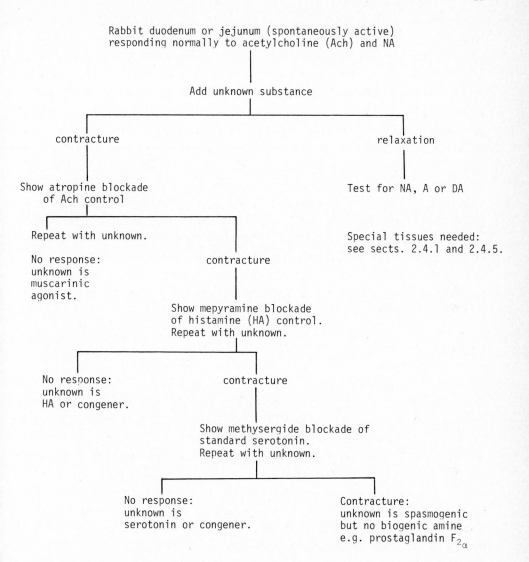

Fig. 2.1. Typical protocol used in identifying unknown bioactive substances in a solution.

For the purposes of bioassay of the biogenic amines (other than toxicity data) the parameters most readily determined are muscle contraction or tension, or relaxation, and these are parameters which are normally distributed about a mean value in a normal or Gaussian curve. In other words, just as populations of animals and their tissues will vary, so does the response of a single tissue at different and often adjacent times. The causes of this variation are not easy to explain. Minor differences in metabolism, receptor sensitivity and so on will produce slightly variant responses. It is assumed, of course, that care is taken to ensure minimal operator error in the bath concentrations of fluids, the recording of responses and the measurement of the records. Because of this variation, any single response of a tissue is unreliable, and only after examining a series of responses to the same dose of drug, can an acceptable average or mean response be obtained. The response of tissue to drugs when plotted against the logarithm of the drug concentration almost always assumes an "S" or sigmoid shape. The rising phase of the curve is virtually linear between 25% and 75% of the maximum response, and for the purposes of bioassay, is regarded as linear. The most accurate estimates of drug effects are obtained where the concentrations of drugs producing a given response are averaged (5), since the slopes of the curves are steeper and less open to interpretive error. However such an approach is profligate in its use of drug, and where limited amounts are available, as in tissue eluates, the most conventional approach is to use a standard concentration of drug and to average the responses to this concentration.

The three point assay

The three point assay is so termed because only three concentrations of drug are used. Two are known and represent different dilutions of a standard solution, while the third is unknown, being the solution under test. The assay assumes that the drug concentrations fall within the linear portion of the concentration-effect curve, and, if necessary, a complete curve may need to be plotted to ensure that this provision is valid. The unknown concentration of drug dose should produce a response that is intermediate between the responses produced by the standard doses. Thus the unknown concentration in the tissue bath is "bracketed" by known ones, and the assay is referred to as a bracketing assay. The standard concentrations are obtained by adding different volumes of a known drug solution(s) to the tissue bath, and these doses are identified as s_1 and s_2, s_2 being greater than s_1. The volume of the unknown drug solution used is u. If x is considered the volume of known solution of drug that theoretically produces the same response as u, then the concentration in the unknown solution can be calculated. The responses of tissue produced by the doses are S_1, S_2 and U respectively.

Since the relationship between log dose and response is linear in this range, it is possible to use simple proportional algebra to relate the known and unknown concentrations as follows:

The difference between the responses to doses of drug is proportional to the difference between logarithms of the doses, i.e. $S_2 - S_1 \propto \log s_2 - \log s_1$ or $\log \left(\frac{s_2}{s_1}\right)$

and $U - S_1 \propto \log x - \log s_1$ or $\log \left(\frac{x}{s_1}\right)$

therefore $\dfrac{\log \left(\frac{x}{s_1}\right)}{\log \left(\frac{s_2}{s_1}\right)} = \dfrac{U - S_1}{S_2 - S_1}$

$$\log \left(\frac{x}{s_1}\right) = \frac{U - S_1}{S_2 - S_1} \log \left(\frac{s_2}{s_1}\right)$$

$$x = \text{antilog} \left\{ \frac{U - S_1}{S_2 - S_1} \log \left(\frac{s_2}{s_1}\right) \right\} s_1$$

Finally the concentration of drug in the unknown solution can be obtained by dividing x by u and multiplying the concentration of the known drug solution by this ratio:

conc of u $= \dfrac{x}{u}$ multiplied by conc of s.

The values S_1, S_2 and U are, of course, measured responses which are subject to the variation previously described. Greater accuracy is therefore encouraged by the repetition of the assay a number of times and using the means of the responses in the calculation.

The more often a response is measured, the more accurate is the mean of the measurements. A balance has to be struck between practicality and desirability. While the number of doses of solution under test may be limited by the small volume of available solution, a minimum number of four additions should be attempted. It is important, too, to note that the doses should not be added always in the same order. Residual effects such as sensitization or desensitization can be circum- vented by changing the sequence in which doses are added, e.g.

$s_1 \longrightarrow u \longrightarrow s_2$, then

$s_1 \longrightarrow s_2 \longrightarrow u$, then

$u \longrightarrow s_1 \longrightarrow s_2$

Such a randomization pattern is termed a Latin square.

The four point assay

In any assay procedure, the accurate establishment of the true slope of the concentration-effect line is essential, and largely dependent on the attempts to
overcome the natural variation within and between tissues. To this end the slope
of a line determined for a larger number of points is normally more accurately
known than that of one where the number of points used is less. Therefore, an assay
using two concentrations of unknown drug has advantages over one where only one
concentration of unknown drug is used. For reasons that are explained in the
mathematical derivation of the formula, one unknown concentration produces a response either greater or less than those produced by doses of known concentration,
and so the assay cannot be described as a bracketing assay. It is more usually
termed the four-point assay for obvious reasons. Using the same annotation as in
the three point assay, two doses of differing but known concentration s_1 and s_2
are applied to a tissue and the responses S_1 and S_2 are measured. Two dilutions
of solution of unknown concentration, u_1 and u_2, are added to the tissue bath and
the responses are again recorded and measured. It is mathematically convenient
to keep the ratio of s_1/s_2 and u_1/u_2 the same (see calculation below). Again,
accuracy is improved if multiple applications by a Latin square randomized procedure are used, and a final mean value is recorded as the values of S_1, S_2, U_1
and U_2.

Again, if x_1 and x_2 are the notional doses of standard solution which would
produce the responses U_1 and U_2 observed with the drug solution of unknown concentration, then the concentration of the unknown solution can be determined thus.

The slope of a line is ordinate/abscissa or y-axis divided by x-axis. For a
concentration-effect line, the slope is:

$$\frac{S_2 - S_1}{\log s_2 - \log s_1} = \frac{S_2 - S_1}{\log \left(\frac{s_2}{s_1}\right)}$$

Similarly, slope for an unknown solution should be

$$\frac{U_2 - U_1}{\log \left(\frac{u_2}{u_1}\right)}$$

If the two values for slope are summed and averaged, the mean value for slope becomes

$$\frac{S_2 - S_1}{\log \left(\frac{s_2}{s_1}\right)} + \frac{U_2 - U_1}{\log \left(\frac{u_2}{u_1}\right)} \div 2$$

but, since the ratio of the doses of both the standard or test solution applied
were the same, i.e.

$$\frac{s_2}{s_1} = \frac{u_2}{u_1}$$

the mean slope is

$$\left\{ \frac{S_2 - S_1 + U_2 - U_1}{\log \left(\frac{s_2}{s_1}\right)} \right\} \div 2$$

By a similar argument, the differences between the responses at the lower and
higher doses of standard and test drug solutions can be averaged to produce a more
accurate mean difference in response. These should be similar at the upper and
lower dose levels since the dose ratios are similar.

Thus, a mean response difference or y parameter is

$$\frac{U_2 - S_2 + U_1 - S_1}{2}$$

We therefore have a mean slope and a mean ordinate parameter, and a mean abcissa
parameter can be calculated from the two. However since u_1 and u_2 are volumes of
unknown concentration, let us again make use of the theoretical volume of standard
solution x which produces an identical response as a dose of unknown solution and
from this determine the amount of drug required to produce a response. It does not
matter if we compare x and u_1 or u_2, so to provide a comparison with a three point
assay let x produce a response U_1, identical to that produced by u_1.

The distance on the abscissa between $\log x$ and $\log s_1$, i.e.

$$\log x - \log s_1 \text{ or } \log \left(\frac{x}{s_1}\right) = \frac{\text{mean ordinate difference}}{\text{slope}}$$

Therefore

$$\log \left(\frac{x}{s_1}\right) = \frac{U_2 - S_2 + U_1 - S_1}{S_2 - S_1 + U_2 - U_1} \log \frac{s_2}{s_1}$$

$$x = \text{antilog} \left\{ \frac{U_2 - S_2 + U_1 - S_1}{S_2 - S_1 + U_2 - U_1} \log \frac{s_2}{s_1} \right\} s_1$$

and the concentration of u = concentration of s multiplied by $\frac{x}{u_1}$ as before.

2.3 CHOICE OF TISSUES

The tissues used in the assay of biogenic amines are determined by sensitivity
to the amine in question, availability of both tissue and appropriate recording

equipment, and of course, skill and competence of the investigator. Assuming that the last can be acquired, it is appropriate to consider here the general use of tissues as indicators of drug concentrations. Muscular tissues have for long been the most convenient tissues to use since their response is mechanical and can be easily recorded with relative accuracy. However muscles fall into different categories which differ in their speed of contraction, their automatic or myogenic contractile activity and their responses to the various amines. As a consequence some thought must be given to the nature of the parameter to be measured.

Where tension is to be measured independently of spontaneous frequency, the choice is essentially between an isometric measurement, where the tension developed by the tissue is measured in the absence of actual shortening of the tissue; or an isotonic measurement where the tension is kept constant and the shortening of the muscle is recorded. Auxotonic methods, where the load is increased as the tissue shortens, are rarely used in bioassay work. In situ, many muscles perform their physiological role by exerting an isometric tension, but most muscles undergo only an initial isometric phase of contraction, during which tension is increased until it overcomes the resistence or inertia against which the muscle acts. Thereafter, shortening akin to either isotonic or auxotonic contraction occurs. When the speed of contraction is rapid, the viscosity of the cytoplasm opposes the shortening of the muscle fibres and reduces the impact of any drug effect which might otherwise increase the rate of contraction. For this reason, cardiac muscle and skeletal muscle are measured isometrically where electronic force transducers and oscillographs are available. Smooth muscle, despite its slow rate of tension development may be measured isometrically, but is more often measured under isotonic conditions In isometric contraction, however, it is wrong to assume that there is no shortening of the contractile elements in the muscle. The process of sarcomere shortening proceeds within the limits imposed by the system. Although the ends of the whole muscle may be rigidly fixed, there remains an imperfectly identified series elastic component, which stretches in response to the shortening of the sarcomeres. There is no practical method at present to obviate the series elastic component, save in the case of skeletal muscle, where it may be fully stretched during tetanic stimulation.

The classical mechanical measurement of contracting tissues involved the action of the muscle on levers which in turn scratched a trace on the sooted surface of paper attached to a revolving kymograph drum. Cardiac and skeletal muscles were made to act against a spring, and thus worked under auxotonic conditions, while smooth muscles had a simple isotonic leverage system. While there may be some disadvantage in using a kymograph system, in that it lacks some accuracy and may be inconveniently dirty with soot in the laboratory, in most cases the results obtained by such methods remain as acceptable as results obtained by more sophisticated, and thus expensive, methods. A useful laboratory handbook in the setting up of tissues is that written by the staff of the Department of Pharmacology, University of Edinburgh (6) to which further reference may be made by readers who have no basic experience of isolated tissue preparations.

2.3.1 Smooth muscle contraction and relaxation

Smooth muscle is to be found in most viscera and in blood vessels. With such a wide distribution comes an equally wide spectrum of sensitivity to pharmacologically active agents, including the biogenic amines. As a result, smooth muscle in one form or another comprises the vast majority of the test tissues used in their bio-assay. Smooth muscle offers the advantage of response which is slow, is measurable by simple apparatus and is reproducible. Desensitization is often slow to develop. Receptors to biologically active substances are distributed over the cell surface, and thus throughout the tissue. Application of drugs is therefore a simple matter of addition to the bathing medium. The muscle occurs mostly as thin sheets in hollow organs and is easily dissected and set up for experiment.

Not all smooth muscle behaves similarly, however, and a considerable degree of selection is required. Some smooth muscle has spontaneous or myogenic activity. Examples are to be found in the duodenum of the rabbit and the portal vein of the guinea pig. This activity, so often originating from a pacemaker group of cells, can be spread directly from muscle cell to muscle cell, in a wave of contraction, or it may be coordinated by plexuses of nerves found embedded in the layers of muscle, as seen in the intestine. Such activity may confuse measurements of contraction produced by stimulant amines; however the inhibition of this activity in turn provides a possible means of assaying relaxant amines.

Most smooth muscles are, however, quiescent until stimulated either through innervating nerves or by addition of an agonist substance to the bathing fluid. In the assay of any substance, the object is to choose a tissue which is sensitive to the substance at the concentrations available and which will produce a quanti-fiable response. In some tissues, such as the uterus, the density of drug receptors is hormone-dependent (oestrogens in the case of the uterus). It is advanta-geous to ensure that the animal is given adequate doses of hormones prior to sacrifice and removal of the tissues.

The choice of parameter to measure is normally reduced either to contraction or relaxation of a tissue. In the case of contraction, either the mechanical tension produced in longitudinally arranged muscle can be measured directly, or the pressure of fluid contained in a hollow organ such as a blood vessel or bladder can be determined. Here the measurement is simple, reproducible, and involves only the substance under investigation. The alternative, measurement of relaxation, is a very much less satisfactory approach which nonetheless is often unavoidable under certain circumstances. Relaxation of muscle implies prior tension, however produced, and therefore the relaxant drug is acting in physiological antagonism to an existing excitatory process. It is the net balance of the mutually opposing influences that provides the parameter for measurement, and the standardization of the excitatory process is essential, but not always easy. Contraction of bronchial muscle can be induced by exposing it to a concentration of excitatory drug such as acetylcholine; the opposing effect of adrenaline is then gauged. However, chronic exposure to drugs, especially at the elevated concentration likely to be used here, brings on problems such as desensitization. The depression of spontaneous activity of rabbit duodenum can also be used as an assay of relaxant drugs but, in this case the origins of the activity are not under the investigator's control and lend an extra dimension of variability into the already variable system. The conclusion is that assays using smooth muscle relaxation rather than contraction are best avoided if at all possible.

2.3.2 Cardiac muscle contraction and frequency

Cardiac muscle is, of course, restricted to the heart. It should not be assumed however, that the heart is a homogeneous organ, where muscle from one part is indistinguishable by pharmacological inertia from muscle from another part. The whole heart can be used as an experimental preparation. It must be perfused either via the aorta (retrogradely) and the coronary circulation (Langendorff preparation - see p. 116 in ref. 6), or by perfusion of the pulmonary vein (7). Such preparations allow the rate of beating driven by the pacemaker in the sinuatrial node and free from nervous influences, to be measured. While coronary perfusion pressure is also measurable by this method, it is not possible to obtain an unambiguous measurement of the force of muscular contraction, since there are several layers of muscle fibres oriented in different directions, which produce a twisting or wringing of the organ during systole. There is no axis along which muscle tension can be measured when the hollow organ exists in three dimensions, and contraction serves to reduce its volume in each dimension. To measure tension, it is necessary to dissect the heart. The atria, both right and left, can be removed separately or together and provide entirely suitable tissues for experiment. The right atrium contains the pacemaker and is spontaneously contractile in a tissue bath. This can be overcome, if desired, by stimulating the tissue at a rate faster than the spontaneous rate, but it is more usual to set the tissues up in pairs and compare drug effects on frequency in the right atrium with drug effects on tension in the left atrium. Ventricular tissue can also be used. This is much thicker than atrial tissue in most experimental mammals, and some further dissection is necessary to allow adequate oxygenation of the tissue during experimentation. Right ventricular strips can be cut, following the axis of the thickest muscle layer, or papillary muscles can be removed from the left ventricle. Ventricular tissues are stronger than atrial tissue, but the magnitude of the response evoked by drugs is often less. The sensitivity, as defined by the concentration of drug producing

50% of the maximum response (EC_{50}), is the same in most cases, and so for ease
of measurement, atrial tissue is often preferred. The frequency of contraction
is usually more sensitive than the force of contraction, and thus effects of drugs
on muscarinic cholinoceptors and β-adrenoceptors may be detected at lower concentrations of drug by observing changes in spontaneous rate.

In general rat cardiac tissue possesses poor sensitivity to pharmacological
agents. There are, for instance, no histamine receptors in rat left atria, although some are present in the right. EC_{50} values in the rat are higher, by one
order of magnitude or more, than in other species, and so for assay work, cardiac
tissue from the rat is best avoided.

Cardiac tissue does require some degree of careful handling over and above that
applied to other tissues. It is very susceptible to anoxia, and adequate oxygenation is essential during the experiment. In dissecting the tissue, oxygen demand
can be reduced by immersing the tissue in ice-cold bathing fluid, although under
these conditions heart cells cannot extrude calcium ions, which leak inwards across
the cell membrane passively. This rise in free cytoplasmic calcium can cause
damage, probably by activating Ca^{++}-dependent proteases, if the cryoplegia is
prolonged.

In addition, cardiac muscle, more than other types of muscle, exerts its contraction as a function of initial length (Frank-Starling relationship). This must
first be determined after setting up each tissue, and an experimental length,
which is associated with a twitch tension less than maximum (usually 70%), is
determined and used throughout subsequent experimental procedures.

2.3.3 Skeletal muscle contraction

The skeletal muscle of vertebrates can be subdivided into two groups. There
are muscles in which each muscle fibre is innervated by several nerve fibres,
where transmitter release occurs at many points along the muscle, and diffusion
of transmitter occurs to allow the receptors found over the entire muscle surface
to interact and induce contraction directly. Such multi-innervated muscles are
common in vertebrate classes other than the mammals where they are restricted to
extraocular muscles and muscle spindle fibres. Because of the distribution of
receptors, and ease of access to them, they are well suited to experimental
procedures in an isolated organ bath. Muscles where each fibre is innervated by
only one nerve fibre are known as focally innervated muscles. Here the excitation-
contraction coupling process follows on not directly from receptor activation, but
only after an intervening action potential has swept over the muscle from the
localized site of receptor activation, the end plate. Because the receptors are
largely restricted to the end plate region, and because they are not freely accessible to substances in the bathing fluid due to the protective wrapping of Schwann
cell which overlies the nerve terminal in such muscles, it is not possible to use

focally innervated muscle in organ bath experiments where the transmitter substance (acetylcholine in vertebrates) is added to the bathing fluid. The striated
muscles of non-vertebrate species can be used in bioassays, but other than the
muscle of the leech, little use has been made of these alternative preparations.

It would be wrong to assume that only receptors to the endogenous neurotransmitter are to be found in striated muscle. While the contractile response is
normally only evoked by the action of the natural transmitter or an analogue
(ignoring direct electrical stimulation), the magnitude, time to peak amplitude
and rate of relaxation are all susceptible to a number of pharmacological agents.
However the responses are small, and such modification of the contractile parameter
is not a suitable assay for biogenic amines or other substances.

2.3.4 Parameters from non-contractile tissues

While most assay procedures use muscle contraction as the quantifiable parameter
there are alternative biological actions which are used under a variety of circumstances to identify and quantify active substances such as the biogenic amines.
Nervous activity itself, either as membrane potential changes or as rates of dis-
charge in neurones, is commonly used where putative neurotransmitters are under
investigation. However neurophysiology of this type requires expensive apparatus
and considerable expertise in its operation. Such preparations are therefore
beyond the scope of the present chapter.

Glands do form an alternative tissue which is responsive to a variety of pharmacological agents. It is relatively easy to collect the secretion of an exocrine gland
and determine the influence of various substances on the rate of secretion. Such a
technique is not common with vertebrate animals, but it is useful in other species,
particularly where the neurotransmitters involved are unknown or merely putative.
The salivary glands of ticks provide an example where this technique is applied in
investigating the chemical control of salivary secretion in this animal.

2.4 METHODS OF ASSAY

While there are a wide variety of preparations which may be used to test for
and assay biologically active substances, there are a number of commonly used ones
which are described briefly in this section. The test is neither exclusive nor
exhaustive and the choice of tissue may well be influenced by special circumstances,
availability of equipment or material, or even the personal preference or competence
of the investigator. In the last analysis, the choice is determined by sensitivity
of the preparation and the specificity of the response. Inevitably, there is a
degree of overlap. Several substances may influence the same receptor type, or
there may be several types of receptor present, all of which when stimulated by
their agonist ligand produce the same response. In these circumstances a degree
of specificity may be ensured by the use of "specific" blocking drugs which effec-

tively eliminate one type of receptor only from the tissue, leaving the others
relatively free and available for activation. Reference is made to such cases
where appropriate.

2.4.1 Catecholamines and congeners

The three principal catecholamines are adrenaline (A), noradrenaline (NA) and
dopamine (DA). Their metabolites also possess some biological activity, but it
is substantially less than that shown by these three substances. Adrenaline is
secreted by postganglionic sympathetic nerves in amphibia, but in mammals it is
principally found in the medulla of the adrenal gland, whence it is secreted into
the bloodstream, forming an intermediate between a hormone and a neurotransmitter,
and in the terminals of some neurones, the perikarya of which are located in the
caudal medulla oblongata. The distribution of the terminals ranges from the spinal
cord to the thalamus and the hypothalamus (8). Noradrenaline is widely distributed
in mammalian postganglionic nerve fibres, and is also found in the central nervous
system, primarily in the projections of cells lying in the locus coeruleus and in
nuclei of the lateral reticular formation (8). Dopamine is not found widely in the
peripheral nervous system. It is restricted to some neurones in autonomic ganglia
and possibly also in the nerve plexuses of the alimentary tract. However, it is
widely distributed in the central nervous system.

Receptors to the catecholamines are classically divided into adrenoceptors
and DA receptors, although DA is active at adrenoceptor sites. The adreno-
ceptors are in turn subdivided into α_1, α_2, β_1 and β_2 subgroupings which can be
recognized by their differing sensitivity to various synthetic agonist or antag-
onist molecules. They are widely but not uniformly distributed in tissues, thus
providing a rich choice of suitable assay preparations. Receptors which show
specificity towards DA are rare in peripheral tissues, although they are common in
the central nervous system. As a result, the possible assay preparations are
restricted in number, but are available and used when the occasion warrants.

Adrenaline shows equal affinity for both α and β adrenoceptors. Noradrenaline,
however, is more potent on α receptors, and on the β_1 adrenoceptors which are
found in heart muscle, than on β_2 receptors. Most assays therefore use tissues
which possess either α_1 or β_1 receptors. The α_2 receptor, recognized as a pre-
synaptic receptor on nerve terminals, does exist as a postsynaptic receptor in some
tissues, but has not as yet been used in assay work. The pressor effect of both A
and NA in the pithed rat has been advocated as the parameter of choice for bioassay (9). Amounts as low as 1.0 ng can be assayed by this method after extraction
from source material. Adrenaline, as pointed out, will affect the β_2 receptors
of some blood vessel beds, causing vasodilation. This would offset the constrictor
effect in other vessel beds and cause little overall changes in systemic pressure
unless a β-blocking drug is first administered to the animal. In this regard,

propranolol at a dose of 0.5 mg/kg will abolish the dilator effect of A and
leave the pressor effect for assay. This preparation is somewhat demanding
in both skill and experience, and a simple isolated tissue preparation may be
preferred. To this end isolated tissues containing α, β_1 and β_2 receptor types
may be used in smooth or cardiac muscle preparations. Of the vascular tissues,
the tension developed in aortic strips is a convenient and simple preparation of
reasonable sensitivity (6), while the rabbit ear artery provides a slightly more
sensitive though more complicated preparation (10) in that a pressure transducer
and peristaltic pump are required rather than the conventional tissue bath. The
sensitivity of these tissues to catecholamines can be increased by adding 10 μg/l
(5.7×10^{-8}M) serotonin to the bathing or perfusing fluid (11). Serotonin itself
has little constrictor action, and the nature of the sensitizing effect has not
been explained. The assay system can be rendered more sensitive by the use of the
superfusion technique, in which tissues are bathed in a stream of aerated fluid
continuously flowing down their surface. The volume surrounding the tissue is thus
much less, and the dilution of minute amounts of active substances is reduced. By
using this technique, amounts as low as 20 pg of NA can be detected using either
rabbit aorta or iliac artery strips (12).

Since α-adrenoceptors are present in these vascular tissues, the catecholamines A and NA are about equipotent. The isolated atria or right atrium of either
the guinea pig or the rabbit (6,13) offer an alternative in which there is a simila
potency, although the receptor in this case is essentially of the β_1 subtype. The
parameter of choice here is the increase in spontaneous beat frequency, a parameter
which shows greater sensitivity and accuracy than force of contraction.

Where mixtures of these two catecholamines are present, special precautions
must be taken and these are detailed in a later section.

Although DA is active at α and β receptors and can be blocked by the classical
antagonists at these sites, phentolamine and propranolol respectively, it has only
one fiftieth of the potency of NA at the α-adrenoceptor, and one thirtieth of the
potency of either NA or A in increasing the force of cardiac contraction.

If DA alone were the subject of bioassay, it would be possible to use any of
the preparations recommended for A. However, the circumstances of bioassay
are often such that the DA may be present along with other substances, including
the other catecholamines, in a solution extracted or eluated from tissues. In this
case, the more specific the response, the better are the requirements of pharmaco-
metrics met. Dopamine receptors are found in some blood vessels and here DA
produces localized vasodilation (14). Suitable lengths of large blood vessels
are not apparently invested with sufficient DA-specific receptors to recommend
them as isolated tissue preparations (15), and it is necessary to rely on either
whole animal blood pressure (14) or regional blood flow in sensitive tissues such
as the kidney (16,17) as the test parameters. In either case the preparation is

not simple since an anaesthetized animal such as the cat or dog is required, and in the case of regional blood flow, electromagnetic flowmeters or some similar equipment is necessary. It is possible to perfuse the kidney in a manner similar to that used in the Langendorff isolated heart preparation (6) and observe the fall in perfusion pressure caused by DA. However this approach has not been widely reported in the literature. Since receptors to other catecholamines are present in blood vessels, it is necessary to pretreat animals, or bathe isolated tissues, with adequate blocking doses of phentolamine or propranolol in order to see a specific DA-mediated effect.

The most sensitive biological test for DA appears to be the inhibition of the electrically-induced contraction of the cardiac portion of the intestine of the tapestry cockle, *Tapes watlingi*, a bivalve mollusc from Australia (18,19). Other members of the Venus clam family (Veneridae) such as *V. verrucosa* or *Mercenaria mercenaria* would probably be suitable substitutes in areas where *T. watlingi* is not available. A test for specificity at DA receptors is, of course, to employ specific blocking agents. Substances which best meet such a requirement are butyrophenones such as haloperidol, a phenothiazine derivative such as chlorpromazine, or metoclopramide (19).

2.4.2 Acetylcholine and congeners

Acetylcholine (Ach) has been recognized as a bioactive amine for almost a century, and its assay by pharmacometric means is long established.

The preparations used vary in sensitivity and the ultimate choice is dependent on the amount of material available for assay. Acetylcholine receptors are classified as two broad categories, nicotinic and muscarinic. The nicotinic receptors are found in vertebrates in skeletal muscle and in autonomic ganglia. The possible assay preparations are limited, as previously indicated, to multiply innervated muscles. The rectus abdominis muscle of the frog and the biventer cervicis muscle of the chick are suitable in this regard, although their sensitivity is only moderate, even if the cholinesterase present in the tissue is first inactivated by treatment with up to 1 μg/ml diisopropylfluorophosphate (DFP) (6). Under these conditions, a threshold response to 10^{-7}M is barely discernible, and it is necessary to use more sensitive preparations. Intestinal smooth muscle is well supplied with muscarinic receptors and provides a suitable assay preparation which is sensitive to atropine, rather than tubocurarine, as a blocking agent. Guinea pig ileum or rabbit jejunum are both suitable tissues (6), but their sensitivity is only slightly better than the skeletal muscle preparations. However, the time for response and recovery is significantly shorter, and for this reason they are more convenient for routine assays. If greater sensitivity is required, the dorsal muscle of the leech provides a useful tissue which will respond to Ach at concentrations as low as 10^{-9}M if the muscle is pretreated with an anticholinesterase such as DFP (6). The disadvantage

of the leech muscle is that a sensitization period of up to three hours is recommended; thereafter the responses are slow in both developing and recovering.
The most sensitive tissue available is the heart of clam species (20,21). Acetylcholine slows the spontaneous beat of this tissue and reduces the amplitude of the
tension developed at concentrations as low as $3 \times 10^{-11}M$; the effect can be quantified to provide an exquisitely sensitive assay.

A point worth noting in relation to Ach is the ubiquity of various cholinesterase enzymes in tissues. Tissue extracts are therefore prone to be cholinolytic
unless the enzyme is inactivated with physostigmine or DFP. Unlike many preparations used to assay the catecholamines, the response elicited by Ach is the classical contracture or at least development of tension. It commends itself to bioassay procedures for the reasons outlined earlier.

2.4.3 Histamine and congeners

The classical tissue used in the assay of histamine (HA) is the guinea pig
ileum (6) or strips of longitudinal muscle from the ileum (22). Threshold concentrations of $10^{-8}M$ are capable of eliciting a contractile response in guinea pig
tissues since this species is particularly sensitive to the amine. The response
resembles that to Ach in that there is rapid contracture of the tissue, and the
recovery after removal of the drug is rapid. It forms an ideal bioassay system
if the amount of material to be assayed is sufficient. Since Ach produces an
identical response, assays are best carried out in the presence of blocking concentrations ($10^{-6}M$) of atropine, and the specificity of the response can be
determined by using a drug such as chlorpheniramine or mepyramine (pyrilamine)
which will block the H_1 receptors present. While HA itself is equipotent at both
H_1 and H_2 receptors, congeners do show specificity for one or other type. H_2
receptors are recognized as mediating gastric acid secretion in vertebrates;
however, this response is not readily amenable to bioassay. Most blood vessels
have both H_1 and H_2 receptors present and bioassay with these tissues is not
simple and unambiguous. The atria of the guinea pig heart manifest a useful
separation of the two receptor types (23), and only the H_2 receptors are found in
the right atrium. There they mediate a chronotropic effect which provides the
basis of a positive useful biological assay. The threshold concentration is about
$10^{-7}M$ in this preparation. Specificity can be confirmed by using an H_2-blocking
drug such as cimetidine or ranitidine.

Histamine occurs in the tissues of invertebrates as widely as in vertebrates,
and it is certainly a neurotransmitter in the nervous system of gastropod molluscs.
No suitable invertebrate tissue, however, has been developed as a pharmacometric
preparation with the same degree of sensitivity as the clam heart.

2.4.4 Serotonin and congeners

The standard pharmacometric assay for 5-hydroxytryptamine (serotonin; 5-HT) makes use of the fundic strip from rat stomach (6,24). Receptors to serotonin are found throughout the alimentary canal, but the rat is a particularly sensitive animal. The tissue is slow in responding to spasmogens, and a moderately long time cycle of 6 minutes is required. In addition, the tissue must be passively stretched after removal of the drug, since it does not relax spontaneously. Adrenaline can be used as a pharmacological relaxant in preference to mechanical stretching. The threshold concentration of serotonin in this preparation is around 10^{-9}M, and so relatively small amounts of material are sufficient for a satisfactory (if slow) assay. The receptor specificity of the spasmogen present in an unknown solution can be confirmed using methysergide (10^{-7}M) or cyproheptadine (10^{-6}M) as the standard blocking agents. Blockade in this case is slow in developing and must be followed over a 60 minute period. Since the rat stomach, like all alimentary tissues, is richly supplied with muscarinic cholinoceptors, and in all likelihood tissue extracts will contain some Ach, the assay of tissue extracts is best carried out in the presence of atropine. Rat stomach is only weakly sensitive to HA and so protection with antihistamines is unnecessary. Owing to the extreme sensitivity of this readily available mammalian preparation, there has been little interest in developing alternative assays. The clam heart is sensitive to serotonin, responding with increases in amplitude and frequency. It is, however, not as sensitive as the rat stomach strip, with a threshold concentration in excess of 10^{-8}M, and there is little advantage to its adoption (25).

2.4.5 Assay of complex mixtures

It is rare that only one biologically active substance is present in a tissue eluate or extract. Such a situation is more in the nature of a student laboratory exercise than part of routine physiological research. Many active substances coexist in tissues, even within a single cell. Since the biogenic amines, by definition, share a common chemical property, they tend to be extracted as a group during routine chemical extractions (see following chapters). It may be preferable to assay small volumes containing mixtures of the amines in trace amounts, rather than risk losing them in the course of further chemical separation. To this end the superfusion technique (26,27), developed originally to identify active substances in blood has proved to be invaluable. In this technique, the standard bath containing warmed, aerated solution bathing the tissue is replaced by a jacketed vessel, kept warm by circulating water. The tissue is mounted in the vessel as before, but the fluid is trickled from a small orifice, such as a cannula, down over the surface of the tissue and drained through a hole or spout in the base of the vessel. The volume of fluid in contact with the tissue at any time is very small, and economies of material are thus possible. By arranging several such preparations above each other, the same fluid can be brought into contact with a number of different tissues as it drips from the drain of one vessel on to the next lower tissue. The differences in response can then be noted. Such an arrangement is known as cascade superfusion, and is important where the substances in the mixture are labile, or are so alike chemically as to make routine chemical separation and analysis difficult and fraught with the risk of losing the sample in the process. The general strategy to be followed is to use tissues uniquely possessed of one or other receptor type. Adrenaline and NA, for example, are approximately equiactive at α-adrenoceptors, but A is considerably more potent at β_2 receptors. Mixtures of the two can be analyzed using rat fundic strip and chick rectum, both tissues being contracted by the presence of 10 $\mu g/l$ serotonin (28). As solutions containing A perfuse over the tissues, they respond approximately equally. However when NA is superfused, only the stomach responds vigorously. The chick rectum, having predominantly β_2 receptors, is much less sensitive to NA. The concentrations of the two substances present can be determined either mathematically or graphically (29,30).

For the mathematical solution let us continue with the supposition that the mixture contains both A and NA in unknown concentrations.

The concentration of pure A solution (prepared from a standard stock solution) which produces the same response as the mixture can be determined experimentally. Likewise can the equivalent NA solution be determined. If two tissues are chosen which possess predominantly α_1 or β_2 receptors, then the A equivalents in these tissues are A_α and A_β and the NA equivalents are NA_α and NA_β. The ratios of the equivalents are therefore

$$R_\alpha = \frac{NA_\alpha}{A_\alpha} \quad \text{and} \quad R_\beta = \frac{NA_\beta}{A_\beta}$$

The A equivalent of the mixture is the sum of the equivalents of the two substances present, i.e.

$$A_\alpha = A + \frac{NA}{R_\alpha} \quad \text{and} \quad A_\beta = A + \frac{NA}{R_\beta}$$

therefore
$$NA = R_\alpha R_\beta \frac{(A_\alpha - A_\beta)}{(R_\beta - R_\alpha)}$$

and
$$A = \frac{A_\beta R_\beta - A_\alpha R_\alpha}{R_\beta - R_\alpha}$$

Normally R_α is negligible in comparison to R_β

then NA $= (A_\alpha - A_\beta) \, R_\alpha$

and A $= A_\beta - \dfrac{A_\alpha \, R_\alpha}{R_\beta}$

The ratios of R_α and R_β are fixed for any one type of tissue and need not be related to the equivalents of each unknown mixture.

Since A is more potent at the β_2 receptor than NA, it follows that once the ratios of R_α and R_β are known, only the A equivalent need be determined and the above equations used. If, however, A comprises less than 10% of the mixture, its more specific effect is less discernible, and NA may be used as the standard solution for determining NA_α and NA_β. In this case the equations can be derived to give:

A $= \dfrac{(NA_\beta - NA_\alpha)}{R_\beta}$ and   NA $= NA_\alpha - \dfrac{NA_\beta \, R_\alpha}{R_\beta}$

The graphical method of solution is to mark on a linear scale the points A_α and NA_α where the ordinate and abscissa are calibrated as A and NA concentrations. When these two values are joined by a straight line, then the co-ordinates of all points on that line represent the concentrations of A and NA in the mixture, which would have the same effect in tissue α. If a similar exercise is carried out with tissue β and a straight line drawn between A_β and NA_β, the intersection of the two straight lines is that point of equivalence in both tissues, and from its co-ordinates the values of A and NA can be read.

All of the foregoing argument presupposes that the two substances in the mixture are acting on the same receptor in a competitive manner. Such would be the case with A, NA and DA, and in this case it can be ensured by carrying out the determinations in the presence of metoclopramide to block any DA receptors present. Although the other biogenic amines may produce similar responses to each other, they are mediated by different receptors and the contribution of any single amine in a mixture can be determined simply by adding the appropriate specific blocking drug to the perfusate and noting the reduction in response.

2.5   CONCLUSIONS

The examples provided in this chapter are not intended as a comprehensive or definitive list of techniques used in the biological assay of active substances. Instead they are provided as an indication of the ingenuity which has been applied to the investigation of such substances, an ingenuity which still finds expression in various ways. As the list of known or suspected active substances lengthens, so too does the list of ways in which they can be assayed. Comparative pharmacology is in its formative stages, and the appreciation of the sensitivities of tissues from non-vertebrate sources is far from complete. It is reasonable to predict, therefore, that biological assay methodology will remain an integral part of the investigation of biologically active substances for many years to come.

REFERENCES

1 British Pharmacopoeia, 1980, Appendix XIV O HMSO, London, pp. A156-A169.
2 United States Pharmacopeia XX, 1979, U.S.P. Convention, Rockville, pp. 891-900.
3 European Pharmacopoeia vol. II, 1971, Maisonneuve S.A., Sainte-Ruffine, pp. 49-5
4 D. Colquhoun, Lectures on Biostatistics, Oxford University Press, London, 1971, 425 pp.
5 D.R. Waud, Analysis of Dose-Response Curves, in E.E. Daniel and D.M. Paton (Eds. Methods in Pharmacology, vol. 3, Plenum Press, New York, 1975, pp. 471-506.
6 Staff of the Department of Pharmacology, University of Edinburgh, Pharmacologica Experiments on Isolated Preparations, Churchill Livingstone, Edinburgh, 1970, 163 pp.
7 J.R. Neely, H. Liebermeister, E.J. Battersby and H.E. Morgan, Am. J. Physiol., 212 (1967) 804-814.
8 R.Y. Moore and F.E. Bloom, Ann. Rev. Neurosci., 2 (1979) 113-168.
9 D.F. Sharman, in R. Fried (Ed.), Methods of Neurochemistry, vol. 1, Marcel Dekker, N.Y., 1971, p. 102.
10 I.S. de la Lande and M.J. Rand, Aust. J. Exp. Biol. Med. Sci., 43 (1965) 639-659
11 J.R. Vane, Pharmacol. Rev., 18 (1966) 317-324.
12 J. Hughes, Br. J. Pharmacol., 44 (1972) 472-491.
13 J.V. Levy, Isolated atrial preparations, in A. Schwartz (Ed.), Methods in Pharmacology, vol. 1, Appleton-Century, Crofts, New York, 1971, pp. 77-104.
14 T.C. Hamilton, Br. J. Pharmacol., 44 (1972) 442-450.
15 M.J. Kelly, Br. J. Pharmacol., 46 (1972) 575P-577P.
16 L.I. Goldberg, Pharmacol. Rev., 24 (1972) 1-29.
17 L.I. Goldberg, P.H. Volkman and J.D. Kobli, Ann. Rev. Pharmacol. Toxicol., 18 (1978) 57-79.
18 D.F.H. Dougan and J.R. McLean, Comp. Gen. Pharmacol., 1 (1970) 33-46.
19 D.F.H. Dougan, P.T. Mearrick and D.N. Wade, Clin. Exp. Pharmacol. Physiol., 1 (1974) 473-478.
20 E. Florey, Comp. Biochem. Physiol., 20 (1967) 365-377.
21 G.A. Cottrell, B. Powell and M. Stanton, Br. J. Pharmacol., 40 (1970) 866-870.
22 H.P. Rang, Br. J. Pharmacol., 22 (1964) 356-365.
23 S.C. Verma and J.H. McNeill, J. Pharmacol. Exp. Ther., 200 (1977) 352-362.
24 J.R. Vane, Br. J. Pharmacol., 12 (1957) 344-349.
25 G.C. Chong and J.W. Phillis, Br. J. Pharmacol., 25 (1965) 481-496.
26 J.H. Gaddum, Br. J. Pharmacol., 8 (1953) 321-326.
27 J.R. Vane, Br. J. Pharmacol., 23 (1964) 360-373.
28 A.K. Armitage and J.R. Vane, Br. J. Pharmacol., 22 (1964) 204-210.
29 J.H. Gaddum and F. Lembeck, Br. J. Pharmacol., 4 (1949) 401-408.
30 E. Marley and W.D.M. Paton, J. Physiol. (Lond.), 155 (1961) 1-27.

Chapter 3

THIN-LAYER CHROMATOGRAPHY OF BIOGENIC AMINES

R. ANTHONY LOCOCK

Faculty of Pharmacy and Pharmaceutical Sciences, University of Alberta, Edmonton, Alberta   T6G 2N8 (Canada)

3.1   INTRODUCTION

3.1.1   Brief history of the development of thin-layer chromatography (TLC) and its application to the separation of biogenic amines

The origin of the chromatographic technique now familiarly known as thin-layer chromatography (TLC) can be traced to the work of Izmailov and Shraiber (1). These Russian investigators used loose layers of alumina on glass plates for the separation of the constituents of belladonna, digitalis and rhubarb tinctures by a process they called "spot chromatography". This development pre-dates, by six years, the publication by Consden et al. (2) in 1944 concerning paper chromatography, which became an important separation technique in biochemical analysis for the next fifteen years.

The present-day system of TLC was introduced by Kirchner and co-workers (3) in 1951, who used glass strips coated with adsorbents bound by starch, "chromatostrips", for the analysis of terpenes in volatile oils. The general utility of TLC was not recognized until Stahl described the first practical equipment for preparing thin-layers on a standardized basis and applied TLC to many diverse separation problems (4,5). Stahl has edited a comprehensive general reference book on TLC (6).

More recent compilations of TLC information include those of Macek (7) and Kirchner (8). A recent textbook which describes all the practical features of TLC has been written for the beginning investigator by Touchstone and Dobbins (9).

The early TLC separations of biogenic amines were accomplished on cellulose thin-layers using mobile phases which had been adapted by analogy from paper chromatographic methods. The biogenic amines were not derivatized and visualization on the thin-layer was achieved by spray reagents which had been used successfully on paper chromatograms. As the need developed for more sensitive and specific analyses for the biogenic amines in biological samples, silica, alumina and polyamide were used as stationary phases and the free amines or amine derivatives were chromatographed. The resulting separations were analyzed by in situ measurement of absorbance or fluorescence or radioactivity or the separated amines and related compounds were removed from the thin-layer and the absorbance, fluorescence, radioactivity or mass spectrum determined for each

component. Important attributes inherent in all TLC methods are the ability to
carry out simultaneous separations of a number of samples and the wide choice of
detection methodology.

From the late 1950's until the early 1970's TLC was an important analytical
separation technique which used a liquid mobile phase. In recent years, with the
development of more efficient high performance (pressure) liquid chromatographic
(HPLC) methods, it has become less popular. In this chapter, the TLC methods for
analyzing biogenic amines will be reviewed and some of the new developments in
TLC technology will be described. High performance thin-layer chromatography
(HPTLC) has been introduced recently (10) and may offer an alternative to existing
gas-liquid chromatographic and HPLC analytical methods for biogenic amines.

3.1.2 High performance thin-layer chromatography (HPTLC)

HPTLC is a relatively new method which follows the pioneering work of Stahl
and has been referred to as the "second-generation of thin-layer chromatography"
(10). The HPTLC designation was chosen by analogy to the more familiar term
HPLC. HPTLC is faster, more sensitive and efficient, uses less solvent and can
accommodate a larger number of samples per plate than conventional TLC. Two
styles of equipment are used. HPTLC plates may be developed linearly in
miniaturized TLC chambers or in an automated radial U chamber, (see Fig. 3.1)
which may be operated conventionally in a circular or radial fashion whereby
solvent is fed to the center of the plate and migrates outwards or in an anti-
circular mode where the mobile phase is applied to the thin layer along a precise
outer circle, from where it flows over the initial zone toward the center.

HPTLC does offer superior resolution compared to conventional TLC and the
absolute detection limits are at least ten times lower than conventional TLC.
Janchen and Schmutz (11) have evaluated TLC and linear and circular HPTLC by
in situ reflectance measurements for the benzodiazepines, corticosteroids and
polycyclic hydrocarbons. In all cases the circular mode of operation of HPTLC
was advantageous.

In HPTLC the samples are applied to the thin-layer plate in volumes of less
than 1 µl and the resulting spot size should be 1-2 mm in diameter. Special
sample applicators such as controlled motion drum syringes are available (12).
The reduction in spot size creates an increase in sample concentration on the
plate and care must be taken to avoid overloading the HPTLC system. The detectability with HPTLC is in the nanogram to picogram range.

The use of optimized thin-layer materials with superior optical and chromatographic properties is necessary for HPTLC. Such materials are available as
preformed precoated plates. These plates have a smaller particle size (5 µm
average), a narrow particle size distribution and are usually 200 µm thick.
Silica gel has been the most popular stationary phase material for HPTLC.

Recently, chemically bonded stationary phases (RP-2, RP-8, RP-18 and C_{18}) have become available (Whatman Inc., New Jersey, and E. Merck, Darmstadt). These plates, in which hydrocarbon chains are bonded to the silica gel support, will permit the use of "reverse phase" TLC systems for the separation of hydrophobic compounds. Reverse phase thin-layer chromatography (RPTLC) may take place under conditions comparable to HPLC on bonded phase columns (13).

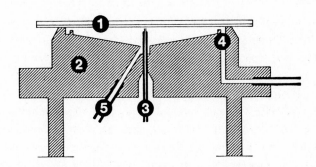

Fig. 3.1 Cross sectional diagram of the CAMAG U-Chamber System. The HPTLC plate (1), measuring 50 x 50 mm, rests with its layer facing down on the U-chamber body (2). Mobile phase is fed to the center of the plate via a platinum-iridium capillary (3) of 0.2 mm internal diameter. An external vapour phase may be passed into the chamber through the circular channel (4) and out through the center base (5) or the gas flow may be reversed. (Courtesy of CAMAG, Muttenz, Switzerland)

A final characteristic common to HPTLC separations is the use of a sophisticated data acquisition system. Quantitative analysis by direct optical scanning of the HPTLC plate with an automatic photodensitometer is possible and avoids the time-consuming and inefficient steps of elution, sample concentration and measurement. Touchstone and Sherma (14) have recently written a comprehensive textbook on densitometry in thin-layer chromatography.

In addition to the book edited by Zlatkis and Kaiser (10), Halpaap and Ripphahn (15,16) and Jupille (17) have also reviewed the principles and performance of HPTLC. The potential separation efficiency by HPTLC is now comparable to that achieved by gas chromatography and HPLC (18).

3.1.3 Multiple development in thin-layer chromatography

If volumes larger than 1 µl are required to be applied to the TLC plate, as is the case for residues and biological extracts, the separation can best be accomplished by multiple development. Multiple development, or unidimensional multiple development, is repeated development of the TLC plate by the same mobile phase in the same direction and for the same distance. Multiple development has been investigated by Thoma (19), who published tables which may be used to determine the number of repeated developments necessary to separate solutes of given R_f values. Programmed multiple development is the repeated development of a TLC plate with the same mobile phase in the same direction for gradually increasing distances (20,21). Each mobile phase advance is followed by solvent removal by an inert gas and/or heating the plate. Each plate development is programmed by a controller which governs all aspects of development and solvent removal. This equipment for HPTLC is available from the Regis Chemical Co. (Morton Grove, Ill.).

An important characteristic of multiple development of the thin-layer plate is the effect of this technique on spot concentrations. As the spots of a two component mixture migrate on a thin layer plate the resolution of the separation will increase as the distance of development of the plate increases. The separation may be expected to double as the development distance doubles. However, this is inefficient since an increase in distance of development also allows increased diffusion of the spots to occur, and doubling the distance of development will quadruple the time of development. Diffusion or spreading of the spot results in decreased sensitivity. A series of short developments (multiple development) of the plate is an attempt to overcome these problems. As development is repeated, the spots are concentrated and an original circular spot assumes an oval or streaked appearance since the developing solvent contacts the lower area of this circular spot first and distorts this to a flat-line shape before reaching the rest of the circular spot. This distortion leads to a "tight" narrow streak of material which for a given separation indicates increased efficiency and sensitivity which is proportional to the amount of spot/area of the spot. Improved resolution results from decreased spot size due to spot reconcentration, and large quantities of sample may be applied to the TLC plate (21). A possible application of multiple development techniques is in the separation of 1-dimethylaminonaphthalene-5-sulfonyl (dansyl or DNS) derivatives of biogenic amines, which as will be discussed later in this chapter, have often required repeated TLC development.

A related development technique uses a unique TLC development chamber (Regis Chemical) that permits plates to be developed continuously over short distances. The SB/CD chamber (short bed/continuous development) allows a mobile phase to move rapidly up the plate, which extends out of the chamber. When the mobile phase front emerges from the chamber it evaporates and this gives a continuous rapid flow of solvent up the plate. The shorter the mobile phase path, the higher the mobile phase velocity. Lower mobile phase solvent strength systems may be used with the SB/CD chamber, which may improve resolution, and slow-moving components in low solvent strength mobile phases may be moved from the spot origin. Figure 3.2 presents an illustration of the application of this technique to the separation of DNS derivatives of noradrenaline (NA) and 5-hydroxytryptamine

(5-HT; serotonin).

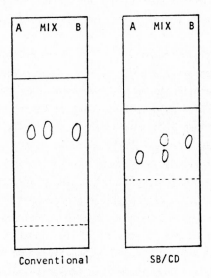

Fig. 3.2 The separation of DNS derivatives of NA (A) and 5-HT (B) on silica gel
by a conventional solvent system (ethyl acetate, cyclohexane, 6:4) and by the
mobile phase with 2 parts hexane added in the SB/CD Chamber. (Courtesy of the
Regis Chemical Co., Morton Grove, Ill.)

3.1.4 <u>Recent developments in TLC technology</u>

Recent developments in the technology of TLC systems are biennially reviewed
by Zweig and Scherma (13,18,22). A cumulative bibliography of the TLC literature
covering the years 1973 to 1977 is available (23). The Camag Bibliography
Service (24) enables one to quickly survey recent TLC literature.

3.2 TLC AS APPLIED TO THE BIOGENIC AMINES

Thin-layer chromatographic systems for biogenic amines and related compounds
are presented here in four tables. Table 3.1 lists methods which have been used
for underivatized biogenic amines. Table 3.2 gives methods for 3-O-methyl
catechol derivatives formed by the action of the enzyme catechol-O-methyl
transferase and radiolabelled S-adenosyl-L-methionine. Table 3.3 presents a
bibliography of TLC systems for DNS derivatives of biogenic amines, and
Table 3.4 is a list of systems for other derivatives of biogenic amines. In
each table, the references are listed chronologically, i.e. lower numbers in
each table refer to early work from the literature while higher numbers refer
to more recent studies.

3.2.1 <u>TLC systems for underivatized biogenic amines</u>

TLC systems for underivatized biogenic amines (25-63) are presented in

Table 3.1. Early systems were adapted from paper chromatographic techniques; therefore cellulose has been used most frequently as the stationary phase (26,28,29-36,41,43,45,47,49,50,51). The next most popular stationary phase is silica gel.

A typical TLC separation of biogenic amines using cellulose as a stationary phase is described by Schneider and Gillis (28). These workers demonstrated that TLC had "definite application as a convenient and reliable investigational tool in biochemical or pharmacological research involving catecholamines" (28). Schneider and Gillis in 1965 (28) studied the biosynthesis of NA from its precursors dihydroxyphenylalanine (DOPA) and dopamine (DA). In a later study (1976) which also used cellulose as a stationary phase for the separation of DA metabolites, Stout et al. (51) found that TLC gave poor recovery (30 - 50%) and poor precision in quantitation of labelled radioisotopes of the catecholamines. The investigators developed HPLC methods which were systems of choice for the quantitative analysis of DA metabolites in physiological fluids (51).

In a recent publication (63), the TLC of underivatized DA metabolites on silica gel is described, and this serves to illustrate the general application of TLC in biochemical analysis. After incubation in appropriate media with ^{14}C-DA or ^{14}C-NA, bovine retinas were centrifuged, and the supernatant was applied to silica precoated plates. Chromatography in two mobile phase systems was carried out and small amounts of carrier DA, NA, 3-methoxytyramine (3-MTA), dihydroxyphenylacetic acid (DOPAC) and homovanillic acid (HVA) were added to the spots on chromatograms or chromatographed in parallel. The standards were identified by spraying with ninhydrin. Autoradiograms of the chromatograms were prepared. Substances were eluted from the silica gel with methanol, water (1:1). Eluates were dried and counted in a scintillation counter. The percentage of radioactive substances was calculated. Recovery in the procedure was found to be 85% (63).

Visualization of the biogenic amines after thin-layer chromatography has been accomplished by several methods. Ninhydrin is often used for this purpose. Diazotized p-nitroaniline and potassium ferricyanide followed by ethylenediamine are also common spray reagents for catecholamines. de Potter et al. (29) found that p-nitroaniline reagent offered the most constant sensitivity for NA and its metabolites. Formulations for all spray reagents are given in the text by Kirchner (8). Aures et al. (36) have studied various detection reagents for imidazoles, indoles and catecholamines and have compared the fluorescence characteristics of indoles and phenols on silica gel after treatment with reagents. Nanogram amounts of biogenic amines could be detected by exposing the thin-layer to o-phthalaldehyde spray or paraformaldehyde gas (36). Cowles et al. (39), using paraformaldehyde and heat, were able to detect 1 ng of serotonin. Osborne (44) lists the minimum detectable

TABLE 3.1

Thin-layer systems for underivatized biogenic amines and related compounds

Compound	Stationary Phase	Solvent System	Detection	Reference
noradrenaline (NA), adrenaline (A), normetanephrine (NMN), metanephrine (MN), 4-hydroxy-3-methoxymandelic acid (VMA), dopamine (DA), 5-hydroxy-tryptophan (5-HTP), 5-hydroxy-tryptamine (5-HT)	silica	1) n-butanol, pyridine acetic acid (70:20:10) 2) isopropanol, ethyl acetate, ammonia (35:45:20)	1) ninhydrin 2) 2,6-dichloroquinone chloroidimide 3) diazotized p-nitro-aniline	(25)
NA, A, histamine (HA), amphetamine, β-phenylethylamine (PEA), ephedrine, isoprenaline	1) cellulose 2) cellulose treated with diazomethane 3) silica	n-butanol, acetic acid, water (4:1:5)	ninhydrin	(26)
NA, A, NMN, MN, VMA, 3,4-dihydroxymandelic acid (DOMA), DA, 4-hydroxy-3-methoxyphenyl-acetic acid (HVA), 3,4-dihydroxy-phenylacetic acid (DOPAC)	polyamide	isobutanol, acetic acid, cyclohexane (80:7:10)	1) diazotized p-nitroaniline 2) ethylenediamine	(27)
3,4-dihydroxyphenylalanine (DOPA), NA, DA	cellulose	2 dimensional development: First) methanol, n-butanol, benzene, water (4:3:2:1) Second) acetone, t-butanol, formic acid, water (180:180:1:39) Both systems contain 0.01% EDTA	potassium ferricyanide, ethylenediamine: fluorescence under long wave UV	(28)
NA, A, NMN, MN, VMA, DOMA	cellulose	n-butanol saturated with 3N HCl and other systems	1) potassium ferricyanide, ethylenediamine 2) diazotized p-nitro-aniline	(29)

TABLE 3.1 (continued)

Compound	Stationary Phase	Solvent System	Detection	Reference
NA, A, DA	cellulose alumina silica	1) n-butanol, acetic acid, water (4:1:5) organic layer 2) n-amyl alcohol, acetic acid, water (4:1:5) 3) methanol, acetone, triethylamine (50:50:1) 4) dimethylformamide 5) phenol, water (8:2)	potassium ferricyanide, sodium hydroxide	(30)
DOPA, NA, A, NMN, MN, DA, amphetamine, ephedrine, isopropylarterenol, nordefrin, phenylephrine	cellulose silica alumina	1) phenol containing 15% 0.1N HCl 2) n-butanol, acetic acid, water (4:1:5) organic layer	1) potassium ferri-cyanide, sodium hydroxide 2) ninhydrin	(31)
NA, NMN, 4-hydroxy-3-methoxy-phenylethyleneglycol (MOPEG), 3,4-dihydroxyphenylethylene-glycol (DOPEG), DOMA, VMA	cellulose	n-butanol, 5N acetic acid (100:35)	diazotized p-nitro-aniline; scintillation counting of H^3-NA and metabolites	(32)
A, adnamine, adrenochrome, diadrenaline ether	cellulose, methylcellulose, silica, alumina	phenol containing 15% 0.1N HCl	potassium ferricyanide, sodium hydroxide	(33)
DOPA, NA, DA	cellulose	n-butanol, ethanol, 1N acetic acid (35:10:10)	potassium ferricyanide, ethylenediamine: fluorescence	(34)
DOPA, NA, NMN, MN, DA, amphetamine, ephedrine, isopropylarterenol, nordefrin, phenylephrine	cellulose	n-butanol, acetic acid, water (4:1:5) organic layer	1) diazotized p-nitro-aniline (amphetamine) 2) ninhydrin (DOPA, DA, ephedrine) 3) potassium ferricyanide, sodium hydroxide (others)	(35)

TABLE 3.1 (continued)

Compound	Stationary Phase	Solvent System	Detection	Reference
HA and various related imidazole derivatives; 5-HT and related indoles; catechol-amines and biogenically related compounds	silica cellulose	1) n-butanol, acetic acid, water (15:3:5) 2) n-butanol, pyridine, acetic acid, water (15:2:3:5) 3) n-butanol, acetic acid (1:1) 4) ethanol, diethyl ether, ammonia, water (10:10:1:4) 5) 8% NaCl in water 6) chloroform, methanol, ammonia (12:7:1) 7) ethyl acetate, acetic acid, water (15:15:10) 8) ethyl acetate, n-propanol, 10% ammonia (4:3:1) 9) n-butanol saturated with 0.1N HCl, 1st direction: isopropanol, 5N NH₄OH, water (8:1:1) 2nd direction	1) o-phthaldehyde 2) paraformaldehyde gas 3) other fluorescent detector reagents	(36)
NMN, MN, VMA, MOPEG	silica	1) n-butanol, ethanol, water (85:20:30) for VMA 2) n-butanol, ethanol, water, ammonia (100:20:30:1) for NMN, MN, MOPEG	VMA - fast red GG or B, sodium carbonate NMN, MN, MOPEG - 4-aminophenazone, potassium ferricyanide	(37)
N,N-dimethyl-β-phenylethanolamine N,N-diethyl-β-phenylethanolamine	silica	1) ethanol, water, ammonia (40:60:2) 2) benzene, acetone, ammonia (70:30:2)	UV fluorescence quenching Dragendorff's reagent	(38)
5-HT, N-methylserotonin	silica	1) n-propanol, ammonia (19:1) 2) n-butanol, acetic acid, saturated aqueous SO₂ (4:1:5) upper layer	paraformaldehyde and heat	(39)

TABLE 3.1 (continued)

Compound	Stationary Phase	Solvent System	Detection	Reference
DOPA, NA, A, NMN, MN, VMA, MOPEG, DOMA, DA, DOPAC, tryptophan (TP), 5-HTP, 5-HT, tyramine (TA), tyrosine (TYR), adrenochrome, homogentisic acid, phenylpyruvic acid, triiodotyrosine	polyamide	1) isobutanol, acetic acid, cyclohexane (80:7:10) 2) isopropanol, ammonia (4:1) 3) n-butanol, acetic acid, water (4:1:1)	1) diazotized p-nitro-aniline 2) diazotized sulfanilic acid 3) ethylenediamine sequential combinations of these detection agents	(40)
DOPA, NA, A, MN, VMA, DA, homovanillic acid (HVA), TA	cellulose	1) n-butanol, acetic acid, water (5:1:3) 2) ethyl acetate, acetic acid, water (5:1.5:3) 3) ethyl acetate, n-butanol, acetic acid, water (3:2:1:3) organic layer	Folin phenol reagent	(41)
NA, A, DA and the corresponding tetrahydroisoquinolines formed by condensation with formaldehyde and acetaldehyde	silica	sec-butanol, formic acid, water (15:3:2) in a nitrogen atmosphere	potassium ferricyanide, ferric chloride	(42)
DOPA, NA, A, NMN, MN, VMA, MOPEG, DOPEG, DOMA, DA, 3-methoxy-tyramine (3-MTA), HVA, DOPAC, octopamine (OA), TA, TYR	cellulose	2 dimensional development: First) 1-butanol, methanol, 1N formic acid (60:20:20) Second) chloroform, methanol, 1N ammonium hydroxide (60:35:5)	diazotized p-nitro-aniline	(43)
DOPA, NA, A, DA, 5-HTP, 5-HT	polyamide	1) methyl acetate, isopropanol, ammonia (9:7:5) 2) butanol, chloroform, acetic acid (4:1:1)	paraformaldehyde, UV light	(44)

TABLE 3.1 (continued)

Compound	Stationary Phase	Solvent System	Detection	Reference
DOPA, NA, NMN, VMA, DA, TP, 5-HTP, 5-HT, tryptamine (T), TYR, bufotenin, dimethyl-tryptamine, indoleacetic acid, 5-hydroxyindoleacetic acid, 5-methoxyindoleacetic acid, melatonin	cellulose	1) n-butanol, 5N acetic acid (100:35) 2) butan-2-one, acetone, 2.5N acetic acid (40:20:20)	diazotized p-nitro-aniline for catechol-amines p-dimethylaminocinnamic aldehyde for indoles	(45)
TP metabolites and related compounds	silica	2 dimensional development: First) propanone, 2-propanol, water, ammonia (50:40:7:3) Second) chloroform, methanol, acetic acid, water (65:10:20:5)	p-dimethylaminobenzalde-hyde	(46)
NA, A, DA	cellulose	n-butanol, acetic acid, water (4:1:1)	potassium ferricyanide, ethylenediamine	(47)
OA, TA, and 21 catecholamines	silica	chloroform, methanol, acetic acid, water (60:25:15:5)	UV, radioscanning	(48)
DOPA, NA, NMN, VMA, MOPEG, DOMA, DA, 3-MTA, HVA, DOPAC	cellulose	1) n-butanol, ethanol, 1N acetic acid (35:10:10) 2) methyl ethyl ketone, formic acid, water (24:1:6) 3) isopropanol, 5N ammonium hydroxide, water (8:1:1) 4) ethyl acetate, acetic acid, water (5:1.5:3)	potassium ferricyanide short wavelength UV	(49)
DOPA, NA, NMN, MOPEG, DA, HVA, OA, TA, TYR ([3H]-labelled)	cellulose	1) ethyl acetate, acetic acid, water (75:26:45) 2) n-butanol, 6N HCl, water (80:12:10) 3) chloroform, methanol, 1N ammonium hydroxide (60:35:5)	scintillation counting of [3H]-tyrosine and metabolites	(50)

48

TABLE 3.1 (continued)

Compound	Stationary Phase	Solvent System	Detection	Reference
NA, DA, 3-MTA, HVA	cellulose	n-butanol, acetic acid, water (100:20:60)	scintillation counting of radioactivity	(51)
NA, A, NMN, MN, isoprenaline, 3-0-methyl isoprenaline	silica impregnated with di-sodium tetra-borate	1) toluene, ethanol (1:1) for isoprenaline and 3-0-methyl isoprenaline 2) toluene, ethanol, water (10:10:1) for NA, A, and the corresponding 3-0-methyl catecholamines	1) diazotized p-nitro-aniline 2) potassium ferri-cyanide, ferrichloride	(52)
NMN, VMA, MOPEG, DOPEG, DOMA, H³-NA and corresponding metabolites	ECTEOLA-cellulose	n-butanol, ethanol (absolute), 0.5N acetic acid (35:7.5:20)	ethylenediamine, fluorescence	(53)
HA	silica	acetone, ammonia (95:5)	1) ninhydrin 2) Pauly's reagent	(54)
NA, NMN, VMA, MOPEG, DOPEG, DOMA	silica impregnated with sodium tetra-borate	n-butanol, ethanol, Tris buffer 10 mmol pH 8.0 (2:1:1)	ferric chloride and potassium ferricyanide	(55)
DA, TA, N-methyltyramine, gramine, hordenine, p-hydroxy-benzoic acid, p-hydroxyphenyl-acetic acid, 3,4-dihydroxy-phenylacetic acid, p-hydroxy-mandelic acid	silica	1) methanol, acetic acid (9:1) 2) ethanol, 2-butanone, ammonia (48:40:12) 3) n-butanol, ammonia (4:1) systems 1-3 for bases 4) toluene, ethyl formate, formic acid (5:4:1) 5) toluene, chloroform, acetone (40:25:35) 6) chloroform, acetic acid, water (60:35:5) 7) chloroform, acetic acid (9:1) 8) benzene, dioxane, acetic acid (5:4:1) 9) n-dibutyl ether, n-hexane, acetic acid (75:10:15)	1) Folin Ciocalteau reagent followed by Na₂CO₃ for phenols 2) ninhydrin 3) Dragendorff's reagent	(56)

TABLE 3.1 (continued)

Compound	Stationary Phase	Solvent System	Detection	Reference
HA and 19 other amines	silica	1) n-butanol, acetone, water (2:2:1), then chloroform, methanol, ammonia (12:7:1) 2) chloroform, methanol, ammonia (2:2:1) 3) methanol, ammonia (20:1) 4) acetone, ammonia (95:5)	1) ninhydrin 2) o-phthalaldehyde 3) fluorescamine 4) o-diacetylbenzene	(57)
NA, amphetamine, nordefrin, mescaline, methoxamine, norephedrine, norfenefrine, phentermine	silica	ethyl acetate, methanol, formic acid (69:30:1)	o-phthalaldehyde then fluorodensitometric estimation	(58)
melatonin and indole metabolites of N-acetylserotonin	silica	1) chloroform, methanol, acetic acid (90:10:1) 2) chloroform, methanol (90:10)	UV and scintillation counting	(59)
DOMA, DA, 3-MTA, DOPAC, TYR	silica	1) n-butanol, methanol, 1N formic acid (60:20:10) for amines 2) n-butanol, methanol, 1N formic acid (60:20:20) for acids	UV 254nm and scintillation counting	(60)
3-MTA and 4-methoxydopamine	silica	n-butanol, ethyl acetate, ammonia (60:20:20)	1) UV fluorescence 2) p-nitrobenzene di-azonium tetrafluoro-borate	(61)
HA, TP, T, agamatine, argenine, cadaverine, diaminopropane, phenylalanine, putrescine	ion exchange	1) Na citrate 2H$_2$O, NaCl (19.6:134.4) Na$^+$ = 2.5M and (19.6:58.4) Na$^+$ = 1.2M	cadmium, ninhydrin	(62)
NA, DA, 3-MTA, HVA, DOPAC	silica	1) n-butanol, water, acetic acid (12:5:3) 2) n-butanol, pyridine, acetic acid (15:2:4.5)	ninhydrin and autoradiography	(63)

amounts of biogenic amines and other related compounds after formaldehyde treatment of polyamide thin-layers. Six ng of DA, 5 ng of 5-HT and 7 ng of NA are the values for the detection limit of these neurotransmitters (44). Lieber and Taylor (57) have compared the specificity and sensitivity of four visualization reagents for histamine (HA). Ninhydrin was the most sensitive and fluorescamine and o-diacetylbenzene reagents were potentially the most specific (57).

3.2.2 TLC systems for 3-O-methyl derivatives of catecholamines

Table 3.2 lists TLC systems which have been used to separate 3-O-methyl derivatives of the catecholamines (64-69) and their metabolites (70). In radio-enzymatic procedures for analysis of catecholamines, adrenaline (A) and NA were converted to their respective metanephrines and DA was converted to 3-MTA by catechol O-methyl transferase in the presence of radiolabelled S-adenosyl-L-methionine. Martin et al. (69) then acetylated the products before TLC. Silica gel is used most often as the stationary phase, and the mobile phases are similar to mobile phases used for underivatized catecholamines. The references listed in Table 3.2 are for radioenzymatic determinations of biogenic amines. This topic is covered in Chapter 9 in this volume. TLC is used as a tool in radioenzymatic methods to obtain separation and specificity. Quantitation is achieved by the determination of the radioactivity of the separated derivatives of the catecholamines.

3.2.3 TLC systems for DNS derivatives of biogenic amines

Thin-layer chromatographic systems for l-dimethylaminonaphthalene-5-sulfonyl (dansyl; DNS) derivatives of biogenic amines are given in Table 3.3 (71-87). Also included in Table 3.3 are TLC systems for analogous dialkylaminonaphthalene-sulfonyl derivatives where the alkyl groups are methyl, ethyl, propyl, butyl and pentyl (88) and chromatographic systems for mixed DNS-acetyl derivatives (89).

Seiler (90) has reviewed the use of fluorescent derivatives of biogenic amines for thin-layer separations, and the formation of fluorescent derivatives for fluorometric determination or detection especially associated with chromatographic or electrophoretic separation has been reviewed by Seiler and Demisch (91). The DNS derivative of low-molecular weight compounds is the most important fluorescent derivative (84,91) and has been the derivative most studied by TLC. The reaction of DNS-Cl with primary amines, secondary amines and hydroxyl groups takes place at alkaline pH to give stable fluorescent compounds. An extensive description of the DNS reaction has been published by Seiler and Wiechmann (92). The limit of visualization for DNS derivatives of monoamines is approximately 10 - 20 ng; however, the destructive oxidation on silica gel gives irreproducible results at these low concentrations (86). An important use of TLC of the DNS derivatives of biogenic amines has been for the isolation of materials for mass

TABLE 3.2

Thin-layer systems for the separation of 3-0-methyl derivatives of catecholamines formed by the action of catechol-0-methyl transferase and labelled S-adenosyl-L-methionine (^{14}C-SAM, ^{3}H-SAM)

Catecholamines	Mobile Phases*	Detection	Reference
NA, A	n-butanol, formic acid, water (15:215:1.5)	short wave UV for location; elution, oxidation to vanillin, scintillation counting H^3 and C^{14}	(64)
NA, A	isopropanol, n-butanol, water, formic acid (60:20:19:1)	short wave UV, elution of spots corresponding to NMN and NM, oxidation to vanillin, scintillation counting	(65)
NA, A, DA	chloroform, ethanol, ethylamine 70% (16:3:2)	UV short wave for location; 3-MTA derived from 0-methylation of DA is eluted and radioactivity measured. NMN and MN are oxidized to vanillin and radioactivity counted after extraction	(66)
NA, A, DA	two dimensional development: First) chloroform, methanol, ammonium hydroxide (120:70:10) Second) ethanol, diethyl ether, water, ammonium hydroxide (150:90:48:12)	iodine vapour, removal of spots and scintillation counting	(67)
NA, A, DA	t-amyl alcohol, benzene, 40% methyl-amine (6:2:3)	UV, removal of spots, scintillation counting	(68)
NA, A, DA	toluene, methanol, water, ethyl acetate (10:5:5:4) upper phase (30 ml) plus ethyl acetate 4 ml, methanol 2 ml, butan-2-one 6.5 ml	concentrated ammonia, Folin-Ciocalteau reagent, scintillation counting	(69)
A, DOPEG, DOMA, DA, DOPET, DOPAC	2 dimensional development on cellulose: First) n-butanol, methanol, 1N formic acid (60:20:20) Second) chloroform, methanol, 1N ammonium hydroxide (60:35:5)	scintillation counting	(70)

* Stationary phase is silica unless otherwise noted.

TABLE 3.3

Thin-layer chromatographic systems for 1-dimethylaminonaphthalene-5-sulfonyl (DNS) derivatives of biogenic amines and related compounds.

Compound	Mobile Phase*	Detection	Reference
NA, A, DA, DOPEG, DOPET, N-acetyldopamine, α-methyl-dopamine, α-methylnoradrenaline, N-methyladrenaline, isoproterenol	2 dimensional development in 2 systems: First) chloroform - 2 times Second) butyl acetate, cyclohexane, ethyl acetate, triethylamine (50:50:20:20) - 2 times and First) diisopropyl ether, then butyl acetate, triethylamine (100:20) - 2 times Second) triethylamine, diisopropyl ether (100:20) - 2 times	UV fluorescence	(71)
TA	1) chloroform, n-butyl acetate (5:1) then 2) ethyl acetate, cyclohexane (3:2)	elution and scintillation counting	(72)
bufotenin	2 dimensional development: First) ethyl acetate, n-butyl acetate (5:1) then benzene, methanol (9:1) Second) triethylamine, chloroform (1:5)	extraction of the spot cochromatographing with DNS-bufotenin and mass spectrometry	(73)
TA, OA, synephrine	For separation of DNS derivatives of TA and synephrine: 1) chloroform, n-butyl acetate (4:1), elution of TA then 2) ethyl acetate, cyclohexane (3:2) For DNS derivative of OA 1) benzene, acetic acid, ethyl acetate (10:1:1) elution of OA then 2) ethyl acetate, cyclohexane (1:2)	elution and scintillation counting	(74)
T	2 dimensional development: First) nitroethane, cumene, diisopropyl ether (5:7:9) Second) pentan-1-ol, propan-2-ol, aqueous ammonia (45:35:10)	elution and scintillation counting	(75)
o-, m-, p-TA	2 dimensional development: First) cyclohexane, n-butyl acetate (8:3) - 3 times Second) n-butyl acetate, cyclohexane,	repeat TLC for separation of mono and di-DNS-TA isomers elution and mass spectro-	(76)

TABLE 3.3 (continued)

Compound	Mobile Phase	Detection	Reference
NMN, MN, HA, 5-HT, T, OA, TA, and some related amines	carbon tetrachloride, ethyleneglycol mono-methyl ether (85:15)	long wave length UV	(77)
some noncatecholic biogenic amines eg. TP, o-, m- and p-TA, OA and related compounds	1) ethyl acetate, cyclohexane (2:1) 2) benzene, triethylamine (8:1)	qualitative and quantitative mass spectrometry of pure derivatives	(78)
amines, amino acids and catecholamines	2 dimensional development on polyamide: For the amines and amino acids - First) water, formic acid (100:3) Second) benzene, acetic acid (9:1) Several different systems for catecholamines	autoradiography of ^{14}C-DNS derivatives	(79)
amino acids, DOPA, NA, A, DA, 5-HT, T, TA, putrescine	2 dimensional development on polyamide or silica systems in different combinations were used for amines 1) heptane, butanol, acetic acid (10:10:1) 2) benzene, methanol, cyclohexane (88.5:1.5:10) 3) heptane, butanol, formic acid (10:10:1) 4) ethyl acetate, methanol, acetic acid (20:1:1) 5) water, formic acid (92:1) 6) benzene, acetic acid (9:1)	UV, internal, external and radioactive standards, mass spectrometry	(80)
amino acids and 5-HT	2 dimensional development on polyamide: First) formic acid, water (3:100) Second) benzene, acetic acid (9:1) sometimes followed by ethyl acetate, methanol, acetic acid (20:1:1) in the same direction	UV monitoring, spot removal and scintillation counting or autoradiography (^{14}C-DNS derivatives), fluorescence spectrophotometry, mass spectrometry	(81)

54

TABLE 3.3 (continued)

Compound	Mobile Phase	Detection	Reference
m-, p-TA, T, β-phenylethyl-amine	1) chloroform, n-butyl acetate (4:1) 2) benzene, triethylamine (12:1) and (8:1) 3) benzene, methanol (10:1) 4) carbon tetrachloride, triethylamine (5:1) 2 or 3 systems used successively elute and transfer zones after each development	UV 365nm, quantitative mass spectrometry	(82)
m-, p-TA, T, β-phenylethyl-amine	1) chloroform, n-butyl acetate (4:1) 2) benzene, triethylamine (8:1) and (12:1) 3) carbon tetrachloride, triethylamine (5:1) 2 or 3 successive unidimensional separations	UV 365nm, quantitative mass spectrometry	(83)
HA, 5-HT, TA, phenethylamine, other amines and amino acids	Systems for the amines 1) ethyl acetate, cyclohexane (3:2) 2) benzene, triethylamine (5:1) and 2 dimensional development for the others First) benzene, cyclohexane, methanol (85:15:2), 2 times Second) diethyl ether, cyclohexane (3:1)	extraction and quantitative fluorimetry	(84)
NA, A, NMN, DA	For the separation of DNS amino acids from DNS amines: methyl acetate, 2-propanol, 25% ammonia (90:60:40) then two dimensional chromatography First) toluene, pyridine, acetic acid (150:15:0.5) 3 times Second) benzene, triethylamine, ammonia (100:20:0.1) then ascending paper chromatog-raphy on Whatman SG 81, paper in either 1) benzene, cyclohexane, methanol (90:10:2) or 2) benzene, ethyl acetate (65:35)	UV measurement or radioactivity	(85)

TABLE 3.3 (continued)

Compound	Mobile Phase	Detection	Reference
trace amines i.e. noncatecholic arylalkylamines, nonhydroxylated indolalkylamines, phenylethylamine, isomers of tryptamine and tyramine	unidimensional development in two or three systems containing mixtures of two or three of the following: toluene, chloroform, ethyl acetate, cyclohexane, methanol, benzene, triethylamine, n-butyl acetate, in various proportions depending on the particular DNS derivative	UV 365nm	(86)
OA from TA	2 dimensional development in two systems: 1) First) cyclohexane, benzene, methanol (2:17:1) Second) cyclohexane, ethyl acetate (1:1) 2) First) chloroform, triethylamine (3:1) Second) benzene, acetone (9:1)	in situ fluorescence	(87)
T, TA, phenylethylamine and O-alkylated TA derivatives**	1) benzene, triethylamine (5:1) 2) cyclohexane, ethyl acetate (3:1) 3) chloroform, ethyl acetate (4:1) 4) benzene, triethylamine (20:1) for O-alkyl derivatives	UV then mass spectrometry	(88)
NM, A, NMN, MN, m- and p-OA, phenylethanolamine, N-methylphenylethanolamine, m- and p-synephrine***	1) chloroform, ethyl acetate (6:1) and (2:1) 2) carbon tetrachloride, triethylamine (10:1) and (5:1) 3) chloroform, n-butyl acetate (5:2) and (5:3) 4) carbon tetrachloride, triethylamine, methanol (10:2:1) 5) benzene, triethylamine (5:2)	UV then mass spectrometry	(89)

* The stationary phase is silica unless otherwise noted.
** Dialkylaminonaphthalenesulfonyl derivatives where the alkyl groups are methyl, ethyl, propyl, butyl and pentyl.
*** DNS and acetyl derivatives are used.

spectrometry (73,76,78,80-83,93). Using a deuterium labelled analogue as an internal standard which is carried through all the steps of extraction, derivatization and TLC, as little as 100 pg of an endogenous amine in original tissue may be quantified by direct probe mass spectrometry (93).

DNS derivatives have been used in radiochemical assays for biogenic amines. In such assays, the tissue extract containing the amine of interest is incubated with radiolabelled dansyl chloride. The DNS derivatives are separated by TLC, the appropriate spots are then scraped from the TLC plates, and the radioactivity counted (75,79,85).

The TLC of DNS derivatives has nearly always employed silica gel as stationary phase. Polyamide has been used (79-81) and may offer some increase in sensitivity. However, polyamide thin-layers have a low capacity (84). Seiler and Knodgen (84) have evaluated the performance of HPTLC precoated silica plates for the separation of DNS derivatives. HPTLC was more rapid and sensitive than conventional TLC with regular silica plates. The sensitivity of the method allowed the determination of picomole quantities with good reproducibility when fluorometric methods were used (84).

A major disadvantage to the DNS derivatization and subsequent separation by TLC and quantitation is that the original DNS reaction itself is not "clean", especially with multifunctional catecholamines such as NA. As can be seen by examination of the mobile phases listed in Table 3.3, multiple development in one or two dimensions with the same or different mobile phases is the rule rather than the exception for the separation of DNS derivatives of biogenic amines. The use of internal standards does not always solve the quantitation problem since the DNS reaction may be altered by extraneous compounds (79). In an attempt to solve this situation, DNS-Cl is used in stoichiometric excess (100x) to completely derivatize all possible amines, but the problem remains to remove excess DNS-Cl (94).

3.2.4 TLC systems for other derivatives of biogenic amines

TLC systems for other derivatives of biogenic amines are included in Table 3.4.

The acetyl derivatives (95,99,102) of the catecholamines are substances with more hydrophobic character and thus mobile phases which are less polar than those used for non-acetylated amines may be used for TLC on silica. Acetylation also increases the stability of the catecholamines and protects them from degradation by oxidation at alkaline pH (98,102,105). An example of a recently reported TLC system for acetylated catecholamines is that of Gelijhens and de Leenheer (102). After separation of free urinary catecholamines on a boric acid gel column, they acetylated the column eluate and spotted a dichloromethane solution of the acetylation mixture on a HPTLC plate. After development, the plates were sprayed with ethylene-diamine and potassium ferricyanide spray reagent, then heated, and

TABLE 3.4

Thin-layer systems for other derivatives of biogenic amines

Amines	Derivative	Mobile Phase*	Detection	Reference
NA, A, other catechol-amines and 5-HT	acetyl	1) chloroform, methanol (9:1) 2) cyclohexane, chloroform, methanol, acetic acid (3:5:1:1)	1) phosphoric acid, phosphomolybdic 2) vanillin, sulfuric acid	(95)
NA, A, DA and pharmaceutical phenyl-ethylamines	tetraphenylborate	n-butanol, acetic acid, water (4:1:5) (compounds move as their parent bases)	ferric chloride, potassium ferricyanide	(96)
m-OA, p-OA, PEOH	N-methyl deriva-tives by phenyl-ethanolamine-N-methyl transfer-ase and ³H-SAM, followed by dansylation	1) chloroform, n-butyl acetate (5:2) 2) toluene, triethylamine, methanol (50:5:1) 3) cyclohexane, ethyl acetate (25:35)	uv detection, elution and scintillation counting	(97)
NMN	N-methyl deriva-tive by phenyl-ethanolamine-N-methyl transfer-ase and ³H-SAM	t-amyl alcohol, benzene, methylamine (60:20:30)	short-wave uv, then elution and scintillation counting	(98)
NA, A, adnamine, adrepine	acetyl	1) chloroform, acetone (1:1) 2) cyclohexane, acetone (3:7) 3) chloroform, pyridine (9:1)	1) ammonia 2) phosphoric acid, phospho-molybdic acid	(99)
primary catecholamines, their 3-0-methyl derivatives and related phenylethylamines	fluorescamine	1) ethyl acetate, n-hexane, methanol, water (60:20:25:10) 2) chloroform, isopropanol, water (2:8:1) 3) n-butanol, acetic acid, water (5:2:3) 4) benzene, dioxane, acetic acid (2:5:1)	spray with 70% perchloric acid observe fluorescence	(100)

TABLE 3.4 (continued)

Amines	Derivative	Mobile Phase*	Detection	Reference
5-HT, T	^{3}H-methyl derivatives of TP metabolism from ^{3}H-SAM and rat pineal gland	2 dimensional development: First: chloroform, methanol, acetic acid (94:4:3) Second) chloroform, methanol, ammonia 25% (60:35:5)	UV, then scintillation counting	(101)
NA, A, DA	acetyl	acetone, dichloromethane, acetic acid, methanol (20:80:4:1)	fluorescence	(102)
HA	1-N-methyl HA by histamine-N-methyl transferase and ^{3}H-SAM	chloroform, methanol, 70% ethylamine (32:6:4)	iodoplatinate	(103)
N$^\alpha$-methylhistamine	N$^\alpha$, N$^\tau$ dimethyl-histamine by histamine-N-methyl transferase and ^{3}H-SAM	ethanol, water, ammonia (50:50:10)	radioactivity	(104)

* The stationary phase is silica.

the resulting fluorescence was measured with a TLC scanner. Recovery values for
urine samples spiked with ^{14}C-labelled A, NA and DA were 74.9 ± 3.9, 81.8 ± 3.7
and 84.4 ± 2.1 percent respectively (102).

Nakamura and Pisano (100) used fluorescamine as a fluorigenic labelling reagent
both before and after TLC development in two systems for primary catecholamines
and their 3-O-methyl derivatives. The mobile phase described in Table 3.4 was
used when the compounds were derivatized at the origin on the TLC plate. The
detection limit varied from 5 to 800 pmole (3-MTA and 6-hydroxydopamine), and
the sensitivities of the methods were comparable to those of other fluorimetric
methods (100).

Radioenzymatic procedures have been employed for the analysis of other amines
in addition to the catecholamines. Such assays have usually involved the replacement of hydrogen on an alcohol, phenol or amine by a radiolabelled methyl group.
In many of these assays, the resultant O-methylated and/or N-methylated products
have been isolated by means of selective organic extractions; TLC has been
employed as a means of confirming the structures of the final derivatives (106-
110), but is not a routine part of the assay procedure itself. However, some
radioenzymatic procedures have utilized TLC as an integral part of the procedure
to isolate the radiolabelled products, and examples are given in Table 3.4
(97,98,101,103,104).

3.3 SUMMARY AND CONCLUSIONS

From the many TLC systems listed in Tables 3.1 - 3.4 it may be concluded that
TLC is an important technique for the analysis of biogenic amines.

TLC plays an adjunctive role as a separation technique when biogenic amines
are analyzed by fluorescent, mass spectrometric, radioenzymatic and radiochemical
methods. TLC has been described as a laborious procedure, but the new technology
associated with HPTLC, eg. preformed multichannel plates, can improve the efficiency and specificity of radioenzymatic assays (111).

The need for pretreatment of samples, the lack of specificity of detection
methods, and the length of time for analysis have not made conventional TLC a
method of choice for the measurement of low concentrations of biogenic amines
(112). The new technology of HPTLC, the convenience of scanning densitometers,
the increasing sophistication of sample application and TLC plate development
(113) have given renaissance to the analytical technique of TLC.

REFERENCES

1 N.A. Izmailov and M.S. Shraiber, Farmatsiya, 3 (1938) 1-7. (For an English
   translation see N. Pelick, H.R. Bottinger and H.K. Mangold, in J.C. Giddings
   and R.A. Keller (Eds.), Advances in Chromatography, Dekker, New York, 1966,
   Vol. 3, 85 pp.).
2 R. Consden, A.H. Gordon and A.J.P. Martin, Biochem. J., 38 (1944) 224-232.

60

3 J.G. Kirchner, J.M. Miller and G.E. Keller, Anal. Chem., 23 (1951) 420.
4 E. Stahl, G. Schröter, G. Kraft and R. Renz, Pharmazie, 11 (1956) 633-637.
5 E. Stahl, Chem. Ztg., 82 (1958) 323-329.
6 E. Stahl (Ed.), Thin-Layer Chromatography, 2nd edition, Springer-Verlag, New York, 1969, 1041 pp.
7 K. Macek (Ed.), Pharmaceutical Applications of Thin-Layer and Paper Chromatography, Elsevier, New York, 1972, 743 pp.
8 J.G. Kirchner, Thin-Layer Chromatography, 2nd edition, Wiley-Interscience, New York, 1978, 1137 pp.
9 J.C. Touchstone and M.F. Dobbins, Practice of Thin-Layer Chromatography, Wiley-Interscience, New York, 1978, 383 pp.
10 A. Zlatkis and R.E. Kaiser in A. Zlatkis and R.E. Kaiser (Eds.), HPTLC High Performance Thin-Layer Chromatography, Elsevier, New York, 1977, pp. 9-13.
11 D. Jänchen and H.R. Schmutz, J. High Resol. Chromatogr., 2 (1979) 133-140.
12 CAMAG NANOMAT sample applicator for TLC and HPTLC, CAMAG Homburgerstr. 24 CH-4132 Muttenz, Switzerland.
13 G. Zweig and J. Sherma, Anal. Chem., 52 (1980) 276R.
14 J.C. Touchstone and J. Sherma, Densitometry in Thin Layer Chromatography: Practice and Applications, Wiley, Chichester, England, 1979, 764 pp.
15 H. Halpaap and J. Ripphahn, Chromatographia, 10 (1977) 613-623.
16 H. Halpaap and J. Ripphahn, Chromatographia, 10 (1977) 643-650.
17 T.H. Jupille, CRC Crit. Rev. Anal. Chem., 6 (1977) 325-359.
18 G. Zweig and J. Sherma, Anal. Chem., 50 (1978) 50R.
19 J.A. Thoma, J. Chromatogr., 12 (1963) 441.
20 J.A. Perry, T.H. Jupille and L.J. Glunz, Anal. Chem., 47 (1975) 65A-74A.
21 T.H. Jupille and J.A. Perry, Science, 194 (1976) 288-293.
22 G. Zweig and J. Sherma, Anal. Chem., 48 (1976) 66R.
23 D. Janchen, Thin-Layer Chromatography Cumulative Bibliography, 4: 1973-1977, Camag: Muttenz, Switzerland, 1977, 282 pp.
24 D. Janchen (Ed.), CAMAG Bibliography Service, Thin Layer Chromatography, No. 37 (1976) C3.
25 E. Schmid, L. Zicha, J. Krautheim and J. Blumberg, Med. Exptl., 7 (1962) 8-14.
26 A.H. Beckett and N.H. Choulis, J. Pharm. Pharmacol., 15 (1963) 236T.
27 R. Segura-Cardona and K. Soehring, Med. Exptl., 10 (1964) 251-257.
28 F.H. Schneider and C.N. Gillis, Biochem. Pharmacol., 14 (1965) 623-626.
29 W.P. de Potter, R.F. Vochten and A.F. de Schaepdryver, Experientia, 21 (1965) 482-483.
30 N.H. Choulis, J. Pharm. Sci., 56 (1967) 196-199.
31 N.H. Choulis, J. Pharm. Sci., 56 (1967) 904-906.
32 J. Giese, E. Ruther and N. Matusek, Life Sci., 6 (1967) 1975-1982.
33 D.J. Roberts and K.J. Broadley, J. Chromatogr., 27 (1967) 407-412.
34 G.A. Johnson and S.J. Boukma, Anal. Biochem., 18 (1967) 143-146.
35 N.H. Choulis and C.E. Carey, J. Pharm. Sci., 57 (1968) 1048-1050.
36 D. Aures, R. Fleming and R. Håkanson, J. Chromatogr., 33 (1968) 480-493.
37 J.M.C. Gutteridge, Clin. Chim. Acta, 21 (1968) 211-216.
38 H. Möhrle and R. Feil, J. Chromatogr., 34 (1968) 264-265.
39 E.J. Cowles, G.M. Christensen and A.C. Hilding, J. Chromatogr., 35 (1968) 389-395.
40 J.D. Sapira, J. Chromatogr., 42 (1969) 134-136.
41 A. Vahidi and D.V.S. Sanakar, J. Chromatogr., 43 (1969) 135-140.
42 G. Cohen and M. Collins, Science, 167 (1970) 1749-1751.
43 R.M. Fleming and W.G. Clark, J. Chromatogr., 52 (1970) 305-312.
44 N.N. Osborne, Experientia, 27 (1971) 1502-1503.
45 P. Bauman, B. Scherer, W. Kramer and N. Matussek, J. Chromatogr., 59 (1971) 463-466.
46 C. Haworth and T.A. Walmsley, Anal. Lett., 5 (1972) 35-43.
47 S. Takahashi and L.R. Gjessing, Clin. Chim. Acta, 36 (1972) 369-378.
48 E.S. Markianos and I.E. Nystrom, Z. Klin. Chem. Klin. Biochem., 13 (1975) 273-276.
49 X.O. Breakfield, J. Neurochem., 25 (1975) 877-882.

50 B. Wexler and R. Katzman, Exptl. Cell. Res., 92 (1975) 291-298.
51 R.W. Stout, R.J. Michelot, I. Molnar, C. Horvath and J.K. Coward, Anal. Biochem., 76 (1976) 330-341.
52 R.J. Head, R.J. Irvine and J.A. Kennedy, J. Chromatogr. Sci., 14 (1976) 578-579.
53 M. Vohra and S. Jayasundar, J. Pharm. Pharmacol., 28 (1976) 810.
54 D.E. Schutz, G.W. Chang and L.F. Bjeldanes, J. Assoc. Off. Anal. Chem., 59 (1976) 1224-1225.
55 R.J. Head, J.A. Kennedy, I.S. De La Lande and G.A. Crabb, J. Chromatogr. Sci., 16 (1978) 82-85.
56 E. Meyer and W. Barz, Planta Med., 33 (1978) 336-344.
57 E.R. Lieber and S.L. Taylor, J. Chromatogr., 160 (1978) 227-237.
58 G. Gubitz, Chromatographia, 12 (1979) 779-781.
59 I. Nir and N. Hirschmann, Experientia, 35 (1979) 1426-1427.
60 A.M. Di Giulio, A. Groppetti, S. Algeri, F. Ponzio, F. Cattabeni and C.L. Galli, Anal. Biochem., 92 (1979) 82-90.
61 J.N. Bidard and L. Cronenberger, J. Chromatogr., 164 (1979) 139-154.
62 S. Pongor, J. Kramer and E. Ungar, J. High Resol. Chromatogr., 3 (1980) 93-94.
63 N.N. Osborne, J. Neurochem., 36 (1981) 17-27.
64 K. Engelman and B. Portnoy, Circ. Res., 24 (1970) 53-57.
65 P.G. Posson and J.D. Peuler, Anal. Biochem., 51 (1973) 618-631.
66 M. Da Prada and G. Zürcher, Life Sci., 19 (1976) 1161-1174.
67 N. Ben-Jonathan and J.C. Porter, Endocrinol., 98 (1976) 1497-1507.
68 J.D. Peuler and G.A. Johnson, Life Sci., 21 (1977) 625-636.
69 I.L. Martin, G.B. Baker and S.M. Fleetwood-Walker, Biochem. Pharmacol., 27 (1978) 1519-1520.
70 M.E. Bardsley and H.S. Bachelard, in V. Neuhoff (Ed.), Proceedings of the European Society for Neurochemistry, Verlag Chemie, Stuttgart, 1978, Vol. 1, 531 pp.
71 N. Seiler and M. Wiechmann, J. Chromatogr., 28 (1967) 351-362.
72 A.A. Boulton and L. Quan, Can. J. Biochem., 48 (1970) 1287-1290.
73 S. Axelsson, A. Bjorklund and N. Seiler, Life Sci., 10 (1971) 745-749.
74 A.A. Boulton and P.H. Wu, Can. J. Biochem., 50 (1972) 261-267.
75 S.R. Snodgrass and A.S. Horn, J. Neurochem., 21 (1973) 687-696.
76 Axelsson, A. Bjorklund and N. Seiler, Life Sci., 13 (1973) 1411-1419.
77 G.C. Boffey and G.M. Martin, J. Chromatogr., 90 (1974) 178-180.
78 D.A. Durden, B.A. Davis and A.A. Boulton, Biomed. Mass Spectrom., 1 (1974) 83-95.
79 N.N. Osborne, Microchemical Analysis of Nervous Tissue, Pergamon, New York, 1974, pp. 58-121.
80 H. Dolezalova, E. Giacobini and M. Stepita-Klauco, Int. J. Neurosci., 5 (1973) 53-59.
81 B.E. Leonard and N.N. Osborne, in N. Marks and R. Rodnight (Eds.), Research Methods in Neurochemistry, Plenum, New York, 1975, Vol. 3, pp. 443-462.
82 J.M. Slingsby and A.A. Boulton, J. Chromatogr., 123 (1976) 51-56.
83 A.A. Boulton, S.R. Philips, D.A. Durden, B.A. Davis and G.B. Baker, Adv. Mass Spectrom. Biochem. Med., 1 (1976) 193-205.
84 N. Seiler and B. Knödgen, J. Chromatogr., 131 (1977) 109-119.
85 M. Recasens, J. Zwiller, G. Mack, J.P. Zanetta and P. Mandel, Anal. Biochem., 82 (1977) 8-17.
86 D.A. Durden, in N. Marks and R. Rodnight (Eds.), Research Methods in Neurochemistry, Plenum, New York, 1978, Vol. 4, pp. 205-250.
87 M. Kopun and M. Herschel, Anal. Biochem., 85 (1978) 556-563.
88 B.A. Davis, Biomed. Mass Spectrom., 6 (1979) 149-156.
89 D.A. Durden, A.V. Juorio and B.A. Davis, Anal. Chem., 52 (1980) 1815-1820.
90 N. Seiler, J. Chromatogr., 143 (1977) 221-246.
91 N. Seiler and L. Demesch, in K. Blau and G.S. King (Eds.), Handbook of Derivatives for Chromatography, Heyden, London, 1977, pp. 346-390.
92 N. Seiler and M. Wiechman, in A. Niederwieser and G. Pataki (Eds.), Progress

in Thin-Layer Chromatography and Related Methods, Ann Arbor-Humphrey Science, Ann Arbor, 1970, Vol. 1, pp. 95-144.

93 D.A. Durden and A.A. Boulton, Techniques of Metabolic Research, B 214, 1-25 (1979).

94 T.J. Danielson, Personal Communication.

95 D. Waldi, Mitt. Deut. Pharm. Ges. (1962) 125-128.

96 S. Hauptmann and J. Winter, J. Chromatogr., 21 (1966) 338-340.

97 T.J. Danielson, A.A. Boulton and H.A. Robertson, J. Neurochem., 29 (1977) 1131-1135.

98 N.D. Vlachakis and V. DeQuattro, Biochem. Med., 20 (1978) 107-114.

99 J.E. Forrest and R.A. Heacock, J. Chromatogr., 44 (1969) 638-640.

100 H. Nakamura and J.J. Pisano, J. Chromatogr., 154 (1978) 51-59.

101 M.G.M. Balemans, F.A.M. Bary, W.C. Legerstee and J. van Benthem, Experientia, 34 (1978) 1434-1435.

102 C.F. Gelijhens and A.P. de Leenheer, J. Chromatogr., 183 (1980) 78-82.

103 M.J. Brown, P.W. Ind, P.J. Barnes, D.A. Jenner and C.T. Dollery, Anal. Biochem., 109 (1980) 142-146.

104 F. Nilam and I.R. Smith, Brit. J. Pharmacol., 72 (1981) 505P.

105 I.E. Hughes and J.A. Smith, J. Pharm. Pharmacol., 30 (1978) 124-125.

106 P.B. Molinoff and J. Axelrod, Science, 164 (1969) 428-429.

107 K.M. Taylor and S.H. Snyder, J. Neurochem., 19 (1972) 1343-1358.

108 J.M. Saavedra and J. Axelrod, J. Pharmacol. Exp. Ther., 182 (1972) 363-369.

109 J.M. Saavedra, M. Brownstein and J. Axelrod, J. Pharmacol. Exp. Ther., 186 (1973) 508-515.

110 J.F. Tallman, J.M. Saavedra and J. Axelrod, J. Neurochem., 27 (1976) 465-569.

111 D.P. Henry, in E. Usdin, I.J. Kopin and J. Barchas (Eds.), Catecholamines: Basic and Clinical Frontiers, Pergamon, New York, 1979, Vol. 1, pp. 859-861.

112 A.M. Krstulovic, in J.C. Giddings, E. Grushka, J. Cazes and P.R. Brown (Eds.), Advances in Chromatography, Dekker, New York, 1979, Vol. 17, 292 pp.

113 D. Rogers, Amer. Lab., 11, No. 5 (1979) 77-79.

Chapter 4

FLUORESCENCE TECHNIQUES FOR DETECTION AND QUANTITATION OF AMINES.

JUDITH M. BAKER and WILLIAM G. DEWHURST

Neurochemical Research Unit, Department of Psychiatry, University of Alberta,
Edmonton, Alberta T6G 2G3 (Canada)

4.1   INTRODUCTION
4.1.1 Theory and Instrumentation

It is well known that organic molecules absorb light energy, which results
in various interatomic bonds being raised to a higher energy level. This energy
may be dissipated in a number of ways, one of which is the emission of light. The
term fluorescence commonly refers to the phenomenon whereby light from the ultraviolet (UV) spectrum is absorbed and emitted in the visible spectrum. Thus, under
normal circumstances, emitted radiation is of a longer wavelength (lower energy)
than that absorbed. The number of photons of light emitted is proportional to the
number of molecules involved, that is the concentration of fluorescent substance(s)
present. The fluorescence intensity observed for a particular substance is determined by the difference in energy between the excited and ground states and the
relative importance of other methods of energy dissipation, such as collisional
deactivation. One important point to remember is that the chemical and physical
properties (eg. pKa, dipole moment, interatomic distances) of the excited state
are very different from those of the ground state. These changes may cause special
difficulties in the measurement of certain compounds (see section 4.1.2).

The basic instrumentation for the measurement of fluorescence is illustrated
in Figure 4.1.

Very simply, such an instrument will consist of a light source capable of
producing UV radiation which passes through the primary filter system so that
only light of a particular wavelength is directed onto the sample cuvette. Light
emitted from the sample passes through a secondary filter system so that only
light of a specified wavelength reaches the detector. Detection systems normally
consist of a photomultiplier tube, an amplifier and an indicator or recorder giving
a readout in arbitrary fluorescence units. Early instruments had filters which
had to be inserted and changed manually, whereas many newer models have gratings
which can be adjusted by an external dial, allowing the experimenter much more
freedom in wavelength selection. Instruments are also available which will scan
either the excitation or emission wavelength, thus providing spectra of responses
for the substance in question. The light source and detector are normally placed
at a 90° angle to each other to decrease interference from the source at the detec-

64

tor. This arrangement provides an increase in sensitivity of approximately 1000 times that achieved by measurement of absorbed light (absorption spectrophotometry).

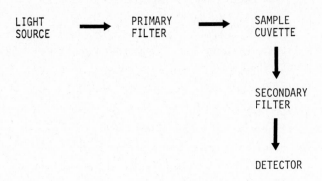

Fig. 4.1. Instrumentation for fluorescence measurement.

Analytical methods based on the measurement of fluorescence achieve specificity for the following reasons:

1. Not all compounds absorb UV light.
2. Not all compounds fluoresce after light absorption.
3. Wavelengths for maximal excitation (λ_{ex}) and emission (λ_{em}) are set for the particular compound of interest and may not correspond to those of other substances present. Even if λ_{ex} is the same for two substances, λ_{em} may be sufficiently different so that no interference exists.
4. Many methods incorporate prepurification steps.

Certain characteristics of the sample itself may limit the usefulness of fluorescence as a method for quantitation. Some substances undergo photochemical decomposition; the light used for excitation causes or accelerates chemical changes in the compound of interest. Such reactions are usually more significant at higher excitation energies (shorter λ_{ex}). Therefore in cases where photochemical decomposition occurs it is important to choose the longest wavelength possible for excitation. Changes in measured fluorescence may also occur due to variations in sample viscosity; generally an increase in viscosity results in an increase in fluorescence. Finally, quenching of fluorescence may occur due to such factors as temperature, dissolved oxygen, and impurities (see section 4.1.2). Limitation of quenching may require strict control over sample processing, purity of chemicals used and cleanliness of glassware.

Fluorescence may be used as a measurement or detection tool in three situations. Certain compounds are naturally fluorescent and may be measured directly in solution. Such compounds are usually not strongly fluorescent and therefore the sensitivity of methods measuring native fluorescence is limited. More commonly, compounds of interest are converted to fluorescent derivatives. Such reactions usually result in strongly fluorescent products. Furthermore, the specificity of the assay method is increased since some potentially interfering compounds may not form fluorescent derivatives under the conditions used. Finally, certain substances may be detected or quantitated because they quench the fluorescence of other strongly fluorescent compounds. An example of this technique is the fluorescent detectors used in thin layer chromatography, whereby compounds are visualized under UV light as dark spots on a fluorescent background.

4.1.2 Common methodological problems

Since fluorescence is usually measured in solution, solvent effects may be important. Polar molecules will show an increase in dipole moment in the excited state which will be stabilized by the use of a polar solvent, resulting in a decrease in fluorescence intensity. Polar compounds dissolved in polar solvents also show a shift in λ_{ex} and λ_{em} to longer wavelengths (lower energy). This bathochromic shift is more pronounced as the dielectric constant of the solvent increases since the magnitude of the shift depends on the strength of solute-solvent interactions. Normally λ_{em} shifts more than λ_{ex}, resulting in a larger separation between the excitation and emission spectra. This property of changing λ_{em} by altering solvent polarity might be used to advantage in separating the emission spectra of substances of differing polarities. Solvent effects are usually insignificant if either the solute or the solvent is non-polar.

Solvents containing a heavy atom, for example ethyl iodide, usually cause enhancement of phosphorescence at the expense of fluorescence. Thus the sensitivity of an analytical procedure utilizing fluorescence may be increased by avoiding such solvents or removing any excess.

Fluorescence measurements may be affected by the pH of the solution being analyzed. The difference between λ_{ex} and λ_{em} (Stoke's shift) illustrates this point: 5-hydroxyindole at pH 7 has λ_{em} at 330 nm and in strong acid at 550 nm with no change in λ_{ex}. Essentially, the effects of pH occur because the fluorescence intensity and λ_{em} of the ionized species differ from those of the unionized species. Therefore, a pH should be chosen at which virtually 100% of the substance to be measured exists as a single species. Furthermore, since the pKa of the excited state is different from that of the ground state, the suitable pH range may be different from that which would be predicted from consideration of the properties of the ground state. Finally, pH adjustment may be used to increase the sensitivity and specificity of a particular assay by decreasing the formation of the fluorescent derivatives of interfering substances.

Intermolecular hydrogen-bonding may result in a decrease in fluorescence inten-

sity in much the same manner as that resulting from polar solute-solvent interactions. Therefore, sensitivity may be increased by choosing a solvent which does
not hydrogen-bond with the substance of interest. This generalization may not
apply to aromatic carboxyl compounds and nitrogen heterocyclics where hydrogen-
bonding tends to decrease n - π^* energy transitions and increase π - π^* transitions
which are more likely to result in fluorescence emission.

Since the presence of other solutes may result in quenching or interference,
it is usually necessary to employ some prepurification steps before analysis of
complex samples. Oxygen is the most ubiquitous fluorescence quencher and its
presence may necessitate degassing the sample or the inclusion of antioxidants.
Metal ions may also cause difficulties due to fluorescence quenching. Not all
metals are implicated, however; Fe^{2+} is an efficient quencher, while Mg^{2+}, Zn^{2+}
and Cd^{2+} are poor.

Ideally there should be a linear increase in fluorescence intensity as solute
concentration increases. Practically, fluorescence often levels off or decreases
at high concentrations (concentration quenching). Since quantitation is usually
accomplished by comparison with suitable standards it is important that sample
concentrations fall within the linear portion of the fluorescence-concentration
curve. Most literature methods quote linear ranges and samples should be diluted
before being assayed if they are expected to fall outside these limits.

Temperature _per se_ usually has little effect on fluorescence intensity; however
quenching and chemical instability often increase with increasing temperature. An
example of an adverse temperature effect is cited by Gerst _et al_. (1966) who found
changes in the fluorescence of the trihydroxyindole derivatives of adrenaline (A)
and noradrenaline (NA) amounting to 14%/°C due to the chemical instability of the
fluorescent molecules at increased temperatures. Such effects may be especially
important in instruments where the light source can cause an increase in sample
temperature while measurements are being conducted. Samples demonstrating temperature lability should be left in place for the minimum time necessary to obtain a
fluorescence reading.

Further detailed information may be obtained from general reference books, such
as the text by Guilbault (35).

4.2   FLUORESCENCE TECHNIQUES FOR DETECTION OF AMINES
4.2.1   Derivatization

Detection of amines is usually accomplished by formation of derivatives which
fluoresce when exposed to UV light. Such a procedure is often used in combination
with paper or thin-layer chromatography and in the histochemical fluorescence methods described in Chapter 5. Many of the same derivatives used for quantitation are
also used for detection purposes. Procedures for the catecholamines include that of
Gaddum _et al_. (43) for detection of A in biological fluids by addition of sodium

hydroxide and observation under UV light. Shore and Olin (88) reported the identification of NA in various tissue extracts by paper chromatography followed by
spraying with ferricyanide and air oxidation. This procedure produced a fluorescent spot similar in colour and chromatographic characteristics to those of NA
standards. A variety of amines may be detected by the method of Bell and Somerville
(8) in which amines are converted to fluorescent derivatives on paper chromatograms
by reaction with formaldehyde. Some of the compounds detected in this manner are
NA, A, dopamine (DA), 5-hydroxytryptamine (5-HT), tryptamine, 3-hydroxy-4-methoxy-
phenylethylamine, 4-hydroxy-3-methoxyphenylethylamine, 3,4-dimethoxyphenylethylamine,
octopamine and bufotenin.

In addition to the procedure of Bell and Somerville (8) various other methods
have been proposed for detection of indoleamines. Jepson and Stevens (44) prepared
fluorescent derivatives of 5-HT and other tryptamines on paper chromatograms sprayed
with ninhydrin. o-Phthalaldehyde has been used extensively as a spray reagent for
thin-layer chromatograms (68), allowing detection of various indoleamines, especially
bufotenin and 5-methoxy-N,N-dimethyltryptamine. Axelsson and Nordgren (7) have
described a method capable of detecting a variety of indoles in blood plasma using
either a formaldehyde spray or o-phthalaldehyde following thin-layer chromatography.
o-Phthalaldehyde spray (27) combined with two dimensional thin-layer chromatography
has also been used as a detection method for 5-methoxytryptamine in rat brain.

Other amines have been analyzed after formation of fluorescent derivatives.
Histamine has been detected after thin-layer chromatography or electrophoresis by
formation of an o-phthalaldehyde derivative (86). Seiler and Demisch (82) have
reviewed the properties of a number of fluorescent derivatives suitable for detection
of amines including 1-dimethylaminonaphthalene-5-sulphonyl chloride (DANSYL-Cl,
DNS-Cl), 5-di-n-butylaminonaphthalene-1-sulphonyl chloride (BNS-Cl), 6-methyl-
anilinonaphthalene-2-sulfonyl chloride (MNS-Cl), 2-p-chlorosulfophenyl-3-phenyl
indone (Dis-Cl), 4-chloro-7-nitrobenzo[c]-1,2,5-oxadiazole(4-chloro-7-nitrobenzofur-
azan; Nbd-Cl), 9-isothiocyanatoacridine, fluorescamine, pyridoxal, o-phthalaldehyde,
formaldehyde, glyoxylic acid, and p-dimethylaminocinnamaldehyde. These authors also
have delineated several properties which are desirable in a reagent used to form
fluorescent derivatives for detection of amines:
1. The reagent must react quickly under mild conditions.
2. The reagent should form non-polar derivatives which can be isolated and
   concentrated in organic solvents.
3. The excess reagent should be easily removable.
4. The derivatives should have a high fluorescence efficiency and should emit
   at a wavelength long enough to be distinguishable from the bluish background
   due to many solvents.
5. The derivative should absorb and emit at wavelengths consistent with the
   instruments available.

4.3   FLUORESCENCE TECHNIQUES FOR QUANTITATION OF AMINES

4.3.1   Derivatization

4.3.1.1   Catecholamines and metabolites

Since the observation by Gaddum et al. (32) that A exposed to oxygen in the presence of strong alkali formed a fluorescent derivative, numerous fluorescence methods for the determination of catecholamines in body fluids and tissues have been reported. This discussion will deal with the two major types of derivatives formed, namely hydroxyindole compounds and condensation products with ethylene diamine, as well as certain miscellaneous derivatives and more sophisticated methods for simultaneous estimation of several compounds.

The general reaction for the formation of trihydroxyindole derivatives of A and NA is shown in Figure 4.2.   The reaction involves two steps: oxidation to

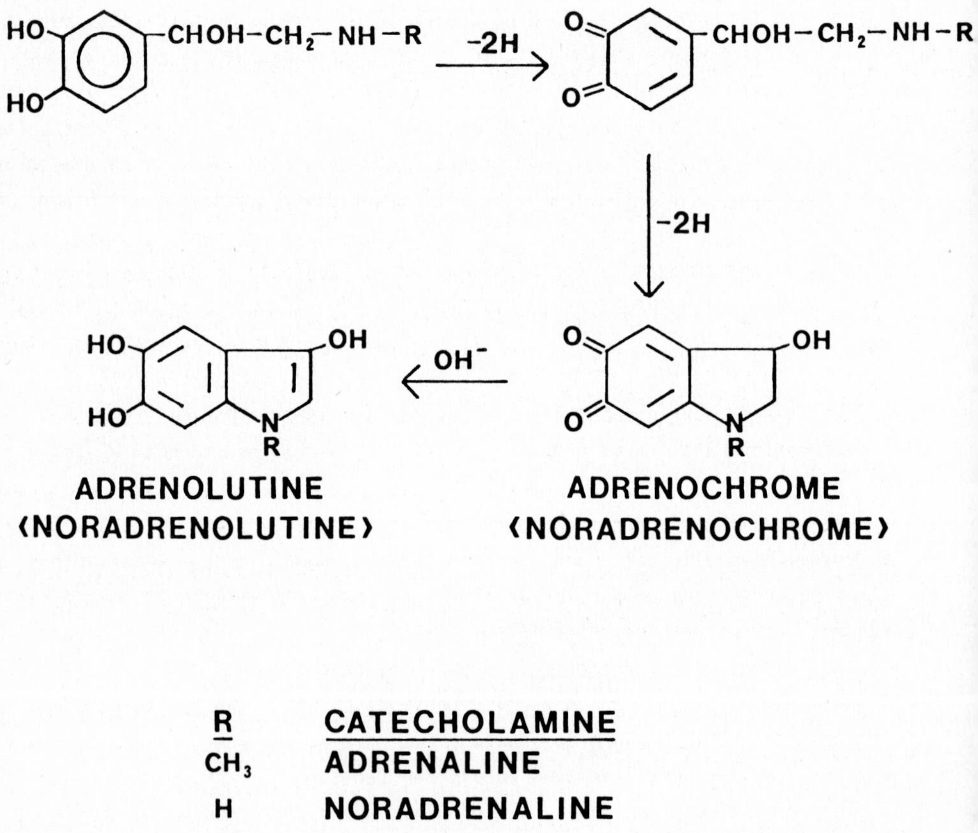

ADRENOLUTINE
〈NORADRENOLUTINE〉

ADRENOCHROME
〈NORADRENOCHROME〉

R	CATECHOLAMINE
CH_3	ADRENALINE
H	NORADRENALINE

Fig. 4.2. Formation of trihydroxyindole derivatives of noradrenaline and adrenaline

form an adrenochrome followed by rearrangement to form an adrenolutine. Since
the "lutines" are easily destroyed in the presence of oxygen, Ehrlen (31) suggested
the addition of a reducing agent to the final solution to stabilize the fluorescence. Ascorbic acid is commonly used for this purpose (31,52,102,18,3), although
other substances have been utilized, including thioglycolic acid (61), β-thio-
propionic acid (74), dimercaptoethanol in sodium sulfite (38) and β-mercaptoethanol
(107). Since ascorbic acid in alkaline solution is itself converted to fluorescent
products, diaminoethane (104) or sodium borohydride (33) must be added to overcome
this problem.

Various oxidizing agents have also been used for the oxidation reaction depicted
in Fig. 4.2. Manganese dioxide was initially utilized by Lund (52) and later by
Cohen and Goldenberg (18). Von Euler and Floding (102) reported the use of potassium
ferricyanide and iodine but concluded that iodine was not suitable since the time
necessary to prepare a faded blank was too long. When iodine is used the derivatives formed are iodinated in the 2-position of the indole ring. These iodinated
derivatives are also reported to be more stable and to allow more complete separation of A and NA than the hydroxyindole derivatives obtained from other methods
(23).

In order to speed up the oxidation reaction various metal ions have been used
as catalysts; Zn^{2+} (102) and Cu^{3+} (38,107) are typical examples.

The detailed work of Anton and Sayre (3,4,5) has provided an excellent illustration of the many complex factors which affect fluorescence measurements, such as
reagent concentrations, cleanliness of glassware, reaction times, and quenching.
In addition Gerst et al. (33) have pointed out that the temperature at which the
fluorescence of trihydroxyindole derivatives is measured may be very important,
resulting in a change of 14%/°C.

Crout (23) has suggested the addition of ethylenediamine tetraacetic acid (EDTA)
to urine before analysis to prevent the formation of a calcium - magnesium phosphate
gel which may quench fluorescence.

Differential estimation of A and NA may be achieved by oxidation at two different
pH values, measurement of fluorescence at two different wavelengths, or a combination
of both. Lund (53) first reported that only A was oxidized at pH 3 whereas both
A and NA were oxidized at pH 6.5. This selective oxidation technique was also
used by von Euler and Floding (102). Price and Price (76) reported that measurement of fluorescence at different wavelengths resulted in a decrease in the volume
of plasma required. This approach was also taken by Cohen and Goldenberg (18),
Bertler et al. (9), von Euler and Lishajko (103), and Haggendal (38). Other authors
have utilized a combination of these two approaches (100,3,107). Finally, a unique
approach was described by Merrills (61) in which thioglycolic acid was employed as a
reducing agent to protect only noradrenolutine while ascorbic acid was utilized to
protect both adrenolutine and noradrenolutine.

In addition, the sensitivity and specificity of these chemical methods has been increased by prepurification using either alumina (52,18,103,3,61) or cation exchange resins (9,100,38). Merrills (61) has reported an automated method for differential analysis of NA and A.

Dopamine has been analyzed by the formation of a dihydroxyindole derivative analogous to the trihydroxyindole derivatives formed from NA and A. The reaction for DA is shown in Figure 4.3. Carlsson and Waldeck (15) first suggested a method

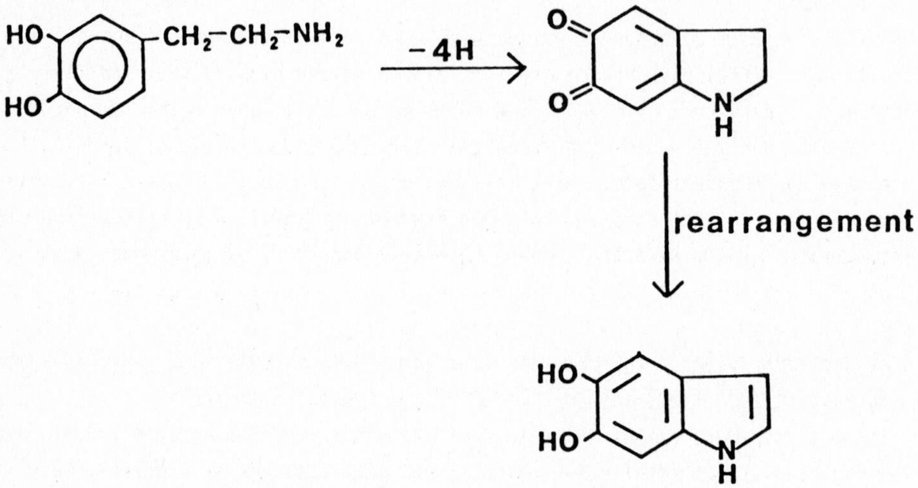

Fig. 4.3. Formation of the dihydroxyindole derivative from dopamine.

for analysis of DA based on this reaction using iodine as an oxidizing agent followed by exposure to UV light to catalyse formation of the final derivative. Under these conditions there was little interference from NA and A but dihydroxyphenylalanine (DOPA) formed the same derivative as DA and thus had to be removed when present in appreciable amounts. Uuspaa (97) suggested oxidation with manganese dioxide and rearrangement with zinc sulfite, a procedure which did not require exposure to UV light. The fluorescence was stabilized by decreasing the pH of the final solution to 5. Anton and Sayre (4) employed sodium periodate as an oxidizing agent and an alkaline sulfite solution to facilitate rearrangement. Noradrenaline and A did not interfere to any extent. After the fluorescence was read at pH 4, 5N HCl was added to decrease the contribution due to DOPA by 53 - 57% and the fluorescence was measured again. A set of simultaneous equations were then used to calculate the concentrations of DA and DOPA present. Atack (6) employed a Dowex 50 column to separate DOPA and then reacted the eluate with potassium ferricyanide and irradiated it with UV light. Fluorescence was measured at 375 nm; at this wavelength interference from NA and A was minimal.

The methoxy derivatives of the catecholamines (normetanephrine, NMN; metanephrine, MN; and 3-methoxytyramine, 3-MTA) have also been analyzed by conversion to hydroxy-indole derivatives. The reaction products of these substances possess spectral characteristics identical with those of the corresponding catecholamines. Analytical procedures must, therefore, include procedures for removal of interference from NA, A or DA as appropriate. Bertler et al. (10) first developed a method for MN and NMN which utilized iodine oxidation at pH 5 and pH 7.2. Only A and NA were oxidized at pH 5 while A and NA as well as MN and NMN were oxidized at pH 7.2. The methoxyamines could then be determined by difference. After oxidation was complete the pH was decreased to 5.3 to minimize interference from DA. Smith and Weil-Malherbe (90) have reported a method for oxidation of MN and NMN at different pH values. No interference from 3-MTA occurred. Prior adsorption on alumina removed interference from A, NA and DA. Haggendal (37) has employed prior oxidation with potassium ferricyanide to remove interference from A and NA since NMN and MN are not oxidized under these conditions. The methoxyamines were then adsorbed on an Amberlite CG 120 column, eluted with HCl and reacted with iodine. This report also suggested that by lengthening the resin column NA, A, NMN, MN and DA could be separated without prior oxidation. Carlsson and Lindqvist (14) have modified the method of Bertler et al. (10) for assaying NMN in brain. Prior oxidation with potassium ferricyanide removed NA and DA and then samples were passed through a Dowex 50 resin column and the eluate was analyzed after iodine oxidation. Brunjes et al. (12) removed A and NA on an alumina column and followed this by potassium ferricyanide oxidation of the eluate. Reaction at low pH allowed estimation of MN; when the pH was raised, NMN was oxidized as well. Fluorescence was also measured at two different wavelengths to allow more complete separation of MN and NMN. Anton and Sayre (5) proposed a method for analysis of MN and NMN in which fluorescent derivatives were formed by periodate oxidation at pH 5 for both MN and NMN and at pH 1.5 for MN. The fluorescence of the mixture was measured at one wavelength whereas the MN derivative alone was analyzed at a different wavelength. If large amounts of NA and/or A were present they could be quantitated by ferricyanide oxidation and MN and NMN estimated by difference.

3-Methoxytyramine has been analyzed by three fluorescence methods. Carlsson and Waldeck (16) employed a Dowex 50 column to separate three fractions containing A and NA, DA and 3-MTA. The final fraction was then oxidized using iodine to give a fluorescent derivative. Guldberg et al. (36) used a Dowex 50 resin column to obtain an eluate containing DA and 3-MTA. Oxidation with potassium ferricyanide in ammonia was apparently specific for 3-MTA, whereas DA could be analyzed by the method of Laverty and Sharman (48). Finally Kehr (47) has proposed chromatographic separation of NA, DA, 3-MTA and 5-HT on a Dowex 50 resin column, followed by oxidation of the 3-MTA eluate by potassium ferricyanide in ammonia.

In addition to their use in analysis of single catecholamines or catecholamine metabolites, the hydroxyindole derivatives have been utilized for simultaneous analysis of various catecholamines and/or related substances (28,93,59,17,49,109, 40). Furthermore, various methods have been proposed for simultaneous analysis of catecholamines and their metabolites by formation of hydroxyindole derivatives and various indole amines by further derivative formation (55,2,63,85,41,21,13,45) or by native fluorescence (46,62,43).

The second reaction commonly used to form fluorescent derivatives of catecholamines and metabolites is oxidation followed by condensation with ethylenediamine. This reaction is shown in Figure 4.4.

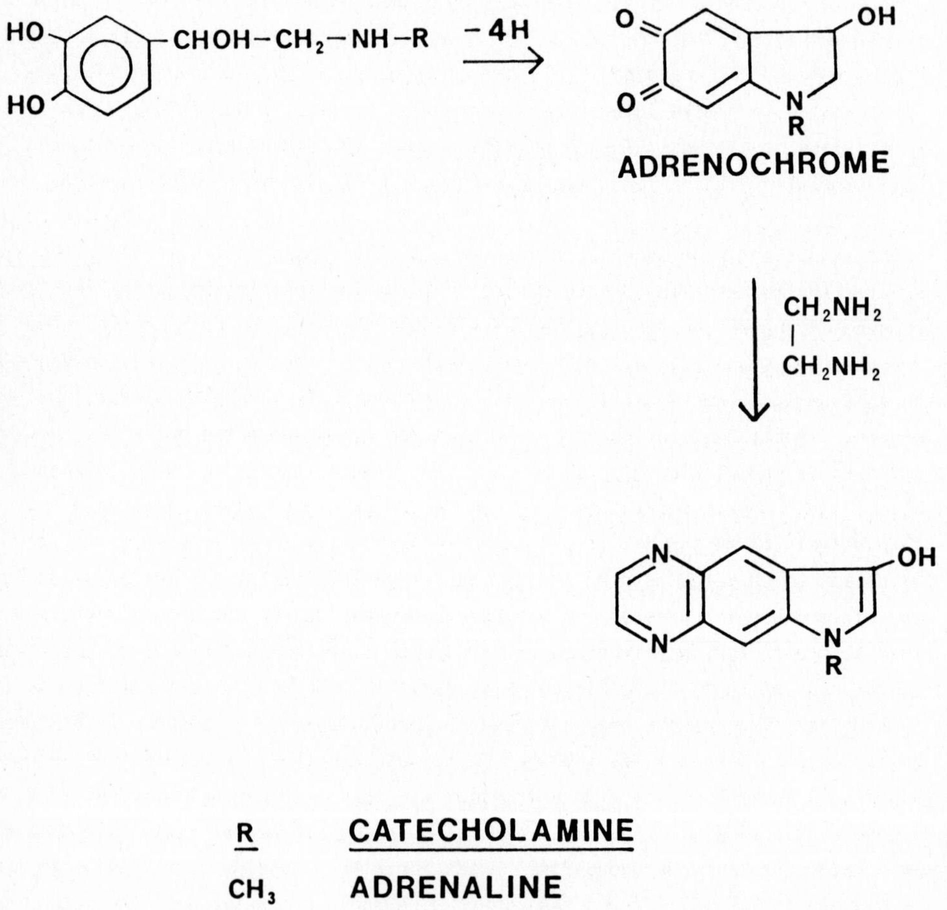
HO
HO
CHOH−CH2−NH−R
−4H
O
O
OH
N
R
ADRENOCHROME
CH2NH2
CH2NH2
N
N
OH
N
R
R
CATECHOLAMINE
CH3
ADRENALINE
H
NORADRENALINE

Fig. 4.4. Ethylenediamine condensation of catecholamines.

Natelson et al. (70) first reported that A in ammonia solution would condense with ethylenediamine, butylamine, aniline or o-phenylenediamine to produce fluorescent derivatives which could be extracted into butyl or amyl alcohol. Weil-Malherbe and Bone (108) suggested the use of this reaction to measure A and NA in blood. Alumina was used to adsorb A and NA from plasma. The catechols were eluted, reacted with ethylenediamine, extracted into isobutanol and fluorescence was measured. These authors have reported that the derivatives formed in this way are more stable than those produced by the trihydroxyindole method. Montagu used this method for the measurement of A and NA in various tissues (64), including brain (65). She also reported the presence of an unknown catechol in brain and suggested it could be DA. In 1961, Weil-Malherbe (106) proposed modifications of his original method to increase its sensitivity and specificity and to allow simultaneous quantitation of A, NA and DA. Even with these changes, however, ethylenediamine condensation was not as specific a procedure as the trihydroxyindole method. Laverty and Sharman (48) and Sharman (84) have proposed a method for DA analysis which involved acetylation, paper chromatography, elution and ethylenediamine condensation and resulted in increased sensitivity and specificity. Crawford and Yates (22) used a similar procedure and found that prior acetylation increased the sensitivity for DA and 3-methoxytyramine (3-MTA), decreased it for DOPA, NA and normetanephrine (NMN) and caused no change for A and metanephrine (MN).

Various authors have criticized the ethylenediamine condensation method. Von Euler et al. (101) suggested that this reaction was not appropriate for analysis of catecholamines in urine since there were many substances present which interfered with the final fluorescence measurement. Valk and Price (98) concluded that the ethylenediamine condensation method agreed well with the trihydroxyindole method for A but overestimated the concentration of NA and thus was not specific enough for plasma analysis. Mangan and Mason (56) reported that NA derivatives were unstable and decayed at a rate proportional to the intensity of the exciting radiation. Finally, Nadeau and Joly (66) suggested that some of the problems associated with this method might be due to the formation of multiple derivatives - nine for A and eight for NA. This criticism was questioned, however, by Weil-Malherbe (105), who stated that under the conditions initially proposed (108) A formed only one product, while NA formed two.

A small number of fluorescence methods for the analysis of catecholamines have been proposed which do not involve either of the two derivatives discussed previously. Bell and Somerville (8) suggested that derivatives formed on paper chromatograms by reaction between formaldehyde and catecholamines could be eluted and their fluorescence measured. Oberman et al. (72) used a DANSYL derivative to quantitate DA in urine, but the method was only sensitive enough to be used in cases of high excretion, such as would occur after DOPA treatment.

4.3.1.2 Indoleamines

Three major fluorescence methods have been used for indoleamines. Early procedures employed condensation of the indoleamines with formaldehyde and oxidation
to form the fluorescent norharman derivative. This reaction is shown for tryptamine (T) in Figure 4.5. Such a reaction has been used to measure T (42,57)

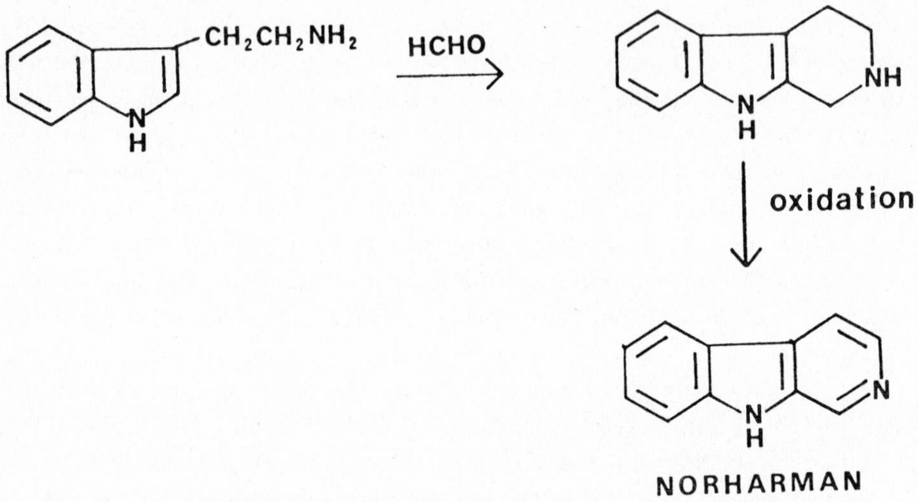

Fig. 4.5. Formation of norharman from tryptamine.

and tryptophan (26).

Condensation with o-phthalaldehyde was initially proposed for measurement of
5-HT and 5-methoxytryptamine (5-MT) by Maickel and Miller (55). Sensitivity for
5-HT was increased by the addition of cysteine (24). This reaction has also been
used to quantitate bufotenin, 5-methoxy-N,N-dimethyltryptamine (68) and 5-MT (77).
The sensitivity of the o-phthalaldehyde condensation reaction is reported to be at
least 2.5 times that obtained using the native fluorescence of 5-HT in strong acid.

The third derivative commonly used for analysis of indoleamines is that formed
by reaction with ninhydrin. Vanable (99) modified the conditions of the method
of Jepson and Stevens (44) so that 5-HT could be determined in solution. Snyder
et al. (91) used the ninhydrin reaction to determine 5-HT in tissues and have state
that this method is 8 times more sensitive than the procedure of using native
fluorescence in strong acid. Quay (77) examined the specificity of this reaction
and found that ninhydrin would form fluorescent derivatives with bufotenin, 5-
hydroxyindoleacetic acid (5-HIAA), N-acetyl-5-hydroxytryptamine and 5-hydroxy-
tryptophan (5-HTP). All these substances required different temperatures and
times of reaction, and thus could easily be differentiated.

4.3.1.3   Other amines

A number of other biogenic amines have been analyzed by formation of fluorescent derivatives. Seiler and Weichmann (83) have reported the use of DANSYL derivatives for fluorescence quantitation of primary and secondary amines, as well as imidazoles and phenols. This reaction is shown in Figure 4.6. Such derivatives may be used

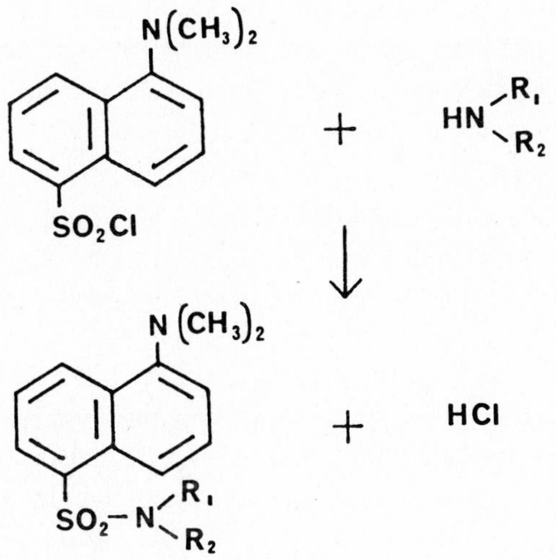

Fig. 4.6. Reaction of DANSYL - Cl with primary and secondary amines.

for direct fluorescence scanning of thin layer chromatograms. A linear quantitation range of 10^{-8} to 10^{-12} moles is reported for this method.

Histamine has been quantified by Shore et al. (87) after reaction with o-phthal-aldehyde in strong alkali. The method was modified by Hakanson et al. (39) to provide increased sensitivity and specificity. Alkon et al. (1) analyzed histamine, methylhistamine, methylimidazoleacetic acid, imidazole acetic acid, 1-methylhistidine, 3-methylhistidine and histidine by using successively N-bromosuccinimide to split the imidazole ring, oxidation, and condensation with o-phenylenediamine to form fluorescent derivatives. Tyramine has been analyzed after reaction with nitro-sonaphthol in urine (71) and tissues (95). β-Phenylethylamine has been measured by formation of fluorescent products with alloxan (11), p-dimethylaminocinnamaldehyde (94) and ninhydrin in the presence of L-leucyl-L-alanine (96). Fluorescence methods also have been suggested for the quantitation of 3,4-dimethoxyphenylethylamine (69) and for mescaline (3,4,5-trimethoxyphenylethylamine) (19).

4.3.2   Native fluorescence

Certain substances are naturally fluorescent and require no derivative formation before quantitation. Analyses using native fluorescence are generally less sensitive

and specific than those involving derivative formation since more interfering
substances may be present and the fluorescence intensity of the compound in question
may not be strong. Native fluorescence in strong acid is a common property of
indoles and has been used for analysis of T and 5-HT (71). Quay (78) used differential solvent extraction and measurement of fluorescence in HCl for the
quantitation of 5-hydroxyindole, 5-methoxyindole, 5-HIAA, 5-HT, 5-HTP, bufotenin,
melatonin, 5-MT, N-acetylserotonin and 5-methoxyindole-3-acetic acid. Similar
methods have been used by Narasimhachari et al. (67) to detect dimethyltryptamine,
5-methoxy-N, N-dimethyltryptamine, and bufotenin and by Cohen and Vogen (20) to
measure injected dimethyltryptamine and diethyltryptamine.

4.3.3   Fluorescence detection combined with another quantitation method
    The property of fluorescence is often utilized in analytical procedures involving
another major instrumental technique. Fluorescence detection may be used with high
performance liquid chromatography (HPLC, see Chapter 10). Davis et al. (25) have
used precolumn derivatization with o-phthalaldehyde followed by HPLC with fluorescence detection to measure NA, DA, 5-HT, NMN, octopamine (OA) and tyramine. A
similar technique has been used for amino acids (50). HPLC combined with detection
of native fluorescence has been used to measure tryptophan, 5-HIAA, indoleacetic
acid and indolepropionic acid (110) and 5-HT (60). Radiolabelled DANSYL-Cl has
been employed to quantify catecholamines and 5-HT (80) and O-methyl-catecholamine
metabolites (81). DANSYL derivatives have also been used for quantitation using
high resolution mass spectrometry (see Chapter 7), a method applicable to many
noncatecholic biogenic amines (29).

4.3.4   Micromethods
    A specialized application of the fluorescence quantitation method is the use of
micromethods developed for the assay of compounds in single cells. Substances
measured include NA, DA, 5-HT, T, A, OA and amino acids as well as many other
compounds. McCaman et al. (58) have reported the determination of 5-HT and DA
in single cells of Hirudo medicinalis using extraction with a specific liquid
cation exchanger and micromodifications of the assay methods of Maickel and Miller
(55) and Shellenberger and Gordon (85). A sensitivity limit of 2 pmol for 5-HT
and 4 pmol for DA was reported.
    Osborne (73) described a modification of the method of Bell and Somerville (8)
for the semi-quantitative estimation of DA, 5-HT and NA in single cells (sensitivities of 6, 5 and 7 ng respectively). The method suffers from the disadvantage
that protein interferes to a greater extent than with the DANSYL-Cl procedures.
    In his detailed 1974 review, Osborne (73) carefully outlined microprocedures
involving reaction with DANSYL-Cl to measure the amines 5-HT, T, NA, A, OA and amine

acids. Such derivatives are highly fluorescent and allow detection of as little
as 5 pmol on thin-layer chromatograms by direct fluorometry. If radiolabelled
DANSYL-Cl is used and measured by autoradiography, as little as 1 pmol may be
detected. While such methods are very sensitive, they are not without problems.
Quantitation is often difficult because of the occurrence of undesirable side
reactions, variability in the degree to which individual compounds react, and
variability due to changes in the ratio of reagent to substrates. Osborne has
stated that the identities and structures of all substances to be dansylated must
be known before quantitation can be carried out. Due to these limitations internal
standards can only be used when the amino acid and amine contents of all samples
are not significantly different. Quantitation can be improved by selectively
isolating the compounds of interest before the dansylation reaction; however, a
decrease in sensitivity of up to a factor of 1000 may result. These micromethods
have been applied to the measurement of single cell content, _in vivo_ synthesis and
turnover of 5-HT in the giant serotonin cells of the cerebral ganglion of _Helix
pomatia_, detection and measurement of T in nervous tissue of rat, mouse, snail
and crab and measurement of NA, A and OA in minute tissue samples.

4.4   COMPARISON OF FLUORESCENCE TECHNIQUES AND OTHER ANALYTICAL METHODS

Quantitative methods using measurements of fluorescence have a number of advantages in comparison with many other analytical techniques. Instrumentation and
reagents are inexpensive relative to those required for techniques such as gas
chromatography or mass spectrometry. No sophisticated training of personnel
is required and large number of samples can be processed daily. In addition,
measurement of fluorescence is a versatile method which can be used for analysis
of any fluorescent molecule or derivative.

These advantages must be considered in relation to the following disadvantages.
Great care must be taken to obtain the necessary sensitivity and specificity. A
good example of the large number of factors which can affect the outcome of any
assay is illustrated in the work of Anton and Sayre (3,4,5). A number of examples
exist whereby values obtained by a fluorescence assay do not agree with more specific methods; for example, Suzuki and Yagi (96) reported phenylethylamine levels
in rat brain of 5 ng/g, as compared to 1.8 ng/g obtained by mass spectrometric
methods (29); Spatz and Spatz (94), and Boulton and Milward (11) reported urinary
phenylethylamine levels of 239 and 47 µg/24 h respectively as compared with 1.7 µg/
24 h (89) using mass spectrometry; Spector et al. (95) have found CNS tyramine
levels of 1-6 µg/g in sharp contrast to the 2 ng/g level determined by mass spectrometry (75).

Several direct comparisons with other methods have been reported. Seiler and
Wiechmann (83) have stated that the use of DANSYL derivatives combined with direct
scanning of thin-layer chromatograms yields a sensitivity comparable with the use

of ^{14}C or ^{3}H tracers (0.005 nmol). Hakanson et al. (39) found that their improved
method for measurement of histamine by reaction with o-phthalaldehyde had a sensitivity equal to that of the radioenzymatic procedure of Snyder et al. (92).
Giacobini (34) compared the sensitivities of various methods (Table 4.1). Only
the three most sensitive methods in this table are applicable to the measurement
of transmitter molecules in single vertebrate cells. In addition, fluorimetry-
cycling [a method whereby substances are subjected to a repeated cycle of enzyme
reactions in order to amplify the resultant fluorescence (51)] does not appear to
have been used for analysis of biogenic amines.

TABLE 4.1
Comparison of the sensitivities of various analytical methods [from Giacobini
(8)].

METHOD	SENSITIVITY (moles)
colorimetric	10^{-9} - 10^{-10}
gas-chromatographic	10^{-10} - 10^{-12}
fluorimetric	10^{-11} - 10^{-12}
radiometric	10^{-12} - 10^{-13}
micro-TLC-DANSYL	10^{-12} - 10^{-14}
gas chromatographic-mass spectrometric	10^{-12} - 10^{-14}
fluorimetry-cycling	10^{-14} - 10^{-16}

4.5   SUMMARY

Fluorescence methods have been used for measurement of a wide range of catechol
amines, indoleamines and other primary and secondary amines of biological interest
Such methods offer reasonable sensitivity and specificity without a large capital
outlay. In addition, large numbers of samples can be processed by relatively inexperienced personnel. Micromethods have also been developed which allow measurement of a number of biogenic amines in single neurons. Needless to say, these
techniques require more careful application than measurement done on larger pieces
of tissue. Despite the limitations discussed in this chapter, fluorescence techniques continue to be very popular for analysis of biogenic amines in biological
samples.

REFERENCES
1   D. Alkon, A. Goldberg, J. Green, P. Levi, and K. Liao, Anal. Biochem., 40
    (1971) 192-199.
2   G.B. Ansell and M.F. Beeson, Anal. Biochem., 23 (1968) 196-206.
3   A.H. Anton and D.F. Sayre, J. Pharmacol. Exp. Ther., 138 (1962) 360-375.
4   A.H. Anton and D.F. Sayre, J. Pharmacol. Exp. Ther., 145 (1964) 326-336.
5   A.H. Anton and D.F. Sayre, J. Pharmacol. Exp. Ther., 153 (1966) 15-29.
6   C.V. Atack, Br. J. Pharmac., 48 (1973) 699-714.
7   S. Axelsson and L. Nordgren, Life Sci., 14 (1974) 1261-1270.

8   C.E. Bell and A.R. Somerville, Biochem. J., 98 (1966) lc-3c.
9   A. Bertler, A. Carlsson and E. Rosengren, Acta physiol. scand., 44 (1958) 273-292.
10 A. Bertler, A. Carlsson and E. Rosengren, Clin. Chim. Acta, 4 (1959) 456-457.
11 A.A. Boulton and L. Milward, J. Chromatogr., 57 (1971) 287-296.
12 S. Brunjes, D. Wybenga and V.J. Johns, Jr., Clin. Chem., 10 (1964) 1-12.
13 R.F. Butterworth, F. Landreville, M. Guitard and A. Barbeau, Clin. Biochem., 8 (1975) 298-302.
14 A. Carlsson and M. Lindqvist, Acta physiol. scand., 54 (1962) 83-86.
15 A. Carlsson and B. Waldeck, Acta physiol. scand., 44 (1958) 293-298.
16 A. Carlsson and B. Waldeck, Scand. J. Clin. Lab. Invest., 16 (1964) 133-138.
17 C.C. Chang, Int. J. Neuropharmacol., 3 (1964) 643-649.
18 G. Cohen and M. Goldenberg, J. Neurochem., 2 (1957) 58-70.
19 I. Cohen and W.H. Vogel, Experientia, 26 (1970) 1231-1232.
20 I. Cohen and W.H. Vogel, Biochem. Pharmac., 21 (1972) 1214-1216.
21 R.H. Cox Jr. and J.L. Perhach Jr., J. Neurochem., 20 (1973) 1777-1780.
22 T.B.B. Crawford and C.M. Yates, Br. J. Pharmac., 38 (1970) 56-71.
23 E. Crout, in D. Seligson (Ed.), Standard Methods in Clinical Chemistry, Vol. 3, Academic Press, N.Y., 1961, 62-80.
24 G. Curzon and A.R. Green, Br. J. Pharmac., 39 (1970) 653-655.
25 T.P. Davis, C.W. Gehrke, C.W. Gehrke Jr., T.D. Cunningham, K.C. Kuo, K.O. Gerhardt, H.D. Johnson and C.H. Williams, Clin. Chem., 24 (1978) 1317-1324.
26 W.D. Denckla and H.K. Dewey, J. Lab. Clin. Med., 69 (1967) 160-169.
27 W.G. Dewhurst and H.R. McKim, Neuropsychobiol., 5 (1979) 156-159.
28 B.D. Drujan, T. Sourkes, S. Layne and G. Murphy, Can. J. Biochem. Physiol., 37 (1959) 1153-1159.
29 D.A. Durden, B.A. Davis and A.A. Boulton, Biomed. Mass Spectrometry, 1 (1974) 83-95.
30 D.A. Durden, S.R. Philips and A.A. Boulton, Can. J. Biochem., 51 (1973) 995-1002.
31 I. Ehrlen, Farm. Rev. (Stockh.), 47 (1948) 242-252.
32 J.H. Gaddum and H. Schild, J. Physiol. (Lond.), 80 (1934) 9P-10P.
33 E.C. Gerst, O.S. Steinsland and W.W. Walcote, Clin. Chem., 12 (1966) 659-669.
34 E. Giacobini, J. Neurosci. Res., 1 (1975) 1-18.
35 G.G. Guilbault, Practical Fluorescence, Marcel Dekker, N.Y., 1973, 664 pp.
36 H.C. Guldberg, D.F. Sharman and P.R. Tegerdine, Br. J. Pharmac., 42 (1971) 505-511.
37 J. Haggendal, Acta physiol. scand., 56 (1962) 258-266.
38 J. Haggendal, Acta physiol. scand., 59 (1963) 242-254.
39 R. Hakanson, A.-L. Ronnberg and K. Sjolund, Anal. Biochem., 47 (1972) 356-370.
40 M. Hamaji and T. Seki, J. Chromatogr., 163 (1979) 329-336.
41 D.R. Haubrich and J.S. Denzer, Anal. Biochem., 55 (1973) 306-312.
42 S. Hess and S. Udenfriend, J. Pharmac. Exp. Ther., 127 (1959) 175-177.
43 R.B. Holman, P. Angwin and J.D. Barchas, Neuroscience, 1 (1976) 147-150.
44 J.B. Jepson and B.J. Stevens, Nature, 172 (1953) 772.
45 T. Karasawa, K. Furukawa, K. Yoshida and M. Shimizu, Japan J. Pharmac., 25 (1975) 727-736.
46 T. Kariya and M.H. Aprison, Anal. Biochem., 31 (1969) 102-113.
47 W. Kehr, Naun.-Schmied. Arch. Pharmac., 284 (1974) 149-158.
48 R. Laverty and D.F. Sharman, Br. J. Pharmac., 24 (1965) 538-548.
49 R. Laverty and K.M. Taylor, Anal. Biochem., 22 (1968) 269-279.
50 P. Lindroth and K. Mopper, Anal. Chem., 51 (1979) 1667-1674.
51 D.H. Lowry, The Harvey Lecture Series, 58 (1963) 1-19.
52 A. Lund, Acta Pharmacol. (Kbh), 5 (1949) 231-247.
53 A. Lund, Acta Pharmacol. Toxicol., 6 (1950) 137-146.
54 R.P. Maickel, R.H. Cox Jr., J. Saillant and F.P. Miller, Int. J. Neuropharmacol., 7 (1968) 275-281.
55 R.P. Maickel and F.P. Miller, Anal. Chem., 38 (1966) 1937-1938.
56 G. Mangan and J. Mason, Science, 126 (1957) 562-563.
57 W.R. Martin, J.W. Sloan, S.T. Christian and T.H. Clements, Psychopharmacologia (Berl.), 24 (1972) 331-346.

58 M.W. McCaman, D. Weinreich and R.E. McCaman, Brain Res., 53 (1973) 129-137.
59 E.G. McGeer and P.L. McGeer, Can. J. Biochem. Physiol., 40 (1962) 1141-1151.
60 H.R. McKim and W.G. Dewhurst, Proc. West. Pharmacol. Soc., 23 (1980) 291-294.
61 R.J. Merrills, Anal. Biochem., 6 (1963) 272-282.
62 G. Metcalf, Anal. Biochem., 57 (1974) 316-320.
63 F.P. Miller, R.H. Cox Jr., W.R. Snodgrass and R.P. Maickel, Biochem. Pharmac.,
    19 (1970) 435-442.
64 K.A. Montagu, Biochem. J., 63 (1956) 559-565.
65 K.A. Montagu, Nature, 180 (1957) 244-245.
66 G. Nadeau and L. Joly, Nature, 182 (1958) 180-181.
67 N. Narasimhachari, B. Heller, J. Spaide, L. Haskovec, H. Meltzer, M. Strahilevi
    and H.E. Himwich, Biol. Psychiat., 3 (1971) 21-23.
68 N. Narasimhachari and J. Plaut, J. Chromatogr., 57 (1971) 433-437.
69 N. Narasimhachari, J. Plaut and H. Himwich, J. Psychiat., 9 (1972) 325-328.
70 S. Natelson, J.K. Lugovoy and J.B. Pincus, Arch. Biochem. Biophys., 23 (1949)
    157-158.
71 J.A. Oates, in J.H. Quastel (Ed.), Methods in Medical Research, Year Book Medic
    Publishers Ltd., Chicago, 1961, pp. 169-174.
72 Z. Oberman, R. Chayen and M. Herzberg, Clin. Chim. Acta, 29 (1970) 391-394.
73 N.N. Osborne, in Microchemical Analysis of Nervous Tissue, Pergamon Press,
    Oxford, 1974, 225 pp.
74 J.F. Palmer, J. Pharm. Pharmac., 15 (1963) 777-778.
75 S.R. Phillips, D.A. Durden and A.A. Boulton, Can. J. Biochem., 52 (1974) 366-37
76 H.L. Price and M.L. Price, J. Lab. Clin. Med., 50 (1957) 769-777.
77 W.C. Prozialeck, D.H. Boehme and W.H. Vogel, J. Neurochem., 30 (1978) 1471-1477
78 W.B. Quay, Anal. Biochem., 5 (1963) 51-59.
79 W.Quay, J. Pharm. Sci., 57 (1968) 1568-1572.
80 M. Recasens, J. Zwiller, G. Mack, J.P. Zanetta and P. Mandel, Anal. Biochem., 8
    (1977) 8-17.
81 C.F. Saller and I.J. Kopin, Soc. Neurosci. Abstr., 6 (1980) 444.
82 N. Seiler and L. Demisch, in K. Blau and G.S. King (Eds.), Handbook of Derivati
    for Chromatography, Heyden, London, 1978, pp. 346-390.
83 N. Seiler and M. Wiechmann, in K. Blau and G. King (Eds.), Progress in Thin-Lay
    Chromatography and Related Methods, vol. 1, Ann Arbor-Humphrey, Ann Arbor, 197C
    pp. 95-144.
84 D.F. Sharman, in R. Fried (Ed.), Methods of Neurochemistry, vol. 1, Marcel Dekk
    New York, 1971, 83-128.
85 M.K. Shellenberger and J.H. Gordon, Anal. Biochem., 39 (1971) 356-372.
86 W.B. Shelley and L. Juhlin, J. Chromatogr., 22 (1966) 130-138.
87 P.A. Shore, A: Burkhalter and V.H. Cohn, J. Pharmac. Exp. Ther., 127 (1959)
    182-186.
88 P. Shore and J. Olin, J. Pharmac. Exp. Ther., 122 (1958) 295-300.
89 J.M. Slingsby and A.A. Boulton, J. Chromatogr., 123 (1976) 51-56.
90 E. Smith and H. Weil-Malherbe, Fed. Proc., 20 (1961) 182.
91 S.H. Snyder, J. Axelrod and M. Zweig, Biochem. Pharmac., 14 (1965) 831-835.
92 S.H. Snyder, R.J. Baldessarini and J. Axelrod, J. Pharmac. Exp. Ther., 153
    (1966) 544-551.
93 T.L. Sourkes and G.F. Murphy, in J.H. Quastel (Ed.), Methods in Medical Researc
    Year Book Medical Publishers Ltd., Chicago, 1961, pp. 147-152.
94 H. Spatz and N. Spatz, Biochem. Med., 6 (1972) 1-6.
95 S. Spector, K. Melman, W. Lovenberg and A. Sjoerdsma, J. Pharmac. Exp. Ther.,
    140 (1963) 229-235.
96 S. Suzuki and K. Yagi, Anal. Biochem., 75 (1976) 192-200.
97 V.J. Uuspaa, Ann. Med. Exptl. Biol. Fenniae (Helsinki), 41 (1963) 194-201.
98 A. DeT. Valk Jr. and H.L. Price, J. Clin. Invest., 35 (1956) 837-841.
99 J. Vanable, Anal. Biochem., 6 (1963) 393-403.
100 A. Vendsalu, Acta physiol. scand., 49 suppl., 173 (1960) 23-32.
101 C. Von Euler, U.S. Von Euler and I. Floding, Acta physiol. scand., 33, suppl. 1
    (1955) 32-38.
102 U.S. Von Euler and I. Floding, Acta physiol. scand., 33, suppl. 118 (1955) 45-5
103 U. Von Euler and F. Lishajko, Acta physiol. scand., 45 (1959) 122-132.

104 U.S. Von Euler and F. Lishajko, Acta physiol. scand., 51 (1961) 348-356.
105 H. Weil-Malherbe, Biochim. Biophys. Acta, 40 (1960) 351-353.
106 H. Weil-Malherbe, in J.H. Quastel (Ed.), Methods in Medical Research, Year Book
     Medical Publishers Inc., Chicago, 1961, pp. 130-146.
107 H. Weil-Malherbe and L.B. Bigelow, Anal. Biochem., 22 (1968) 321-334.
108 H. Weil-Malherbe and A.D. Bone, Biochem. J., 51 (1952) 311-318.
109 B.H.C. Westerink and J. Korf, J. Neurochem., 29 (1977) 697-706.
110 S.N. Young, G.M. Anderson, S. Gauthier and W.C. Purdy, J. Neurochem., 34 (1980)
     1087-1092.

Chapter 5

HISTOCHEMICAL APPROACHES TO THE DETECTION OF BIOGENIC AMINES

JOHN M. CANDY

MRC Neuroendocrinology Unit, Newcastle General Hospital, Westgate Road, Newcastle upon Tyne, NE4 6BE (United Kingdom)

5.1 INTRODUCTION
5.1.1 Validation of histochemical approach

It can be reasonably argued that the impetus for the explosive multidisciplinary interest in the biogenic monoamines over almost the last two decades resulted from the development of an exquisitely sensitive method for their cellular localization (1-3). The availability of detailed maps of the distribution of the monoamine-containing neurones and nerve terminals in the central nervous system (4,5) had a catalytic effect on neurochemistry, pharmacology and psychopharmacology. The fact that the chemical basis of the histochemical method that resulted in the controlled selective transformation of the biogenic monoamines into intensely fluorescent products had been extensively investigated and was well understood (see Section 5.2.1), was instrumental in the wide acceptance of this approach. In addition, microspectrofluorimetric analysis of the excitation and emission spectra allows the reaction products to be identified (see Section 5.2.3).

5.1.2 Outline of history of development of histochemical methods for the detection of the monoamines

Eros (6) and Eranko (7) were the first to report fluorescence after the form-aldehyde fixation of the adrenal medulla. This fluorescence was later shown to be due to reaction of the formaldehyde with 5-hydroxytryptamine (5-HT; serotonin), (8,9) and noradrenaline (NA), (10) respectively. The failure of the aqueous formaldehyde method to detect the monoamines in neurones was later attributed to diffusion of the monoamines, (11,12). The solution to this problem was the discovery by Falck et al. (1) that the catecholamines in a dried protein matrix are converted into intensely fluorescent derivatives when exposed to formaldehyde gas. The subsequent application of this technique to freeze-dried tissues (2,3) determined the methodology used for the histochemical detection of the monoamines for the ensuing decade. Although there were several attempts (13-16) to refine the aqueous formaldehyde method (7), because of its intrinsic simplicity none resulted in a method with comparable sensitivity to that developed by exposure of freeze-dried tissue to formaldehyde gas (formaldehyde-induced fluorescence, F.I.F.). The next major advance in the F.I.F. method was the introduction by Hokfelt and

Ljungdahl (17) of a new way of processing and sectioning tissue, allowing unfixed or lightly fixed tissue to be reproducibly sectioned down to 20 μm at 0-5°C using a vibratome. At about the same time Axelsson et al. (18) were searching for reagents, other than formaldehyde, that were capable of forming highly fluorescent molecules from the monoamines. From a range of carbonyl compounds screened, glyoxyl acid appeared to be one of the most potentially useful reagents. Subsequently, it was shown by Lindvall et al. (19,20) that application of the glyoxylic acid method to sections cut on a vibratome resulted in a remarkable increase in the sensitivity and precision with which the catecholamines could be visualized. The glyoxylic acid method which has been successfully applied to cryostat sections (21-23) is an improvement on the F.I.F. cryostat method (15,24-27). However, even with the introduction of the magnesium-catalyzed glyoxylic acid formaldehyde reaction and the aluminium-catalyzed formaldehyde reaction to cryostat-cut sections (28,29), the cryostat method is not as sensitive as the vibratome technique. The glyoxylic acid method has improved the localization of the monoamines in whole mount preparations (20,30,31) compared with the F.I.F. method (32-34). Interestingly, it has been reported that monoamines can be demonstrated in whole mount preparations with similar sensitivity to that of the glyoxylic acid method after incubation of the tissue at ambient temperature with a mixture of formaldehyde and glutaraldehyde (35). The main advantage of this technique is that the tissue can be subsequently processed for electron microscopy. A major increase in the sensitivity of the F.I.F. method has been made possible by the discovery that aluminium ions increase the fluorescence yield; this modification of the F.I.F. method has made it as sensitive as the glyoxylic acid vibratome procedure (29).

5.1.3 Fluorescence microscopy and microspectrofluorimetry in the detection of the monoamines

Fluorescence microscopy allows the detection of very low concentrations of substances that are fluorescent, or that can be made by chemical modification into fluorescent derivatives. The basic mechanisms of fluorescence are described in Chapter 4 (this volume). Suffice it to say here that the phenomenon of fluorescence involves the absorption of light of short wavelength, its excitation of the fluorophore molecules and its re-emission at a longer wavelength. Thus, fluorescence microscopy involves the illumination of the specimen with light of the appropriate wavelength (for detection of the monoamines the exciting light should be between 370 and 410 nm) and this is achieved by using an excitation filter in the illumination pathway. The light emitted by the specimen is detected after filtering out the shorter wavelength light using a secondary or barrier filter (for the monoamine fluorophores a secondary filter that transmits light above 460 nm should be used).

The ability to both accurately localize and quantitate substances within tissues is obviously of fundamental importance for quantitative histochemistry. Micro-

chemical and cytophotometric techniques have been used which enable the localization and quantitation of a substance in tissue. Cytophotometry involves the
optical quantitation of a substance within a cell and the most commonly used methods
are those of fluorimetry, absorptiometry and reflectiometry. For quantitation,
microfluorimetry has a number of advantages over microdensitometric methods,
including its specificity and extreme sensitivity, allowing both spectral analysis
and fluorescence quantitation of the amine fluorophores. Spectral analysis requires that both excitation and emission spectra are recorded and as these are
often characteristic of a particular fluorophore this approach allows tentative
identification of the amine fluorophore.

5.2 CHEMICAL BACKGROUND TO THE FORMALDEHYDE AND GLYOXYLIC ACID METHOD
5.2.1 Reaction mechanisms in the formaldehyde method

The chemical mechanisms involved have been extensively studied and elucidated
(12,36-40). Fluorophore formation in the formaldehyde reaction involves initially
a Pictet-Spengler type cyclization, which only occurs with primary and secondary
amines. Ring cyclization is produced by an electrophilic attack on the phenyl
nucleus and is promoted by a high electron density at the point of ring closure,
i.e. the 6-position of the phenylethylamine (PEA) molecule (Fig. 5.1, A1) and the
2-position of the indolylethylamine molecule (Fig. 5.1, B1), yielding weak or non-
fluorescent 1,2,3,4-tetrahydroisoquinoline compounds in the case of the PEA derivatives (Fig. 5.1, A2) or 1,2,3,4-tetrahydro-β-carboline compounds (Fig. 5.1, B2)
in the case of an indolylethylamine derivative. In the presence of a protein
matrix, the tetrahydro-derivatives are converted to strong fluorophores. Two
routes of fluorophore formation are possible; 1) autoxidation to the 3,4-dihydro-
isoquinoline (Fig. 5.1, A3) for the PEA derivatives or formation of a 3,4-dihydro-
β-carboline for the indolylethylamine derivatives (Fig. 5.1, B3); 2) a further
acid-catalyzed reaction of the 1,2,3,4-tetrahydroisoquinoline compound to yield
a 2-methyl-3,4-dihydroisoquinoline derivative (Fig. 5.1, A5) or in the case of the
1,2,3,4-tetrahydro-β-carboline a 2-methyl-3,4-dihydro-β-carbolinium compound
(Fig. 5.1, B4). The 3,4-dihydroisoquinolines derived from 3-hydroxylated PEAs
exhibit a pH-dependent tautomerism between two states of the fluorophore and this
is reflected in their fluorescent characteristics. In the pH range 6-10 the
intensely fluorescent quinonoidal form is dominant (Fig. 5.1, A4 and A6).

5.2.2 Reaction mechanisms in the glyoxylic acid method

Primary and secondary PEAs and indolylethylamines have been shown to react with
glyoxylic acid to form strong fluorophores in basically the same manner in which
they react with formaldehyde (41-44). The initial reaction of a PEA or indolylethylamine derivative in an acid-catalyzed Pictet-Spengler condensation results
in the formation, via a Schiff base, of a 1,2,3,4-tetrahydroisoquinoline-1-carboxylic

86

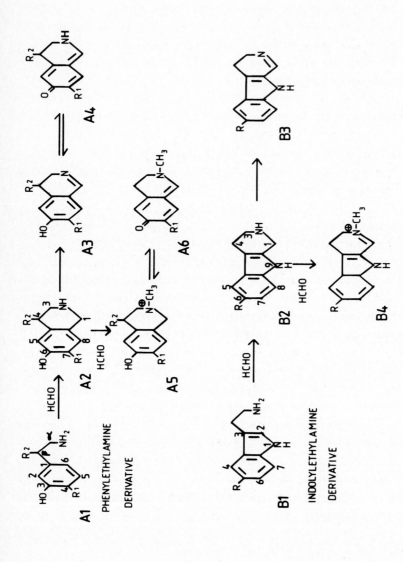

Fig. 5.1. Formation of fluorophores from phenylethylamines and indolylethylamines by formaldehyde. In the case of a phenylethylamine either a 3,4-dihydroisoquinoline derivative (A3) or a 2-methyl-3,4-dihydroisoquinolinium compound (A5) is formed. These derivatives exhibit a pH-dependent tautomerism with their quinonoidal forms (A4 and A6 respectively). For dopamine R_1 = OH, R_2 = H and for noradrenaline R_1 = OH, R_2 = OH. The indolylethylamines yield a 3,4-dihydro-β-carboline (B3) or through a second acid-catalyzed reaction a 2-methyl-3,4-dihydro-β-carbolinium compound (B4). For 5-hydroxytryptamine R = OH and for tryptamine R = H.

acid in the case of a PEA (Fig. 5.2, A2) or of a 1,2,3,4-tetrahydro-β-carboline-1-carboxylic acid (Fig. 5.2, B2) with an indolethylamine. These intermediate compounds are only weakly fluorescent and are transformed into intense fluorophores in similar ways to those outlined for the formaldehyde reaction, i.e. 1) autoxidative decarboxylation to form a 3,4-dihydroisoquinoline (Fig. 5.2, A3) in the PEA reaction sequence and a 3,4-dihydro-β-carboline (Fig. 5.2, B3) in the case of an indolylethylamine or 2) a further acid-catalyzed reaction with glyoxylic acid to yield in the case of a PEA a 2-carboxymethyl-3,4-dihydroisoquinolinium compound (Fig. 5.2, A5) or with an indolylethylamine a 2-carboxymethyl-3,4-dihydro-β-carbolinium compound (Fig. 5.2, B4), which may possibly be decarboxylated to yield a 2-methyl-3,4-dihydro-β-carbolinium derivative (Fig. 5.2, B5). The 3,4-dihydroisoquinoline and 2-carboxymethyl-3,4-dihydroisoquinolinium derivatives formed from 3-hydroxylated PEAs exhibit a pH-dependent tautomerism, i.e. at neutral pH their quinonoidal forms are dominant (Fig. 5.2, A4 and A6 respectively).

A number of studies have used a combined formaldehyde and glyoxylic acid treatment (e.g. 21,28,45); the most likely reaction sequences are shown in Fig. 5.3, using a 3-hydroxylated PEA derivative as an example. The PEA can either react with glyoxylic acid (Fig. 5.3, A1 to A4) or with formaldehyde (Fig. 5.3, A1 to A5 to A10 or A1 to A5 to A8) in the manner already described. In addition the 1,2,3,4-tetrahydroisoquinoline formed after condensation of a PEA with formaldehyde (Fig. 5.3, A5) can further react with glyoxylic acid to form a 2-carboxymethyl-3,4-dihydroisoquinolinium derivative (Fig. 5.2, A6). No further reaction with glyoxylic acid at the 4-position can occur with this derivative however; the 3,4-dihydroxyphenylalanine (DOPA) derivative reacts with glyoxylic acid to form a 2,4-dicarboxymethyl-3,4-dihydroisoquinolinium compound. Thus the fluorophores formed from DOPA and dopamine (DA) in this mixed reaction have marked structural differences and this forms a possible basis for their microspectrofluorimetric differentiation (46). The indolylethylamines and their corresponding amino acids after condensation with formaldehyde almost certainly can react further with glyoxylic acid in an analogous way to the reactions of the DA and DOPA derivatives (46).

5.2.3 Fluorescence and microspectrofluorimetric analysis in the detection of the monoamines

Table 5.1 lists the fluorescence yields relative to those of NA or DA in the F.I.F. reaction and the spectral characteristics of a number of PEA derivatives after reaction in a protein matrix with formaldehyde vapour or glyoxylic acid vapour. It can be seen that the fluorescence yield in both the formaldehyde and glyoxylic acid reactions is strongly influenced by both the nature and position of the substituents on the 3,4-position of the molecule. Also, in all cases where formaldehyde induced fluorescence, the glyoxylic acid method gave a greater fluo-

88

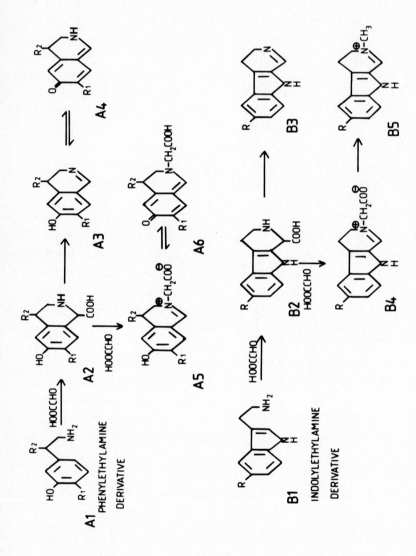

Fig. 5.2. Reaction of phenylethylamine and indolylethylamine derivatives with glyoxylic acid. Phenylethylamines yield a 3,4-dihydroisoquinoline (A3) or in a second acid-catalyzed reaction a 2-carboxymethyl-3,4-dihydroisoquinolinium derivative (A5). These derivatives at neutral pH are in their quinonoidal forms (A4 and A6 respectively). Indolylethylamines can form either a 3,4-dihydro-β-carboline (B3) or a 2-carboxylmethyl-3,4-dihydro-β-carboline (B3) or a 2-carboxylmethyl-3,4-dihydro-β-carbolinium compound (B5).

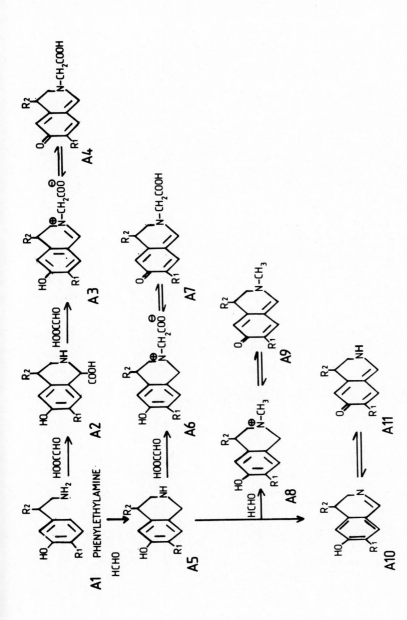

Fig. 5.3. Illustrates the possible reaction pathways for a phenylethylamine in the combined formaldehyde and glyoxylic acid reactions. Derivatives formed by reaction solely with glyoxylic acid include 2-carboxymethyl-3,4-dihydroisoquin-olinium compounds (A3, A4), while derivatives formed by reaction solely with formaldehyde are a 2-methyl-3,4-dihydro-isoquinolinium compound (A9) and 3,4-dihydroisoquinoline compounds (A10, A11). In addition a 2-carboxymethyl-3,4-dihydroisoquinolinium derivative (A6) may be formed after reaction of a phenylethylamine with formaldehyde and glyoxylic acid.

TABLE 5.1

FLUORESCENCE YIELDS AND SPECTRAL CHARACTERISTICS OF A NUMBER OF PHENYLETHYLAMINE DERIVATIVES AFTER FORMALDEHYDE AND GLYOXYLIC ACID TREATMENTS [ADAPTED FROM BJORKLUND et al. (47), LINDVALL et al. (42), BJORKLUND et al. (43), BJORKLUND et al. (51) and LINDVALL and BJORKLUND (20)]

COMPOUND	FORMALDEHYDE VAPOUR TREATMENT			GLYOXYLIC ACID VAPOUR TREATMENT		
	RELATIVE FLUORESCENCE YIELD*	EXCITATION MAX (nm)	EMISSION MAX (nm)	RELATIVE FLUORESCENCE YIELD*	EXCITATION MAX (nm)	EMISSION MAX (nm)
PEA (Phenylethylamine)	0	-	-	8a(0)b	-	-
Phenylalanine	0	-	-	3	-	-
3-Hydroxy-PEA (m-tyramine)	58	385	415 and 510	178	NA	NA
α-Carboxy-3-Hydroxy PEA (m-tyrosine)	NA	NA	NA	480	NA	NA
2-Hydroxy-PEA (o-tyramine)	0	-	-	3	-	-
α-Carboxy-2-hydroxy PEA (o-tyrosine)	NA	NA	NA	23	NA	NA
4-Hydroxy-PEA (p-tyramine)	0	-	-	3	-	-
α-Carboxy-4-hydroxy-PEA (p-tyrosine)	0	-	-	13	NA	NA
4,β-dihydroxy-PEA (octopamine)	0	-	-	10(25)	NA	NA
3-methoxy-PEA	2	-	-	0	-	-
4-methoxy-PEA	0	-	-	0	-	-
3,4-dihydroxy-PEA (dopamine)	109	320 and 410	475	452(810)	330 and 375	460
3-hydroxy-4-methoxy-PEA	227	405	460	3443	NA	NA
3-methoxy-4-hydroxy-PEA	0	-	-	90	NA	NA
3,4-dimethoxy-PEA	0	-	-	250	-	-
N-methyl-3,4,β-trihydroxy-PEA (adrenaline)	47	320 and 410	475	14(45)	335	485
α-Carboxy-3,4-dihydroxy-DOPA	120	320 and 410	475	638(570)	330 and 380	480
α-Carboxy-3,4,β-trihydroxy-PEA (DOPS)	10	380	470	145	NA	NA
3,4-β-trihydroxy PEA (noradrenaline)	100	320 and 410	475	296(445)	330 and 375	460
α-methyl-3,4,β-trihydroxy-PEA (α-methylnoradrenaline)	100	320 and 410	475	330	NA	NA
α-methyl-3,4-dihydroxy-PEA (α-methyl dopamine)	70	320 and 410	475	360	NA	NA

NA, data not available.

* Fluorescence yield relative to NA or DA yield in the F.I.F. method.

a Data from Lindvall et al. (42)

b Data from Bjorklund et al. (43) or Lindvall and Bjorklund (20).

rescence yield. PEA and its amino acid phenylalanine gave no detectable fluorescence in either reaction. Only the 3-hydroxylated PEAs (m-tyramine, m-tyrosine) of
the PEA derivatives with only one substituent on the benzene ring gave a significant
fluorescence yield. 2-Hydroxylated PEA derivatives (o-tyramine, o-tyrosine),
4-hydroxylated PEA derivatives (p-tyramine, p-tyrosine and octopamine), 3-methoxy-
PEA and 4-methoxy-PEA were without significant fluorescence. 3-Hydroxylated PEAs
gave a greater fluorescence yield than 3-methoxylated PEAs, thus m-tyramine
produced a more intensely fluorescent fluorophore than 3-methoxy-PEA. The presence
of a hydroxy or methoxy group in the 4-position in addition to a hydroxy group
at the 3-position (DA, 3-hydroxy-4-methoxy-PEA) greatly increased the fluorescence
yield in comparison to a PEA derivative with a hydroxy group in the 3-position
and no substituent in the 4-position (m-tyramine). The presence of a hydroxy
group in the 3-position was essential for fluorescence in the formaldehyde reaction as 3-methoxy-PEA, 3-methoxy-4-hydroxy-PEA and 3,4-dimethoxy-PEA formaldehyde
derivatives were non-fluorescent. Interestingly, in the glyoxylic acid reaction
3-methoxy-4-hydroxy PEA gave significant fluorescence while the 3,4-dimethoxy-PEA
derivatives did not. Substitution on the nitrogen of the side chain reduces the
fluorescence yield, thus adrenaline (A) produced only half the fluorescence yield
compared with DA or NA in the formaldehyde method and a considerably lower fluorescence yield in the glyoxylic acid reaction. The presence of a carboxy group
on the α-carbon of the side chain resulted in an increased fluorescence yield in
the glyoxylic acid reaction in the case of m-tyrosine and DOPA, while DOPA has a
similar fluorescence yield to DA in the formaldehyde reaction. However, the presence
of a methyl group on the α-carbon (α-methyl-NA and α-methyl-DA) did not significantly affect the fluorescence yield compared to that of the parent molecule.
The presence of a β-hydroxyl group in NA did not affect the fluorescence yield in
the formaldehyde reaction but significantly decreased it in the glyoxylic acid
reaction; however, with α-carboxy-3,4, β-trihydroxy-PEA (DOPS), which has both a
carboxyl group on the α-carbon and a β-hydroxyl group, there was a marked decrease
in the fluorescence yield compared with that of DA in both the formaldehyde and
glyoxylic acid reactions. Table 5.1 shows that the spectral characteristics of
the phenylethylamine derivatives (with the exception of m-tyramine) are similar for
both the formaldehyde and glyoxylic acid methods. Thus the excitation/emission
maxima are at 320 nm and 410 nm/470-480 nm in the case of the formaldehyde-induced
fluorophores and 330 and 375 nm/460-480 nm for the glyoxylic acid-induced fluorophores. The NA and A fluorophores produced have a labile 4-hydroxy group which is
easily split off by acid (42,48-50) yielding a dihydroisoquinoline derivative
which exhibits quite different spectral properties with the main excitation peak
at 320-330 nm. In contrast, the excitation peak for the DA fluorophore after
acid treatment remains unchanged (42,48-50). This change in the excitation spectra
of the NA fluorophore after exposure to acid forms the basis of a method for differ-

TABLE 5.2

FLUORESCENCE YIELDS AND EXCITATION/EMISSION MAXIMA OF SOME INDOLYLETHYLAME DERIVATIVES AFTER FORMALDEHYDE OR GLYOXYLIC ACID TREATMENT [ADAPTED FROM BJORKLUND et al. (47), LINDVALL et al. (42), BJORKLUND (43) and BJORKLUND, LINDVALL and BJORKLUND (20)]

COMPOUND	FORMALDEHYDE VAPOUR TREATMENT			GLYOXYLIC ACID TREATMENT		
	RELATIVE FLUORESCENCE YIELD*	EXCITATION MAX (nm)	EMISSION MAX (nm)	RELATIVE FLUORESCENCE YIELD*	EXCITATION MAX (nm)	EMISSION MAX (nm)
Tryptamine	9	370	495	305[a](280)[b]	370	495
Tryptophan	8	375	435 or 500	375 (240)	370	500
5-Hydroxytryptamine	33	(315)[d],385 and 415	520-530	125 (70)	375	520-540
5-Methoxytryptamine	14	(330),380 and (410)	505[c]	(100)	375	520-540
5-Hydroxytryptophan	3	310, 385 and 415	520-530	130 (30)	(320),380 and 410	520-540
5-Methoxytryptophan	3	(385) and 410	505	NA	NA	NA
6-Hydroxytryptamine	195	385 (420)	505	(600)	385-400	500
5,6-Dihydroxytryptamine	14	310, (380) and 405	500	(190)	405	500
N-Acetyl-5-hydroxytryptamine	NA	NA	NA	115 (20)	370	480
N-Acetyl-5-methoxytryptamine (Melatonin)	0	-	-	80 (40)	(335) and 370	500
N,N-Dimethyltryptamine	0	-	-	115	(340) and 370	430
N,N-Dimethyl-5-hydroxy-tryptamine (Bufotenin)	0	-	-	35	(340) and 370	430
N-Methyl-5-hydroxytryptamine	3	315,(390) and 415	490	80 (20)	370	500

NA, Data not available.
*Fluorescence yield relative to NA or DA yield in the F.I.F. method.
[a]Data from Lindvall and Bjorklund (20).

[b]Data in brackets from Bjorklund et al. (43)
[c]Emission peak maximum 520-530nM at high concentrations.
[d]Figure in brackets indicates the position of a small peak or shoulder in the spectrum.

entiating DA from NA at the cellular level using microspectrofluorimetric analysis (49,50).

Table 5.2 shows the fluorescence yields relative to DA and NA in the F.I.F. method and spectral characteristics of a number of indolylethylamine derivatives in the formaldehyde and glyoxylic acid reactions. Tryptamine (T) and its amino acid tryptophan (TP) produce only a very low fluorescence yield in the formalde- hyde reaction; however, glyoxylic acid converts these compounds into intensely fluorescent derivatives. A hydroxy or methoxy group in the 5-position of T (5-HT, 5-methoxy-T) considerably reduces the fluorescence yield in both the formaldehyde and glyoxylic acid reactions. The presence of a carboxy group on the α-carbon of the side chain results in a decreased fluorescence yield for both 5-hydroxytryp- tophan (5-HTP) and 5-methoxytryptophan (5-MTP) in the formaldehyde reaction but has little effect on the fluorescence yield of 5-HTP in the glyoxylic acid method. The presence of a hydroxy group in the 6-position (6-hydroxy-T) results in a dramatic increase in the fluorescence yield in both reactions. However, hydroxy substitu- tion in the 5,6-position (5,6-dihydroxy-T) drastically reduces the fluorescence yield. Acetylation of the nitrogen of the side chain results in some reduction in fluorescence yield in the glyoxylic acid reaction for both melatonin and N-acetyl- 5-HT, whereas melatonin exhibits no fluorescence in the formaldehyde method. N- Methylated indolylethylamines (N,N-dimethyl-T, bufotenin and N-methyl-5-HT) all exhibit fluorescence in the glyoxylic acid method while they are almost totally non- fluorescent in the formaldehyde method. It is apparent that neither the glyoxylic acid nor formaldehyde methods optimally detects 5-HT.

The indolylethylamine fluorophores produced in the formaldehyde reaction have emission peak maxima that are at a longer wavelength than the phenylethylamine fluorophores formed by condensation with formaldehyde (cf Tables 5.1 and 5.2). Two groups of indolylethylamine fluorophores can be distinguished on the basis of their different emission peak maxima. One group of indolylethylamines (T, TP, 5-methoxy- T, 5-MTP, 6-hydroxy-T, 5,6-dihydroxy-T and N-methyl-5-HT) has emission peak maxima in the range from 490-505 nm. The other group of indolylethylamines has excitation maxima ranging from 520-530 nm and includes 5-HT and 5-HTP as well as 5-methoxy-T at high concentration (Table 5.2). The group of indolylethylamines with emission maxima in the range 490-505 nm can be further split into two groups on the basis of their excitation maxima. The first group is characterized by a single excitation peak in the region of 370 nm, e.g. T. The other group is distinguished by the presence of two excitation peaks with maxima at 380-390 nm and at 405-420 nm, e.g. 5-HT; in addition this group often has a small peak or shoulder in its excitation spectrum at 310-330 nm.

After glyoxylic acid treatment the indolylethylamines can be separated into 3

TABLE 5.3

Methods for the visualisation of the monoamines utilising freeze-drying.

Procedure	Methods utilising freeze-dried paraffin embedded tissue		
	Formaldehyde Reaction Falck and Owman (52)	Aluminium-Catalysed Reaction Ajelis et al. (29)	Magnesium-Catalysed Reaction Lorén et al. (45)
Perfusion Conditions	-	Ice-cold tyrode buffer (pH7) containing 2% glyoxylic acid followed at high pressure (0.8-2 bar/cm^2) by tyrode buffer (pH3.8) containing 4% paraformaldehyde with 150mM Al$_2$(SO$_4$)$_3$.	Either a) 4% paraformaldehyde with 1M MgSO$_4$ in 0.1M phosphate buffer (final pH4-5) or b) 2% glyoxylic acid and 0.5% paraformaldehyde with 1M MgSO$_4$ in 0.1M phosphate buffer (final pH4-5).
Tissue Processing	Tissue rapidly excised and quenched (isopentane or propane-propylene precooled in liquid nitrogen). Freeze-dried at -30°C to -40°C. Treated with para-formaldehyde (50-70% relative humidity) at 80°C for 1h. Embedded in paraffin wax. Sectioned. Sections mounted using non-fluorescent mountant.	Same as formaldehyde reaction.	Same as formaldehyde reaction.

groups on the basis of their excitation spectra. Thus 5-HT, 5-methoxy-T
and 5-HTP have excitation maxima ranging from 520-540 nm and N,N-dimethyl-T
and N,N-dimethyl-5-HT have an excitation peak at a much shorter wavelength
(430 nm), while the remaining indolylethylamines have similar excitation peak
maxima in the range 480-500 nm.

5.3 PRACTICAL ASPECTS OF THE HISTOCHEMICAL DETECTION OF THE MONOAMINES
5.3.1 Tissue preparation

There are a number of methodological approaches available for the visualization
of the biogenic monoamines and several of these methods are summarized in Tables
5.3, 5.4, 5.5 and 5.6. The essential differences between the methodological
approaches are: a) whether the tissue is perfused with a fluorophore-forming agent
or a chemical that will promote fluorophore formation, b) whether or not the tissue
is frozen and c) which fluorophore forming agent is used. A prerequisite of all
the methodologies is the rapid removal or perfusion of the tissue under investigation, as any lengthy post-mortem delay will almost certainly have deleterious
effects on the visualization of the amines.

The question of whether or not to freeze the sample is always relevant. The
method adopted will obviously depend on the equipment available and the type of
investigation being undertaken. If the tissue is to be frozen, then ice crystal
formation must be prevented, as such crystals disrupt the tissue. Rapid freezing
of the tissue limits ice crystal formation and this can be achieved by immersing
the tissue in a propane-propylene mixture that has been precooled to the temperature
of liquid nitrogen or in isopentane precooled in liquid nitrogen until crystals
of isopentane start to appear. The tissue can then be stored almost indefinitely
in liquid nitrogen. Direct immersion of the tissue in liquid nitrogen is not
adequate since nitrogen gas forms around it and slows the rate of cooling. Rapid
freezing of the tissue frequently results in the formation of tissue cracks;
covering the tissue with fine cotton gauze helps to keep the tissue intact during
subsequent processing. The frozen sample can then either be freeze-dried or
sectioned in a cryostat.

Freeze-drying involves the removal of water by sublimation at a temperature
below the freezing point of the tissue and is carried out in vacuo using a cold-
finger or desiccant (52-54). A number of suitable tissue dryers are available
commercially (e.g. from Edwards High Vacuum, U.K., Bergman and Beving Co., Sweden,
and FTS Systems Inc., U.S.A.). Freeze-drying of the tissue is normally carried
out at -30°C to -40°C under a vacuum better than 0.01 torr using phosphorus pentoxide as the desiccant for periods ranging from 2-14 days depending on the size
of the tissue pieces. Large pieces of tissue up to 1cm^2 are adequately freeze-
dried after a period of 7-10 days. At the end of the drying process, to avoid
condensation, the tissue is warmed to room temperature before releasing the vacuum

TABLE 5.4

Cryostat methods for the visualisation of the monoamines.

Methods utilising cryostat sections

Procedure	Formaldehyde Method Watson and Ellison (55)	Glyoxlyic Acid Method Bloom and Battenberg (21)	Aluminium-Catalysed Reaction Ajelis et al. (29)	Magnesium-Catalysed Reaction Lorén et al. (28)
Perfusion Conditions	4% paraformaldehyde with 0.5% MgCl2 in 0.1M phosphate buffer (pH7).	2% glyoxylic acid with 0.5% paraformaldehyde in 0.1M phosphate buffer (pH7.4).	Perfusion not essential. Perfusate tyrode buffer (pH3.8) containing 4% paraformaldehyde with 150mM Al2(SO4)3 at high pressure (0.8-2 bar/cm²).	4% paraformaldehyde in 0.1M phosphate buffer with 60-130mM MgSO4 (final pH 6.2-6.5).
Tissue Processing	Tissue dissected. Frozen using compressed Freon. Sections cut at -20°C, dried in vacuo over phosphorus pentoxide for minimum 1h. Treated with paraformaldehyde (60% relative humidity at 80°C for 1h). Mounted with immersion oil.	Tissue dissected. Frozen using solid carbon dioxide. Sections cut at -20°C. Immersed in ice-cold 2% glyoxylic acid in 0.1M phosphate buffer (pH7.4) for 1min. Dried in warm air for 5 min. Heated 100°C for 10min. Mounted in paraffin oil.	Sections cut. For non-perfused tissue immersed 30-60s in ice-cold tyrode buffer (pH7) containing 75mM Al2(SO4)3. Dried in warm air for 10-15min, further dried in vacuo over phosphorus pentoxide for minimum 2h. Heated 80°C for 10min, then heated with paraformalde-hyde (50% relative humidity) at 80°C for 1h. Sections mounted in paraffin oil.	Sections cut at -30°C. Sections either air-dried or immersed 2% glyoxylic acid in 0.1M phosphate buffer (pH7) before air-drying. Further air-dried in vacuo over phosphorus pentoxide for up to 24h. Heated with paraformalde-hyde (50-70% relative humidity) at 80°C for 1h. Sections mounted in liquid paraffin.

TABLE 5.5

Vibratome techniques for the demonstration of the monoamines.

Procedure	Methods utilising vibratome sections		
	Formaldehyde Method Hökfelt et al. (17)	Glyoxylic Acid Method Lindvall et al. (20)	Magnesium-Catalysed Reaction Lorén et al. (28)
Perfusion Conditions	Ice cold calcium-free tyrode buffer (pH7).	2% glyoxylic acid in Krebs-Ringer bicarbonate buffer (pH7). Perfusion at high pressure.	4% paraformaldehyde in 0.1M phosphate buffer with 60-130mM $MgSO_4$ (final pH6.2-6.5).
Tissue Processing	Sections (20µm) cut at 0-5°C, in warm stream of air. Further dried 15h in vacuo over phosphorus pentoxide. Heated with paraformaldehyde at 80°C for 1h. Mounted using non-fluorescent mountant.	Sections cut at 0-5°C. Incubated in 2% glyoxylic acid in Krebs-Ringer buffer (pH7) for 3-5min. Dried in warm stream of air for 15min, further dried for 15h in vacuo over phosporus pentoxide. Sections heated at 100°C for 6min or treated with dried glyoxylic acid vapour at 100°C at a partial pressure of 300 torr for 2min. Mounted in liquid paraffin.	Sections cut at 0-5°C. Incubated in 0.1M phosphate buffer with 130mM $MgSO_4$ (pH6.2) for 1-2min. Sections dried in warm stream of air for 10-15min. Further dried in vacuo over phosphorus pentoxide for up to 24h. Treated with paraformaldehyde (50-70% relative humidity) at 80°C for 1h. Mounted using non-fluorescent mountant.

TABLE 5.6

Methods for the preparation of whole mounts of tissue for the visualisation of the monoamines.

Procedure	Whole mount methods			
	Glyoxylic Acid Reaction Furness and Costa (59)	Formaldehyde-Glutaraldehyde Aqueous Method Furness et al. (35)	Aluminium-Catalysed Reaction Ajelis et al. (29)	CNS Smears Olson and Ungerstedt (60)
Tissue Processing	Incubated in 2% glyoxylic acid in 0.1M phosphate buffer (pH7) for 30min. Tissue partially air-dried for 3-5min. Heated 100°C for 4min. Mounted using non-fluorescent mountant.	Incubated in mixture 4% paraformaldehyde and 0.5% glutaraldehyde in 0.1M phosphate buffer (pH7) at room temperature for 1h. Mounted in fixative.	Incubated 3-5min in ice-cold tyrode buffer (pH7) containing $Al_2(SO_4)_3$ 75 mM. Dried in warm air stream for 10-15min, further dried in vacuo for minimum 2h. Heated with paraformaldehyde at 80°C for 1h. Sections mounted in paraffin oil.	Slice from tissue punch smeared onto clean, degreased slide. Dried over phosphorus pentoxide 15h. Heated with paraformaldehyde (60% relative humidity) at 80°C for 1h. Mounted in immersion oil.

and removing the tissue. For the aluminium-catalyzed (29) and magnesium-catalyzed
(45) reactions on freeze-dried tissue, the tissue is perfused with glyoxylic acid
or glyoxylic acid and formaldehyde together either with aluminium sulphate or
magnesium sulphate before it is frozen, while for the F.I.F. method (52) the
tissue is not perfused (Table 5.3). Methods of amine localization which use
cryostat sections are summarized in Table 5.4. The tissue is normally perfused
with formaldehyde, glyoxylic acid or a mixture of these reagents with or without
the addition of high concentrations of various cations (21,55,28,29). Sections
10-20 µm thick are normally cut at between -20°C and -30°C.

An alternative approach that does not involve freezing the tissue prior to
embedding and sectioning uses 20-50 µm thick sections that are cut using a
Vibratome (Oxford Instrument Co., U.S.A.) from material that has been perfused with
glyoxylic acid, formaldehyde or a mixture of these substances with or without the
addition of magnesium of aluminium salts (Table 5.5). The sections are cut in
buffer at a temperature of between 0°C and 5°C. Two other approaches that do not
involve freezing the tissue involve the use of whole mounts of peripheral organs
(35,59) or a tissue smear technique (60) (Table 5.6).

5.3.2 Methods of fluorophore formation

In the F.I.F. method (52) and also in the aluminium-catalyzed (29) or magnesium-
catalyzed (45) reactions, the freeze-dried tissue is exposed to formaldehyde gas
in a tightly sealed container at 80°C for 1 h. The formaldehyde gas is generated
from 5 g paraformaldehyde in a 1 litre vessel. A critical factor is the water
content of the paraformaldehyde (52,56,57). The formaldehyde reaction requires
the presence of a certain amount of water; thus if the formaldehyde gas is too
dry the fluorescence is weakly but distinctly localized and if too much water is
present quenching can occur (58). Paraformaldehyde equilibrated to a relative
humidity of between 50 and 70% has been found to be generally satisfactory. This
is achieved by equilibrating paraformaldehyde over a solution of sulphuric acid of
the appropriate density (56). After formaldehyde treatment the freeze-dried tissue
specimens are embedded in paraffin wax in vacuo and block-embedded. The infiltra-
tion of the specimens with molten paraffin wax should be as rapid as possible to
avoid extraction of the fluorophores. The fluorophores are stable in the paraffin
block for several months. Because of quenching of the fluorescence by water (58),
paraffin sections cannot be relaxed in the conventional way on a warm water bath.
Instead the sections have to be gently melted onto slides, and this obviously can
create problems due to folding of the tissue. A non-fluorescent mountant such as
Entellan (E. Merck, W. Germany), Fluormount (E. Gurr, U.K.), liquid paraffin or
immersion oil is used. The sections are then warmed for a short period at 60°C
to dissolve the paraffin wax.

The cryostat methods (21,28,29,55) involve a brief incubation in a buffer

solution containing glyoxylic acid or, in the case of the aluminium-catalyzed reaction, aluminium sulphate followed by drying in a warm stream of air. In the
glyoxylic acid method (21) the sections are heated at 100°C for 10 min and mounted
in liquid paraffin. In the cases of the aluminium-catalyzed (29) and magnesium-
catalyzed (28) reactions the sections are further dried over phosphorus pentoxide
for up to 18 h. After the period of desiccation the sections are heated with
paraformaldehyde (50-70% relative humidity) at 80°C for 1 h and mounted in liquid
paraffin. The formaldehyde cryostat method (55) involves drying the cryostat
sections over phosphorus pentoxide and then heating with paraformaldehyde (60%
relative humidity) at 80°C for 1 h.

In the glyoxylic acid and magnesium-catalyzed vibratome methods (20,28) the
sections are briefly incubated in glyoxylic acid or a solution containing a high
concentration of magnesium sulphate. The sections are air-dried and kept in vacuo
over phosphorus pentoxide for up to 24 h and then heated either with paraformaldeh
(50-70% relative humidity) at 80°C for 1 h or glyoxylic acid at 100°C for 2 min.
The glyoxylic acid vapour treatment is carried out by heating glyoxylic acid
(previously dried for 24 h over phosphorus pentoxide) at 100°C for 1 h. The hot
glyoxylic acid-saturated air is introduced into an evacuated vessel containing
the sections to a partial pressure of 300 torr. The sections are mounted in liquid
paraffin. In the formaldehyde Vibratome method (17) the sections are initially
air-dried then further dried over phosphorus pentoxide for 15 h; this is followed
by heating with paraformaldehyde (50-70% relative humidity) at 80°C for 1 h.

Whole mounts of peripheral organs can be processed by immersing the whole mount
in glyoxylic acid (59), in 4% formaldehyde with aluminium sulphate (29) or in a
mixture of formaldehyde and glutaraldehyde (35). In the glyoxylic acid method the
tissue is partially air-dried and then heated. In the formaldehyde-glutaraldehyde
method no heating above ambient temperature is necessary; the tissue is simply
incubated at room temperature for 1 h. In the aluminium-catalyzed reaction the
whole mounts are thoroughly air-dried before heating with paraformaldehyde.

A simple smear technique was developed by Olson and Ungerstedt (60) which involves drying a tissue smear and then treating it with formaldehyde vapour. This
method has been applied successfully to the visualization of monoamine-containing
nerve terminals in a number of different areas of the rat brain (60).

5.3.3 Fluorescence microscopy and microspectrofluorimetry

Normally the fluorescent specimen is viewed against a dark background, thus
enabling the visual detection of very low concentrations of a fluorophore. The
apparent colour and contrast of the fluorescence will depend on the secondary
filter used and also on the intensity of the fluorescence because the human eye
is unable to appreciate colour properly at low light intensities. Two ways of
illuminating the specimen are employed. One system uses dark ground illumination;

in this case only fluorescence from the specimen and light scattered from the dark ground condenser enters the objective. This system gives a moderate light intensity with low power objectives but with high power objectives the light intensity is very low. Epi-illumination is the method of choice for high power objectives as in this case the objective is also the condenser, thus concentrating the illumination precisely on the field of view. The light source is normally a mercury vapour lamp since this has mercury emission lines at 365 nm and 406 nm which are very close to the excitation maxima of the PEA and indolylethylamine fluorophores. The choice of primary and secondary filters is crucial and depends primarily on the fluorophore. Thus for the NA fluorophore in the F.I.F. method, which has an excitation maximum at 410 nm, the closest mercury emission is at 406 nm. A combination of a wide-band glass filter (e.g. BG12) and a red-absorbing filter (e.g. BG38) has been frequently used as the excitation filter for the NA fluorophore but selective excitation would be achieved by using in addition a narrow band interference filter (e.g. Schott AL406). The emission maximum of the NA fluorophore is at 475 nm, and consequently the barrier filter should be chosen so that its cut-off is below 480 nm (e.g. K460 or K470 (Leitz) or LP478 (Zeiss)).

A possible pitfall in relating the colour of the observed fluorescence to the fluorophore present is highlighted by the reports that in certain cases the fluorescence can appear yellow (indicative of the presence of an indolylethylamine) while microspectrofluorimetric analysis indicates it is typical catecholamine fluorescence at 480 nm (61,62). This discrepancy is probably explained by a shift in the maximum sensitivity of the human eye with increasing light intensity (see 11). Thus microspectrofluorimetric analysis is an essential tool for the unequivocal identification of the fluorophore. The essential features of a microspectrofluorimeter (63) are a stabilized xenon lamp for illumination, excitation and emission monochromators, field diaphragms to restrict the area illuminated and the area from which measurements are to be taken and a sensitive photometer. There are three different microspectrofluorimeter systems that are available commercially and that have monochromators for both excitation and emission. These are the Schoeffel-Leitz Microfluorimeter (Schoeffel Instrument Co., U.S.A.), the Leitz Microspectrograph (E. Leitz, W. Germany) and the Farrand MSA (Farrand Optical Co., U.S.A.).

Excitation and emission spectra are distorted by factors inherent in the instrumentation and these are characteristic for each instrument. For instance, both the excitation and emission spectra are affected by the transmission characteristics of the monochromators and optics. A calibrated tungsten lamp can be used to calibrate the instrument. Correction of the excitation spectrum requires that the intensity of the existing light be monitored since the characteristics of this light vary between lamps and throughout the life of a particular lamp. For recording excitation spectra down to about 300 nm, quartz optics and a quartz con-

denser must be used. In addition the monochromator must have a high transmittance
at this wavelength. Rapid spectral recordings should be made to avoid distortion
of the spectra due to fluorescence fading.

Fluorescence quantitation relies on the use of a fluorescence standard and one
of the problems is that this should have similar fluorescence characteristics to
the fluorophore being investigated and it should also exhibit stable fluorescence.
Ritzen (58) has used uranium glass particles as a fluorescence standard for the
quantitation of the monoamines.

5.4 EVALUATION OF THE FORMALDEHYDE AND GLYOXYLIC ACID TECHNIQUES

5.4.1 <u>Formation and extraction of the fluorophores</u>

Originally, it was thought that for intensely fluorescent amines the F.I.F.
method gave an almost quantitative yield of the fluorophore (37,64) while for the
less intensely fluorescent amines there was a correspondingly lower fluorophore
yield (65,66). However, the fluorophore yield obtained with DA and T, which
produce intense and weak fluorophores respectively in the F.I.F. method, is only
about 10% (40-42). Attempts therefore have been made to increase the fluorophore
yield in the F.I.F. reaction. The Pictet-Spengler reaction is catalyzed by
hydrogen ions (67). Thus protein micro droplets containing monoamines and tissue
have been treated with hot formaldehyde gas in the presence of hydrochloric acid
vapour (47,60). In addition, the effects of a number of metal ions with Lewis
acid properties have been studied (28,45,67,68). Treatment with ozone has also
been tried in an attempt to increase the fluorophore yield from the indoleethyl-
amines (65). Surprisingly, none of these treatments resulted in a dramatic increase in the fluorophore yield using the F.I.F. method (see section 5.4.2). Acid
catalysis is also important in both steps of the glyoxylic acid reaction. Glyoxyli
acid promotes the initial cyclization step and the subsequent presence of a carboxy
group at the 1-position of both the tetrahydroisoquinoline and tetrahydro-β-carboli
derivatives promotes intramolecular catalysis, resulting in the formation of intensely fluorescent derivatives (see section 5.2.2). An important finding was that
if the gaseous reactions with formaldehyde or glyoxylic acid vapour are carried out
under conditions simulating the perfusion-immersion procedure (see section 5.3.2),
then there is an almost quantitative fluorophore yield (20,42).

The F.I.F. reaction requires the presence of a certain amount of water (11,57,
70), although the exact role played by the water is uncertain. It may help to
increase the reactivity of the formaldehyde vapour or favour the formation of the
more fluorescent quinonoidal form of the fluorophore (12). The reproducibility and
sensitivity of the glyoxylic acid method may be due, at least in part, to the fact
that this reaction is not dependent on the presence of water (42).

The fluorophores formed in the formaldehyde and glyoxylic acid methods are not
covalently linked to the tissue but are trapped in a protein matrix (12,42).

Several organic solvents do not readily extract the fluorophores however, and water produces a marked decrease in fluorescence intensity which is probably due to a quenching effect (12). Warm xylene and paraffin wax appear to have an adverse effect on the fluorescence yield (71); it is prudent therefore to keep exposure of the tissue to these agents in the F.I.F. method as short as possible while still allowing good histology.

A major drawback of glyoxylic acid vapour is its poor penetration into tissue, making it impossible to use on freeze-dried tissue. Freeze-drying and paraffin-embedding have several advantages over the vibratome method. These include the ability to be able to store material for long periods and the facility with which serial sections can be cut.

5.4.2 Fluorescence yield and photodecomposition of the fluorophores

There is a dissociation between the amount of fluorophore formed and the fluorescence yield. Thus while the fluorophore yield for DA in a protein microdroplet is only 10% with glyoxylic acid vapour, the fluorescence yield is similar to that of the pure fluorophore at the same concentration as DA (42). Furthermore, Lindvall et al. (72) have found that while for the indoleamines T and 5-methoxy-T in the acid catalyzed F.I.F. method, the formaldehyde ozone reaction and the aluminium-catalyzed reaction there is a 10-20 fold increase in the fluorescence yield compared to the F.I.F. method, this is not accompanied by a corresponding increase in the fluorophore yield; indeed the fluorophore levels are only slightly greater than those observed in the F.I.F. method. It appears that previously unrecognized factors are operating to enhance the fluorescence yield in a protein matrix. These factors could conceivably involve an energy transfer between the intermediary reaction products and the fluorophores (72). Using protein micro-droplets to simulate the perfusion immersion procedure (20) the conversion of DA to a tetra-hydroisoquinoline was almost quantitative; however, under these conditions the fluorescence yield increased by only 50-100% (42). Consequently there does not appear to be a direct relationship between the amount of fluorophore formed and fluorescence yield.

The fluorescence yield for both NA and DA is greater in the glyoxylic acid compared to the F.I.F. method (see section 5.2.3). However both these methods are not as sensitive in detecting 5-HT (see section 5.4.4) as they are for the catecholamines.

The fluorophores formed from the biogenic monoamines undergo photodecomposition when exposed to blue or ultraviolet light (73). The rate of fading is much greater for the 5-HT fluorophore than for the catecholamine fluorophores and this difference can be used to differentiate 5-HT fluorescence from catecholamine fluorescence (73).

5.4.3 Quantitation of the fluorescence

The relationship between fluorescence intensity and amine concentration has been investigated by reacting protein microdroplets containing known concentrations of the amine with hot formaldehyde gas. A linear relationship between fluorescence intensity and monoamine concentration existed up to 4.5×10^{-2}M for the catecholamines (70) and up to 9×10^{-2}M for 5-HT (74); above these concentrations marked deviations from linearity were observed. The possibility of incomplete reaction with formaldehyde was excluded, as the pure catecholamine fluorophore showed the same concentration-dependent quenching. In the case of glyoxylic acid-induced DA fluorescence there was a linear correlation between fluorescence intensity and amine concentration up to 8×10^{-2}M (42). Jonsson (75) has shown that fluorescence quenching normally occurs in nerve terminals in the rat iris. It appears that the fluorescence intensity is proportional to the NA concentrations up to about 30-40% of the endogenous NA level in adrenergic nerves (76). Quantitation of endogenous monoamine fluorescence is therefore fraught with difficulty unless the amine levels are markedly reduced from normal.

5.4.4 Specificity and sensitivity

A method of differentiating the DA fluorophore from those of NA or adrenaline (A) has been developed (49,50). This method involves splitting off the labile 4-hydroxy group on the NA or A fluorophore, using acid treatment and recording the shift in the excitation peak (see section 5.2.3). NA and A can be differentiated on the basis of the reaction conditions required to produce a fluorophore. Adrenalin as it is a secondary PEA, requires more intense reaction conditions (higher reaction temperature and/or longer reaction time) than does NA (76).

A free hydrogen atom on an amino group of the side chain is a prerequisite for the Pictet-Spengler reaction. Therefore of the DA, NA and A metabolites only 3-methoxytyramine (3-MTA), normetanephrine (NMN) and metanephrine (MN) respectively can undergo cyclization with formaldehyde or glyoxylic acid. In the F.I.F. method none of these metabolites produces significant fluorescence, while in the glyoxylic acid reaction strong fluorescence is produced by 3-MTA and weak fluorescence by NMN (43). Of the precursor amino acids neither phenylalanine nor tyrosine produce significant fluorescence but DOPA produces a strong fluorophore in both the glyoxylic acid and formaldehyde methods. DOPA and the catecholamine fluorophores have similar spectral characteristics (see section 5.2.3). There is a possibility of distinguishing between DOPA and the other catecholamine fluorophores using a combined formaldehyde and glyoxylic acid reaction under carefully controlled conditions (46).

5-HT fluorescence can frequently be distinguished from catecholamine fluorescence on the basis of colour. Thus 5-HT gives a yellow fluorescence while the catecholamines exhibit a green fluorescence in keeping with their emission maxima if a

470 nm secondary filter is used. However, since catecholamine fluorescence can under certain conditions appear yellow (see section 5.3.3) it is important to use microspectrofluorimetric analysis to distinguish between 5-HT and the catecholamines. The rapid fading of fluorescence exhibited by 5-HT fluorophores can also be used to help distinguish 5-HT from catecholamine fluorophores. The 5-HT metabolites 5-hydroxyindoleacetic acid and 5-hydroxytryptophol do not exhibit fluorescence in the formaldehyde or glyoxylic acid reactions (43) while the 5-HT precursors TP or 5-HTP produce fluorophores (43). Tryptophan can be distinguished from 5-HT or 5-HTP using microspectrofluorimetric analysis (see section 5.2.3). It is therefore apparent that both the formaldehyde and glyoxylic acid methods can be relatively specific for identifying NA, DA, A or 5-HT in tissue sections.

Dahlstrom et al. (77) have calculated that approximately 10^{-17} moles of NA are contained in a single varicosity of the rat iris. NA can still be detected using the F.I.F. method when the tissue stores are depleted by 90% (75). Thus the sensitivity limit of the F.I.F. method may be even lower than 10^{-18} moles of NA. In the case of the glyoxylic acid method it has been estimated that this method is ten times more sensitive for DA (20), thus allowing the detection of 10^{-19} moles of DA.

The F.I.F. method, although extremely sensitive, does not permit the visualization, particularly in the central nervous system, of the catecholamine systems in their entirety. The increased sensitivity of the glyoxylic acid method however, enables catecholamine-containing axons as well as cell bodies and nerve terminals to be detected.

Because of the very low fluorescence yield from 5-HT fluorophores, there are problems in visualizing 5-HT-containing systems in the brain. This problem has been partially solved by using pharmacological treatments to increase the 5-HT levels. Thus, the sensitivity of the method can be increased either by prior administration of a monoamine oxidase inhibitor (4,5) or a monoamine oxidase inhibitor and L-tryptophan (78).

5.4.5 Comparison with other analytical methods

The fluorescence histochemical methods allow the precise localization of amines in discrete areas of the nervous system. Microdissection procedures, typified by the 'punch' technique (79) in which small amounts of tissue are dissected out and analyzed by methods such as radioenzymic and gas chromatographic-mass spectrometric assays, are relatively crude by comparison. In contrast, radioenzymic procedures and gas chromatography-mass spectrometry and several of the other techniques discussed in this book allow precise quantitation and a measure of specificity not obtainable with histochemical methods. The combination of these techniques has made, and doubtless will continue to make, substantial contributions

106

to our understanding of the monoamine-containing systems in the central nervous system.

REFERENCES

1 B. Falck, N.-A. Hillarp, G. Thieme and A. Torp, J. Histochem. Cytochem., 10 (1962) 348-354.
2 B. Falck, Acta physiol. scand., 56 (1962) Suppl. 197.
3 A. Carlsson, B. Falck, N.-A. Hillarp, Acta physiol. scand., 56 (1962) Suppl. 196
4 A. Dahlstrom and K. Fuxe, Acta physiol. scand., 64 (1964) Suppl. 232.
5 K. Fuxe, Acta physiol. scand., 64 (1965) Suppl. 247.
6 G. Eros, Zbl. Allg. Path. Path. Anat., 54 (1932) 385-391.
7 O. Eranko, Acta anat. (Basel) (1962) Suppl. 17.
8 R. Barter and A.G.E. Pearse, Nature, 172 (1953) 810.
9 R. Barter and A.G.E. Pearse, J. Path. Bact., 69 (1955) 25-31.
10 O. Eranko, Acta endocr., 18 (1955) 174-179.
11 M. Ritzen, Cytochemical identification and quantitation of biogenic monoamines - A microspectro-fluorimetric and autoradiographic study. M.D. Thesis, Stockholm, 1967.
12 G. Jonsson, The formaldehyde fluorescence method for the histochemical demonstration of biogenic monoamines - A methodological study. M.D. Thesis, Stockholm 1967.
13 O. Eranko and L. Raisanen, J. Histochem. Cytochem., 14 (1966) 690-691.
14 A. El Badawi and E.A. Schenk, J. Histochem. Cytochem., 15 (1967) 580-588.
15 A.M. Laties, R. Lund and D. Jacobwitz, J. Histochem. Cytochem., 15 (1967) 535-541.
16 A.V. Sakharova and D.A. Sakharov, Prog. Brain Research, 34 (1971) 11-25.
17 T. Hokfelt and A. Ljungdahl, Histochemie, 29 (1972) 325-339.
18 S. Axelsson, A. Bjorklund and O. Lindvall, J. Histochem. Cytochem., 20 (1972) 435-444.
19 O. Lindvall, A. Bjorklund, T. Hokfelt and A. Ljungdahl, Histochemie, 35 (1973) 31-38.
20 O. Lindvall and A. Bjorklund, Histochemistry, 39 (1974) 97-127.
21 F.E. Bloom and E.L.F. Battenberg, J. Histochem. Cytochem., 24 (1976) 561-571.
22 S.J. Watson and J.D. Barchas, Psychopharm. Commun., 1 (1975) 523-531.
23 J.C. de La Torre and J.W. Surgeon, Histochemistry, 49 (1976) 91-93.
24 B. Hamberger and K.A. Norberg, J. Histochem. Cytochem., 12 (1964) 48-49.
25 T. Spriggs, J. Lever, P. Rees and J. Graham, Stain Technol., 41 (1966) 323-327.
26 J.S. Nelson and P.L. Wakefield, J. Neuropath. Exp. Neurol., 27 (1968) 221-223.
27 G.F. Placidi and D. Masuoka, H. Histochem. Cytochem., 16 (1968) 491-492.
28 I. Loren, A. Bjorklund and O. Lindvall, Histochemistry, 52 (1977) 223-239.
29 V. Ajelis, A. Bjorklund, B. Falck, O. Lindvall, I. Loren and B. Walles, Histochemistry, 65 (1979) 1-15.
30 J.B. Furness and M. Costa, Histochemistry, 41 (1975) 335-352.
31 T. Waris and S. Partanen, Histochemistry, 41 (1975) 369-372.
32 T. Malmfors, Acta physiol. scand., 64 (1965) Suppl. 248, 1-93.
33 C. Sachs, Acta physiol. scand., (1970) Suppl. 341.
34 J.B. Furness and T. Malmfors, Histochemie, 25 (1971) 297-309.
35 J.B. Furness, M. Costa and A.I. Wilson, Histochemistry, 52 (1977) 159-170.
36 H. Corrodi and N.-A. Hillarp, Helv. Chim. Acta, 46 (1963) 2425-2430.
37 H. Corrodi and N.-A. Hillarp, Helv. Chim. Acta, 47 (1964) 911-918.
38 H. Corrodi and G. Jonsson, Acta Histochem., 22 (1965) 247-258.
39 G. Jonsson, Acta Chem. Scand., 20 (1966) 2755-2762.
40 A. Bjorklund, B. Falck, O. Lindvall and L.A. Svensson, J. Histochem. Cytochem., 21 (1973) 17-25.
41 A. Bjorklund, O. Lindvall and L.A. Svensson, Histochemie, 32 (1972) 113-131.
42 O. Lindvall, A. Bjorklund and L.-A. Svensson, Histochemistry, 39 (1974) 197-227.
43 A. Bjorklund, B. Falck and O. Lindvall, in P.B. Bradley (Ed.), Methods in Brain Research, John Wiley and Sons, London, 1975, pp. 249-294.
44 L.-A. Svensson, A. Bjorklund and O. Lindvall, Acta Chem. Scand. B, 29 (1975)

341-348.
45 I. Loren, A. Bjorklund, B. Falck and O. Lindvall, Histochemistry, 49 (1976) 177-192.
46 O. Lindvall, A. Bjorklund, B. Falck and L.-A. Svensson, Histochemistry, 46 (1975) 27-52.
47 A. Bjorklund, A. Nobin and U. Stenevi, J. Histochem. Cytochem., 19 (1971) 286-298.
48 H. Corrodi and G. Jonsson, J. Histochem. Cytochem., 13 (1965) 484-487.
49 A. Bjorklund, B. Ehinger and B. Falck, J. Histochem. Cytochem., 16 (1968) 263-270.
50 A. Bjorklund, B. Ehinger and B. Falck, J. Histochem. Cytochem., 20 (1972) 56-64.
51 A. Bjorklund, B. Falck and U. Stenevi, Prog. Brain Res., 34 (1971) 63-73.
52 B. Falck and C. Owman, Acta univ. Lund., 2 (1965) 1-23.
53 L. Olson and U. Ungerstedt, Histochemie, 22 (1970) 8-19.
54 A.G.E. Pearse, Histochemistry Theoretical and Applied, Churchill Livingstone, Edinburgh, 1980.
55 S.J. Watson and J.P. Ellison, Histochemistry, 50 (1976) 119-127.
56 B.T. Hamberger, T. Malmfors and C. Sachs, J. Histochem. Cytochem., 13 (1965) 147-150.
57 B. Hamberger, Acta physiol. scand. (1967) Suppl. 295.
58 M. Ritzen in A.A. Thaer and M. Sernetz (Eds.), Fluorescence Techniques in Cell Biology, Springer-Verlag, Heidelberg, 1973, pp. 183-189.
59 J.B. Furness and M. Costa, Histochemistry, 41 (1975) 335-352.
60 L. Olson and U. Ungerstedt, Brain Res., 17 (1970) 343-347.
61 K.-A. Norberg, M. Ritzen and U. Ungerstedt, Acta physiol. scand., 67 (1966) 260-270.
62 G. Jonsson, Acta Histochem., 26 (1967) 1-11.
63 F.W.D. Rost, in A.G.E. Pearse, Histochemistry Theoretical and Applied, Churchill Livingstone, Edinburgh, 1980, ch. 11, p. 379.
64 H. Corrodi, G. Jonsson and T. Malmfors, Acta Histochem., 25 (1966) 367-370.
65 A. Bjorklund, B. Falck and R. Hakanson, Acta physiol. scand., (1968) Suppl. 318.
66 A. Bjorklund and U. Stenevi, J. Histochem. Cytochem., 18 (1970) 794-802.
67 W.M. Whaley and T.R. Govindachari, in R. Adams (Ed.), Organic Reactions, Vol. 6, John Wiley, New York, 1951, pp. 151-206.
68 I. Loren, A. Bjorklund, B. Falck and O. Lindvall, Acta physiol. scand. (1977) Suppl. 452, 15-18.
69 A. Bjorklund, B. Falck, O. Lindvall and I. Loren, J. Neurosci. Methods, 2 (1980) 301-318.
70 M. Ritzen, Exp. Cell Res., 44 (1966) 505-529.
71 A. Bjorklund and B. Falck, J. Histochem. Cytochem., 16 (1968) 717-720.
72 O. Lindvall, B. Bjorklund, B. Falck and I. Loren, Histochemistry, 68 (1980) 169-181.
73 G. Jonsson, Histochemie, 8 (1967) 288-296.
74 M. Ritzen, Exp. Cell Res., 45 (1967) 178-194.
75 G. Jonsson, J. Histochem. Cytochem., 17 (1969) 714-723.
76 G. Jonsson, Prog. Histochem. Cytochem., 2 (1971) 1-36.
77 A. Dalstrom, J. Haggendal and T. Hokfelt, Acta physiol. scand., 67 (1966) 287-294.
78 G.K. Aghajanian and I.M. Asher, Science, 172 (1971) 1159-1162.
79 M. Palkovits, Brain Research, 59 (1973) 449-450.

Chapter 6

GAS CHROMATOGRAPHIC ANALYSIS OF AMINES IN BIOLOGICAL SYSTEMS

GLEN B. BAKER, RONALD T. COUTTS AND DONALD F. LEGATT

Neurochemical Research Unit, Department of Psychiatry and Faculty of Pharmacy
and Pharmaceutical Sciences, University of Alberta, Edmonton, Alberta T6G 2G3
(Canada)

6.1   INTRODUCTION

Gas chromatography (GC) is an analytical procedure used to separate mixtures of
organic compounds for the purposes of identification and/or quantitation. GC
separations are usually performed on solutions of compounds in inert solvents and
are accomplished on columns containing a non-volatile liquid (the stationary
phase). The GC column is contained in the oven of a gas chromatograph; the oven
temperature is variable, normally over a 20-300° range, depending on the volatility
of the solutes and the stability of the stationary phase. The components of a
mixture in solution are carried through the column by an inert carrier gas, and
separate from one another according to their partition coefficients between the
carrier gas and the stationary phase. As each component elutes from the column,
it is detected and displayed as a peak on a recorder. The interval between the
time of injection and the apex of the recorded peak is called the retention time
of the compound. This value is characteristic of, although not unique to, the
compound giving rise to it under the GC conditions used.

Chromatographic separations can be performed isothermally (column temperature
kept constant) or by temperature programming. In the latter case, the column
temperature is altered at preset rates during the analysis; this may permit
improved resolution and a great reduction in analysis time when investigating
solutions containing a wide range of components.

6.1.1   GC columns

Most GC columns are made of glass. Stainless steel, nickel, copper and aluminum
columns, and glass-lined metal columns are also available. Glass columns are used
frequently in biological studies because of their inertness; degradation of organic
compounds occurs more readily on heated metal columns. Both packed and capillary
columns are used. The former are typically 1-2 m in length, with a 2-4 mm internal
diameter, and are packed with an inert solid support which has previously been
coated with the stationary phase. Numerous stationary phases are employed in GC
analyses, and the amount of stationary phase used to coat a solid support can vary
greatly. In practice, however, only a few stationary phases are used routinely

(1), and 1-5% w/w of a stationary phase on a solid support is the common concentration range. Most support materials are prepared from diatomaceous earth (kieselguhr) which is porous and has a large surface area. Support materials are available in different densities; low density supports can be loaded with more stationary phase than high density supports.

Capillary columns are made of glass or fused silica and are generally 10-100 m in length and 0.25 - 0.50 mm in internal diameter and coated with a thin layer of the stationary phase. Two types of glass capillary columns are commonly used. In wall-coated open tubular (WCOT) columns, a thin liquid phase is film coated directly onto the inside surface of the capillary. Support-coated open tubular (SCOT) columns have the inside surface of the capillary coated with a thin layer of support material which is coated with the liquid phase. Capillary columns generally permit much better resolution of components of a mixture than can be attained using packed columns. However, packed columns are relatively inexpensive and robust, and readily prepared in the laboratory. Ordinary glass capillary columns are fragile but flexible fused silica columns are virtually indestructible. In addition, time-consuming column straightening is not required with the flexible fused silica columns. Generally, capillary columns must be purchased from commercial sources. Comprehensive information about the properties of the various types of capillary columns is available in the catalogues of several commercial suppliers. The potential users of such columns would be well advised to read such information carefully before purchase to ensure that they are obtaining the column best suited to their needs.

6.1.2   Detectors

A variety of techniques have been used to detect compounds eluting from a gas chromatograph. Four different detectors are routinely available: the thermal conductivity detector (TCD), the flame ionization detector (FID), the electron-capture detector (ECD) and the nitrogen-phosphorous detector (NPD). The mass spectrometer can also be used as a sophisticated detector (MSD) in GC analysis.

6.1.2.1   Thermal conductivity detector

The TCD is durable, easy to operate and, because of its lack of selectivity, is capable of detecting a wide range of organic compounds. TC detection is nondestructive, that is the eluted compounds can be collected for further investigation. The TCD also produces a linear response over a wide range of sample amounts. However, the TCD is rarely used in neurochemical studies because of its relatively poor sensitivity. The limit of sensitivity for a TCD is around 1 μg.

6.1.2.2   Flame ionization detector

The FID has probably been the most widely used detector in biological investigations. During operation, the effluent from the GC column is mixed with hydrogen, passed through a metal jet and burned in an atmosphere of air. This produces ions which are collected on a charged collector. The strength of the electric current produced is directly proportional to the amount of compound combusted. Only compounds which combust with ionization in a hydrogen/air flame produce an FID response, and this includes most organic compounds. Detection sensitivity of an FID is approximately proportional to the number of carbon atoms in the molecule being detected and is linear over a wide sample range. Quantities of sample as low as 1 ng can often be detected.

6.1.2.3   Electron-capture detector

The ECD is a relatively selective detector which has the potential to detect as little as 1 pg of an organic compound containing an electrophoric substituent. An ECD usually contains a radioactive (^{63}Ni or ^{3}H) source which emits relatively high energy β-particles. The particles collide with carrier gas molecules (normally 95% argon/5% methane) and produce a small current (the standing current). When an electrophoric compound elutes from the GC column, it reduces the strength of the standing current which returns to its original levels when the sample has left the detector. The change in current is amplified and appears as a peak on the recorder. The sensitivity of the recorder varies greatly according to the ability of a compound to absorb electrons, which causes the reduction in standing current strength. Compounds which contain halogen atoms, ketone or nitro groups or other electrophoric groups are particularly well suited for analysis by ECD.

6.1.2.4   Nitrogen-phosphorous detector

The NPD is also a relatively selective detector and is extremely sensitive to most compounds which possess nitrogen- and/or phosphorous-containing functions. In its operation, the effluent from the GC column is mixed with a smaller volume of hydrogen; this mixture enters an electrically heated detector chamber which contains an alkali source (often a rubidium salt). A low-temperature plasma is formed and produces a minute electric current of a magnitude which is proportional to the amount of compound reaching the detector. This current is amplified and recorded. The NPD is approximately 30,000 times more sensitive towards nitrogen and 60,000 times more sensitive towards phosphorous than it is towards carbon, and has been used for detection of low picogram quantities of nitrogen- and phosphorous-containing compounds.

6.1.2.5   Mass spectrometric detection

A mass spectrometer can be utilized as a sensitive, specific detector in GC analysis. The eluate is first passed through a separator which removes most of the carrier gas, then is directed into the mass spectrometer where the eluted compound is bombarded with high energy electrons. The compound is initially ionized, then undergoes fragmentation into positive, negative and neutral fragments. Either the positively or negatively charged fragments are converted into an electric current (total ion current) which is amplified and recorded. The strength of the electric current is proportional to the amount of compound entering the mass spectrometer. Instead of measuring total ion current, the mass spectrometer can be programmed to amplify the current produced by an ion of a particular mass-to-charge ratio. A single (or selected) ion current is produced. Thus, if two or more compounds elute simultaneously from the GC column, an ion which is formed from only one of the eluting compounds would be selected for amplification, and recorded. Combined GC-MSD is discussed in detail in Chapter 8 of this volume.

6.1.3 Derivatization for gas chromatographic analysis

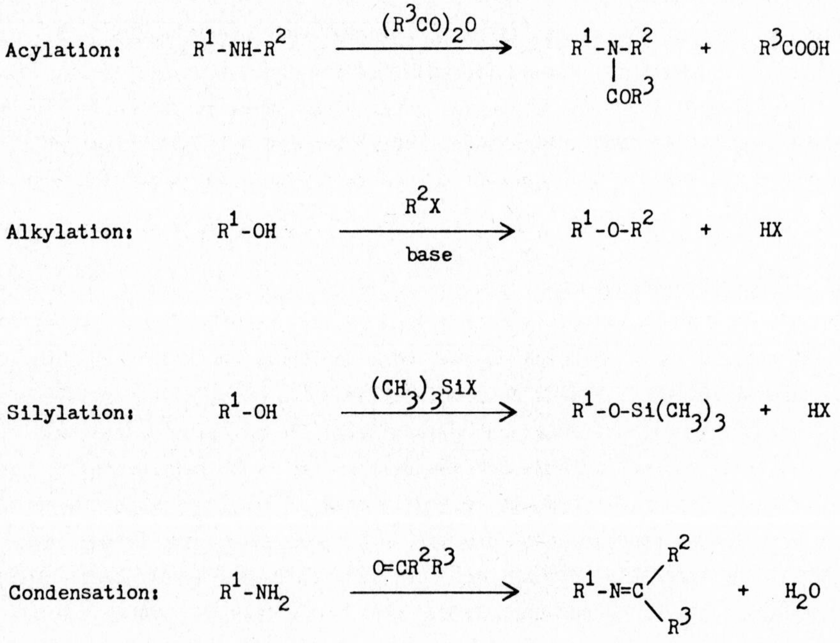

Acylation: $\quad R^1\text{-NH-}R^2 \xrightarrow{(R^3CO)_2O} R^1\text{-N-}R^2 \quad + \quad R^3COOH$
$\quad\quad\quad\quad\quad\quad\quad\quad\quad\quad\quad\quad\quad\quad\quad\quad\quad |$
$\quad\quad\quad\quad\quad\quad\quad\quad\quad\quad\quad\quad\quad\quad\quad COR^3$

Alkylation: $\quad R^1\text{-OH} \xrightarrow[\text{base}]{R^2X} R^1\text{-O-}R^2 \quad + \quad HX$

Silylation: $\quad R^1\text{-OH} \xrightarrow{(CH_3)_3SiX} R^1\text{-O-Si}(CH_3)_3 \quad + \quad HX$

Condensation: $\quad R^1\text{-NH}_2 \xrightarrow{O=CR^2R^3} R^1\text{-N=C} \overset{R^2}{\underset{R^3}{\diagdown\!\!\diagup}} \quad + \quad H_2O$

Fig. 6.1. General reactions for the principal types of derivatization used for GC analysis.

Derivatization of compounds for GC analysis is generally performed for a variety of reasons: a) to increase volatility; b) to increase stability; c) to reduce the polarity of the compound since polar compounds (acids, phenols and some alcohols and amines) often chromatograph poorly; d) to improve extraction efficiency from aqueous media (e.g. acylation of phenolic amines); and e) to introduce a functional group which is particularly sensitive to selective detectors such as ECD, NPD or MSD.

Most derivatives are formed by replacing the active hydrogen atom of polar compounds (e.g. NH, OH, SH) by chemical procedures such as acylation, alkylation, silylation and condensation (1,2). General reactions for each type of derivatization are illustrated in Figure 6.1. Acyl anhydrides, acyl halides or activated acyl amide reagents are usually used for acylation of OH, SH and NH groups. In alkylation reactions, replacement of active hydrogens of OH (including COOH), SH and NH (including CONH) groups yields ethers, esters, thioethers, thiolesters, N-alkylamines or N-alkylamides. The most common type of silylation employed is replacement of an active hydrogen on OH, SH or NH by a trimethylsilyl group. The various trimethylsilylating reagents available and their silyl donor strength and reactivity have been reviewed (2). Boronates, oximes and hydrazones are among products formed for GC analysis by condensation reactions; such reactions involve the joining of two molecules with loss of water. The reactions described above account for the majority of derivatizations used in GC studies, but other products such as phosphoryl and sulfonyl derivatives, ketals, ureas, oxazolidines and oxazolidinones (2) have also been employed. Texts and articles are available describing derivatization procedures in considerable detail (1-15). Some examples of commonly used derivatizing reagents are shown in Fig. 6.2.

6.2   ANALYSIS OF SPECIFIC AMINES
6.2.1   Catecholamines and their O-methyl metabolites

Most of the GC analytical procedures for these amines in biological systems has utilized electron-capture detection (ECD). The amine portion of these compounds can be derivatized readily for gas chromatographic analysis. These amines also have in their structures one or two phenol groups and, in the case of noradrenaline (NA), adrenaline (A), normetanephrine (NMN) and metanephrine (MN), an alcohol moiety, which may be derivatized. For the catecholamines, extraction from tissue frequently involves absorption on alumina or isolation using an ion-exchange resin, followed by elution with acid, removal of the liquid phase under vacuum, and reaction with a perfluoroacylating reagent. Solvent extraction procedures, ion-exchange resins and ion-pair extraction have been utilized for isolation of the O-methylated amine metabolites of the catecholamines from tissue and biological fluids.

114

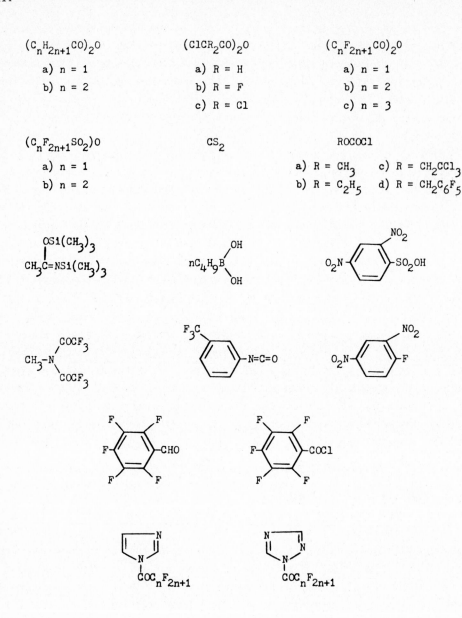

Fig. 6.2 Structures of some typical derivatizing reagents used in gas chromatography.

Trifluoroacetic anhydride (TFAA), pentafluoropropionic anhydride (PFPA) and heptafluorobutyric anhydride (HFBA) are frequently used for derivatization for subsequent GC analysis. Structures of the resultant derivatives are shown in Fig. 6.3. Greer et al. (16) analyzed urinary 3-MT, NMN and MN in patients with

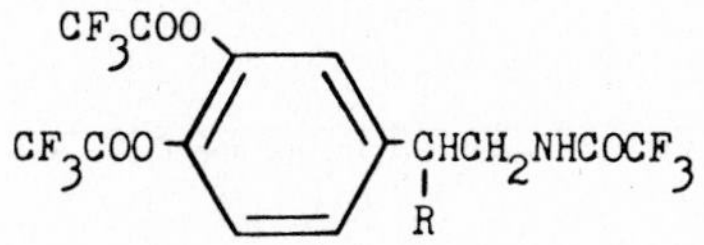

DA : R = H

NA : R = OCOCF$_3$

Fig. 6.3. Products formed when DA (a) and NA (b) are reacted with trifluoroacetic anhydride under anhydrous conditions.

neuroblastoma and pheochromocytoma by converting the amines to their trifluoro-acetyl (TFA) derivatives and quantitating by GC-FID. Kawai and Tamura (17) employed trifluoroacetylation for analysis of catecholamines by GC-ECD in extracts from urine and tumors. A similar procedure was used by Imai et al. (18) to measure catecholamines in a variety of rat tissues and serum. These authors were able to confirm the presence of A in rat brain. In the procedure developed by Wong et al. (19), urinary catecholamines were converted to their pentafluoropropionyl (PFP) derivatives for quantitation by GC-ECD. Martin and Ansell (20) developed a procedure which permitted the analysis of NA, dopamine (DA) and 5-hydroxytryptamine (5-HT) in rat brain using GC-ECD. The amines were converted to their TFA derivatives prior to analysis by GC. A number of other reports appeared in the literature in which perfluoroacylation and GC-ECD analysis were used to measure catecholamines and their O-methylated amine metabolites in body fluids and tissues (21-27). Maruyama and Takemori (28) also reported a GC-FID procedure for estimation of NA and DA in brain; TMS-imidazole was employed to derivatize the catecholamines in this case. Gyllenhaal et al. (29) recently used methyl chloroformate derivatization and GC-NPD for quantitation of catecholamine standards in aqueous solution (Fig. 6.4); the derivatization was performed directly in the aqueous phase, and it is conceivable that the procedure could be applied to biological samples. Mixed derivatives, formed by reaction of the catecholamines or their O-methylated metabolites with a combination of derivatizing agents, have also been utilized for quantitation of these compounds in tissues and body fluids. Kawai and Tamura (30) converted the hydroxyl groups of the catecholamines into trimethylsilyl ethers and the primary amines into Schiff bases (secondary amino groups remained unchanged);

this technique was used in conjunction with GC-FID for analysis of catecholamines in bovine adrenal medulla. Sharman (31) employed acetylation in aqueous medium followed by perfluoroacylation under anhydrous conditions to measure glycol metabolites of NA in brain tissue. Sharman suggested that the method could be modified for analysis of NA and NMN. In a method developed for the assay of catecholamines in brain tissue, Arnold and Ford (19) etherified the alcohol group of NA prior to perfluoroacylation of the amine and phenol functions. In the procedure of Haeffner et al. (32), mixed TFA-TMS derivatives were used for identification in urine of DA,

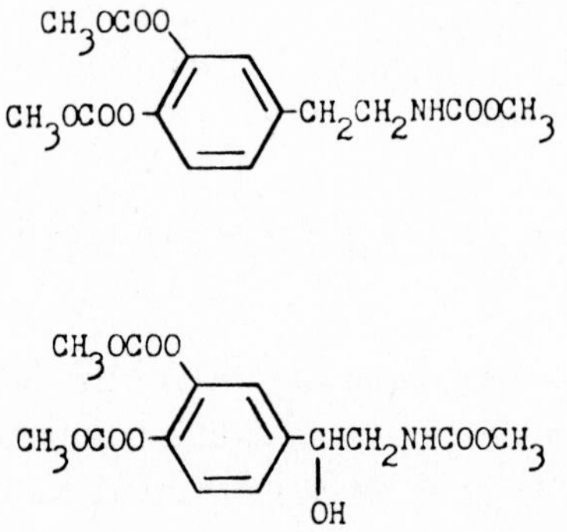

Fig. 6.4. Products formed by reaction of catecholamines with methyl chloroformate under aqueous conditions.

NA, NMN and MN by GC-FID. Lhuguenot and Maume (33) formed the pentafluorobenzylidine-TMS derivatives of NA and DA for quantitation of these amines in rat adrenal extract. Nelson et al. (34) etherified alcohol groups and then applied perfluoroacylation; this provided for the simultaneous measurement of 3-MT, NMN and MN in urine. Doshi and Edwards (35) employed N-2,6-dinitro-4-trifluoromethylphenyl, O-trimethylsilyl derivatives for analysis of catecholamines in rat brain (Fig. 6.5) In another recent report, LeGatt et al. (36) acetylated 3-methoxytyramine (3-MT) and NMN and extracted the resultant N-acetyl, phenolic O-acetyl derivatives into ethyl acetate. Prior to removal of the ethyl acetate under a stream of nitrogen, the samples were hydrolyzed with ammonium hydroxide to specifically hydrolyze the acetylated phenol group. Subsequent reaction with TFAA led to the formation of an N-acetyl, N-TFA, O-TFA compound in the case of 3-MT; a similar compound was formed with NMN, in which the alcohol group was also trifluoroacetylated. This technique was used for quantitation of these two O-methylated amines in brain tissue and human urine (36,37) and has since been applied to the simultaneous analysis of a number of amines (see section 6.2.3). Recent findings in our laboratory (38) indicate that pentafluorobenzoylation followed by trifluoroacetylation may also provide a convenient means for simultaneous measurement of NA and DA in tissue and body fluids. In the last three methods mentioned above, the initial derivatization procedure in each case (reaction with 2,6-dinitrophenyl-4-trifluoromethylphenyl chloride, acetic anhydride or pentafluorobenzoyl chloride) can be performed in aqueous medium.

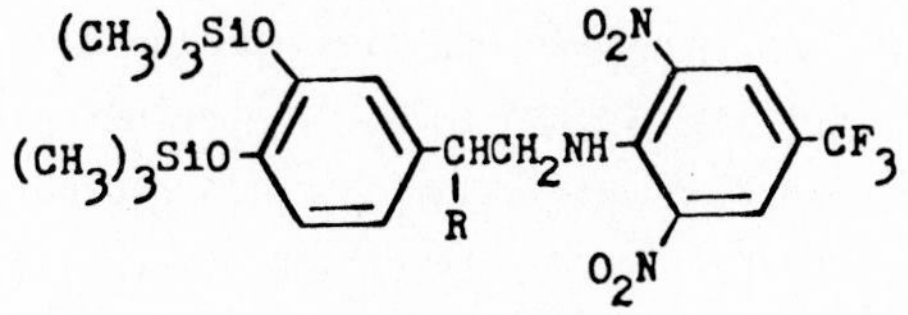

$$R = H \quad (DA)$$
$$R = OSi(CH_3)_3 \quad (NA)$$

Fig. 6.5. N-2,6-dinitro-4-trifluoromethylphenyl, O-trimethylsilyl derivatives of catecholamines used by Doshi and Edwards (35).

There are other reports in the literature on the formation of mixed derivatives of catecholamines and their O-methylated amine metabolites (39-41). In many cases, these studies have been performed with standards only and have not been applied to biological samples. However, the potential GC user should be aware of them since most should be adaptable to analysis in biological systems.

While formation of mixed derivatives may provide certain GC advantages such as increased sensitivity, improved peak shape, altered retention time or separation from another compound which would otherwise interfere in the analysis, there may also be increased likelihood of multiple product formation.

Brief mention should be made here about boronate derivatives. Biogenic amines with the correct structural requirements (e.g. catecholamines, NMN, MN, octopamine (OA) and phenylethanolamine (PEOH) can be converted to cyclic boronates by treatment with an alkylboronic acid. The resultant products have good sensitivity on on GC-ECD (42,43) but the technique has not been employed extensively for analysis of amines in biological systems.

6.2.2 5-Hydroxytryptamine (5-HT; serotonin)

GC has not been a particularly popular technique for the analysis of 5-HT in

biological systems. Although there are numerous reports on the derivatization of
5-HT for GC (44-50), much of this work has been done with standards or has been
utilized for analysis in biological media by mass spectrometric techniques.

Martin and Ansell (20) used GC-ECD for analysis of 5-HT in rat brain tissue;
the tri-TFA derivative was formed after extraction of 5-HT from the tissue homogenate. Baker et al. (51) employed extraction with a liquid ion-exchanger, acetylation under aqueous conditions and reaction with PFPA to analyze 5-HT in rat brain;
the resultant cyclized derivative (46,49,51,52) has excellent sensitivity and
stability. This method has now been modified for concomitant analysis of 5-HT and
T using GC-ECD and a capillary column (37,53). Another possibility for analysis of
5-HT (and a number of other indolealkylamines) is propionylation in aqueous brain
homogenate, omitting the preliminary extraction with a liquid ion-exchanger. When
the resultant N,O-propionyl derivative of 5-HT was further reacted with a perfluoroacylating reagent and gas chromatographed, two peaks were always observed (54),
presumably because of the formation of two structural isomers.

The formation of mixed derivatives of 5-HT has been reported (55-58) and potentially these should be applicable to GC analysis in biological extracts.

6.2.3 Trace Amines: phenylethylamine (PEA), tyramine (TA), octopamine (OA), phenylethanolamine (PEOH) and tryptamine (T)

GC-FID has been used for analysis of PEA and T in putrefying human tissue (59)
and for measurement of concentrations of PEA in dog plasma following infusions of
PEA (60). Several GC-FID techniques have been reported for the investigation of
levels of p-TA and related amines in foodstuffs (61-65). Using a mixed TFA-TMS
derivative and GC-FID, Haeffner et al. (32) identified p-TA in urine. A number of
GC assays utilizing FID or ECD have been reported for analysis of PEA in brain and
urine (66-69), and many of these have been reviewed and compared by Schweitzer and
Friedhoff (70). Martin et al. (71) utilized perfluoroacylation and GC-ECD for
identification of T perfusates of dog brain. Reynolds and colleagues (72,73)
analyzed PEA in urine and brain as its pentafluorobenzoyl derivative. Martin and
Baker (74,75) acetylated PEA in an aqueous solution, extracted with ethyl acetate
and reacted the N-acetyl-PEA with PFPA; the procedure permitted the quantitation of
PEA in a single control rat brain. This method has now been modified for analysis
of PEA in urine (37). A similar procedure (acetylation followed by perfluoroacylation) has been applied to the analysis of tryptamine (T) in urine (76) and
brain (53). In the case of T, a cyclization occurs during the reaction with PFPA
(46,49,51), and reports by other workers on the derivatization of similar compounds
(52) suggests that a spirocyclic derivative is formed.

Acetylation followed by perfluoroacylation has also proven useful for analysis
of phenolic arylalkylamines such as tyramine (TA) and octopamine (OA). Extraction
of such compounds from aqueous solutions is often difficult because of their am-

photeric nature. Reaction with acetic anhydride under aqueous conditions results in acetylation of amine and phenol functions (but not alcohol) (77-82) and the

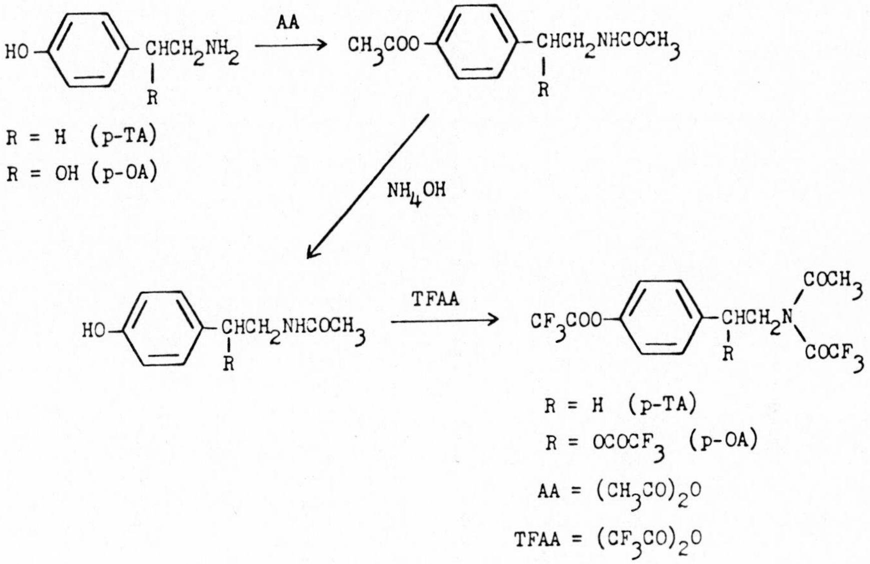

R = H (p-TA)
R = OH (p-OA)

NH$_4$OH

TFAA

R = H (p-TA)
R = OCOCF$_3$ (p-OA)

AA = (CH$_3$CO)$_2$O

TFAA = (CF$_3$CO)$_2$O

Fig. 6.6. Derivatization of p-TA and p-OA for analysis by GC-ECD. Phenol and amine groups are acetylated in aqueous medium, the acetylated phenols are specifically hydrolyzed with ammonium hydroxide, and the resultant N-acetyl compounds are perfluoroacylated under anhydrous conditions.

resultant neutral compounds are readily extracted into organic solvents such as ethyl acetate (83-86). These derivatives can be further reacted with TFAA (or a similar perfluoroacylating reagent) under anhydrous conditions to form N-acetyl, N-TFA, O-acetyl compounds, with alcohol functions also being perfluoroacylated. These final derivatives can be utilized for analysis by GC-ECD (86) but we have found it more suitable to further modify the derivatization process by hydrolyzing the acetylated phenol group prior to the final derivatization, freeing it for reaction with the TFAA. This results in a derivative with increased volatility and sensitivity (see Figure 6.6). Separation of derivatized p-TA from an interfering substance in brain and urine was also achieved (84,87). By combining this procedure with the use of capillary GC-ECD, it is now possible to separate PEA, m- and p-TA, p-OA, NME and 3-MT in a single run (37,86) (see Fig. 6.7). The modified procedure has not proven useful for analysis of T and 5-HT, but it is possible to analyze these substances in the same sample of homogenate or body fluid as the other amines by removing a portion of the ethyl acetate extract prior to hydrolysis (37,54). This

Urine sample or supernatant from
brain homogenate in perchloric acid

Make slightly alkaline.
Shake with di-(2-ethylhexyl)
phosphoric acid
(2.5% v/v in chloroform).

Retain bottom layer

Elute with 0.5 N HCl.

Retain top layer

Neutralize.
Acetylate with acetic anhydride.
Extract with ethyl acetate.

Retain organic layer

Shake with 1/10 the volume
of 10 N NH$_4$OH.
Neutralize with conc. HCl.

Retain organic layer

Evaporate to dryness under N$_2$.

Residue

React with trifluoroacetic
anhydride.
Partition briefly between
cyclohexane and saturated sodium
borate solution.

Retain the cyclohexane layer

Aliquot on GC-ECD

Fig. 6.7. A procedure for the simultaneous analysis of PEA, m- and p-TA, p-OA, NME
and 3-MT. Benzylamine, 3-phenylpropylamine, tranylcypromine or 2-(4-chlorophenyl)-
ethylamine may be used as internal standards.

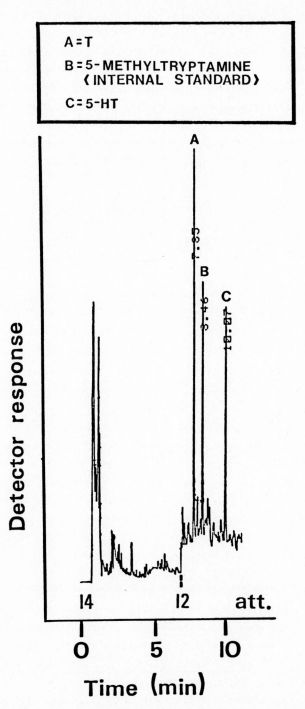

Fig. 6.8. Peaks obtained from T and 5-HT in an extract of human urine. The amines were acetylated under aqueous conditions and reacted with pentafluoropropionic anhydride under anhydrous conditions.

portion can be evaporated to dryness and reacted with PFPA to provide for quantitation of T and 5-HT (53) (see Figure 6.8 for GC trace).

The derivatives of the trace amines produced by the combination of acetylation and perfluoroacylation are sensitive and remain stable for several days under refrigeration. However, care must be taken to exclude water when collecting the samples by the partitioning process at the termination of the perfluoroacylation reaction, since the presence of water in the samples for long periods of time facilitates degradation. In the case of the derivatives of amines with an alcohol group in the side chain, oxazoline derivatives (83) form on standing over a period of time (Fig. 6.9).

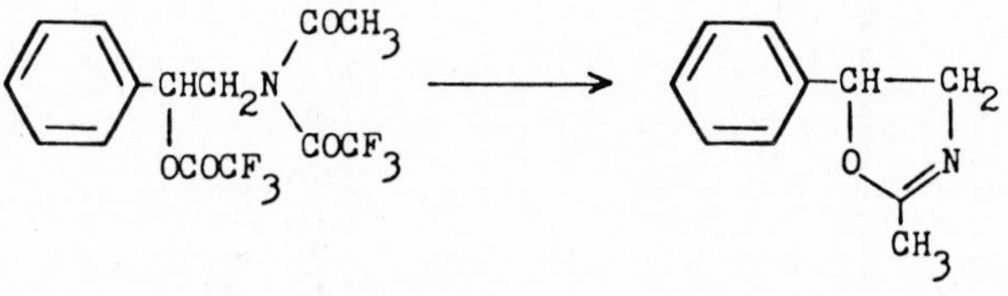

Fig. 6.9. Formation of an oxazoline derivative from derivatized PEOH.

In the assay procedures just described above, preliminary extraction must be performed to separate any endogenous N-acetylated amine from the amine itself. Otherwise the N-acetylated compound would give the same final derivative as the amine when carried through the procedure and give a misleadingly high figure for the levels of the amine. We have found that extraction with a liquid ion-exchanger followed by elution with HCl is a useful way to effect this separation (88). The HCl phase can then be basified and acetylated directly.

Two other derivatization procedures which can be conducted under aqueous conditions may have potential use for analysis of trace and other amines. Pentafluorobenzenesulfonyl chloride and pentafluorobenzoyl chloride have both been used in our laboratories to derivatize PEA standards directly in aqueous medium. The aqueous phase containing the amine is shaken with a solution of the derivatizing reagent in a mixture of ethyl acetate and acetonitrile. Acetonitrile acts as a phase transfer agent in the procedure. Similar phase-transfer reactions have been utilized for GC analysis of other amines and phenols (89-91), and these techniques may prove useful for assaying a variety of important amines in biological systems.

GC-NPD has been little used for analysis of trace and other low molecular weight arylalkylamines, although the technique may be employed more extensively in the future. Tranylcypromine, an antidepressant related closely in structure to PEA, has been assayed in plasma samples after conversion to its HFB derivative (92). Measurement of the urinary levels of N-methyl-T and N,N-dimethyl-T using trifluoro-acetylation and GC-NPD has also been reported (93). The properties of various derivatives of standards of PEA and related analogues on GC-NPD has been the object of a recent study (94). Narasimhachari and Friedel recently reported the use of GC-NPD for measurement of PEA and T in plasma, urine, cerebrospinal fluid and brain (95); amines were reacted with carbon disulfide to form isothiocyanate derivatives. It seems possible that reagents such as dimethylthiophosphinic chloride (96) and di-ethylchlorophosphate (97) may be used increasingly for analysis of biogenic amines in biological systems by GC-NPD. Jacob et al. (96) found a detection limit of 500 fg for N,N-dimethylthiophosphinylaniline. In preliminary studies (98) we have found that PEA may be reacted with diethylchlorophosphate under aqueous conditions; the structure of the final derivative was confirmed by mass spectrometry, but in-vestigations on the applicability of the derivatization procedure to measurement of PEA in tissues and body fluids have not yet been carried out.

6.2.4 Histamine

Mita et al. (99) have developed an assay for histamine (HA) in which the amine is extracted from basified aqueous solution and derivatized with HFBA and then with ethyl chloroformate. The detection limit on GC-FID was claimed to be 20 ng. This technique was applied to the analysis of HA and 1-methyl-HA in urine, whole blood and leukocytes by GC-MS (100,101). Mahy and Gelpi (102) formed the penta-fluoroacylated derivatives of HA, histidine and their 1-methyl metabolites, and the detection range found on GC-FID was 10-30 ng.

Edwards and colleagues have reported derivatization with 2,4-dinitrobenzene sulfonic acid, a procedure useful for quantitation of a number of phenylethylamines (69,103), to be unsuitable for analysis of HA. However, Doshi and Edwards (104) dis-covered that 2,6-dinitro-4-trifluoromethylbenzenesulfonic acid (DNTS) reacts readily with catecholamines, histamines and related amines to yield products with very suitable properties for GC-ECD. These workers have applied the procedure to anal-ysis of HA and 1-methyl-HA in urine samples.

Formation of the trifluoroacetyl (40), trimethylsilyl (105) and heptafluoro-butyryl (106) derivatives of HA for GC have been reported, but these compounds have been found to be unsuitable for quantitation of HA because of excessive tailing (99).

Gas chromatography with NPD (107) has been applied to the simultaneous analysis of HA and its basic metabolites in biological samples. Although procedural details are lacking, the general method is described. HA and metabolites are first ex-tracted from the biological material with an ion-exchange resin. After elution and lyophilization, the sample is reacted with HFBA and the derivatives purified

on a silicone acid column. Acetylation of the ring NH group, if present, precedes
GC analysis. Sensitivity was reported to be in the picomolar range.

A procedure has recently been developed in our laboratories which provides a
rapid, sensitive assay for HA and 1-methyl-HA in brain tissue using GC-ECD (108).
Briefly, this involves extraction with a liquid ion-exchanger, elution with HCl,
basification of the eluate and reaction with pentafluorobenzoyl chloride in ethyl
acetate containing a small amount of acetonitrile as a phase-transfer agent. The
procedure works very well for HA and 1-methyl-HA, with values obtained in rat hypo
thalamus for these two amines which are in good agreement with literature reports.
The reaction with HA is shown in Figure 6.10.

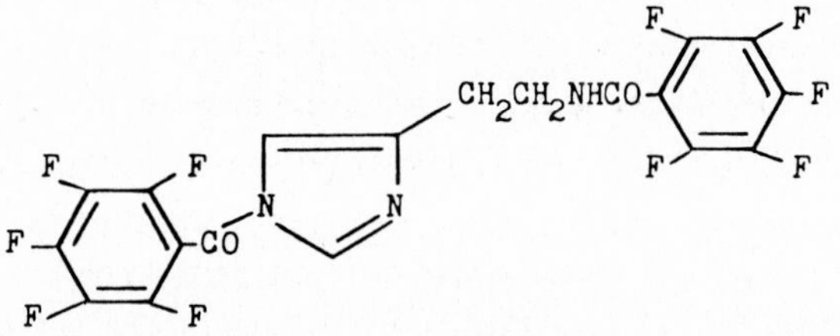

Fig. 6.10. Derivative formed by reaction of HA with pentafluorobenzoyl chloride
under aqueous conditions.

6.3 ADVANTAGES AND LIMITATIONS OF GC COMPARED TO OTHER ANALYTICAL METHODS

GC is a relatively inexpensive technique (particularly when compared to mass
spectrometry) which can be utilized for analysis of a variety of amines. In many
cases, the simultaneous assay of several amines can be achieved. Although this
chapter has dealt primarily with analysis of biogenic amines, several metabolites
of these amines and numerous other biologically important substances and drugs
can also be readily measured (109 for review). Although the sensitivity of GC is
generally lower than that obtained with techniques such as mass spectrometry and
radioenzymic procedures, we have observed that GC-ECD is sufficiently sensitive to
allow quantification of even the 'trace' amines in nerve tissue. In the case of
analysis of TA, GC-ECD analysis permits separation of meta and para isomers (84,
86), which is an advantage over the radioenzymic procedure reported for this amine
(110). Thermolability is an inherent problem in GC but this does not appear to be
an imposing problem for analysis of biogenic amines.

GC may be criticized for its relative lack of specificity compared to mass
spectrometric methods, but this is a criticism which may be directed at a number

of analytical techniques. In fact, structures of final derivatives utilized in GC may be readily confirmed by attaching the GC column to a mass spectrometer and analyzing the column eluate. It is also important to employ GC-MS to confirm structures of derivatives and to follow reaction sequences during development of assays for GC-ECD. Following this confirmation, we feel that GC-ECD can be used alone for routine analysis; this would seem to be substantiated by our results for brain levels of the 'trace' amines, HA and 5-HT, which are in good agreement with values obtained using mass spectrometric procedures.

Interference peaks arising from tissue, solvents and reagents may be more of a problem with GC than with other techniques. However, it has been our experience that this problem can be circumvented by using high quality chemicals, deionized glass-distilled water and glass-distilled organic solvents. The use of capillary columns may also be an important aid in overcoming interference by extraneous peaks.

GC technology is constantly being updated, and the wide variety of derivatizing reagents, columns, packing materials and accessories (e.g. automatic samplers) now available make it an extremely versatile technique. In addition to the reviews and handbooks referred to in this chapter, much information can be gleaned from catalogues of companies selling GC equipment, and the potential user will find study of these sources invaluable when setting up a GC apparatus for analysis.

In summary, GC is a useful tool which continues to play an important role in the analysis of amines in biological systems.

6.4 ACKNOWLEDGEMENTS

Some of the work reported in this chapter was supported by grants from the Medical Research Council of Canada, the Alberta Mental Health Research Fund and the Alberta Heritage Foundation for Medical Research.

REFERENCES

1   K. Blau and G.S. King (Eds.), Handbook of Derivatives for Chromatography, Heyden, London, 1978, pp. 104-151.
2   D.R. Knapp, in Handbook of Analytical Derivatization Reactions, John Wiley & Sons, New York, 1979, 741 pp.
3   B.J. Gudzinowicz in Gas Chromatographic Analysis of Drugs and Pesticides, Marcel Dekker, Inc., New York, 1967, 605 pp.
4   A.C. Moffat and E.C. Horning, Anal. Lett., 3 (1970) 205-216.
5   E. Anggard and G. Sedvall, Anal. Chem., 41 (1969) 1250-1256.
6   F. Karoum, F. Cattabeni, E. Costa, C.R.J. Ruthven and M. Sandler, Anal. Biochem., 47 (1972) 550-561.
7   A.C. Moffat, E.C. Horning, S.B. Matin and M. Rowland, J. Chromatogr., 66 (1972) 255-260.
8   S.B. Matin and M. Rowland, J. Pharm. Sci., 61 (1972) 1235-1240.
9   E. Gelpi, E. Paralta and J. Segura, J. Chromatogr. Sci., 12 (1974) 701-709.
10 E.L. Arnold and R. Ford, Anal. Chem., 45 (1973) 85-89.
11 J.J. Franken and M.M.F. Trijbels, J. Chromatogr., 91 (1974) 425-431.
12 R. Mussini, F. Marcucci and S. Garattini, in L.L. Iversen, S.D. Iversen and

126

S.H. Snyder (Eds.), Handbook of Psychopharmacology, Vol. 1, Plenum Press, New York, 1975, pp. 25-61.
13 S. Ahuja, J. Pharm. Sci., 65 (1976) 163-182.
14 A.D.R. Harrison, in E. Reid (Ed.), Methodological Developments in Biochemistry, Vol. 5, North Holland, Amsterdam, 1976, pp. 11-13.
15 J.A. Perry and C.A. Freit, in K. Tsuji and W. Morozowich (Eds.), GLC and HPLC Determination of Therapeutic Agents, Part 1, Marcel Dekker, Inc., New York, 1978, pp. 137-208.
16 M. Greer, T.J. Sprinkle and C.M. Williams, Clin. Chim. Acta, 21 (1968) 247-253.
17 S. Kawai and Z. Tamura, Chem. Pharm. Bull., 16 (1968) 1091-1094.
18 K. Imai, M. Sugiura and Z. Tamura, Chem. Pharm. Bull., 19 (1971) 409-411.
19 K.P. Wong, C.R.J. Ruthven and M. Sandler, Clin. Chim. Acta, 47 (1973) 215-222.
20 I.L. Martin and G.B. Ansell, Biochem. Pharmac., 22 (1973) 521-533.
21 L.M. Bertani, S.W. Dziedzic, D.D. Clarke and S.E. Gitlow, Clin. Chim. Acta, 30 (1970) 227-233.
22 K. Imai, M.-T. Wang, S. Yoshiue and Z. Tamura, Clin. Chim. Acta, 43 (1973) 145-149.
23 M.-T. Wang, K. Imai, M. Yoshioka and Z. Tamura, Chem. Pharm. Bull., 22 (1974) 970.
24 M.G. Bigdeli and M.A. Collins, Biochem. Med., 12 (1975) 55-65.
25 M.-T. Wang, K. Imai, M. Yoshioka and Z. Tamura, Clin. Chim. Acta, 63 (1975) 13-19.
26 C.D. Kilts, J.J. Vrbanac, D.E. Rickert and R.H. Rech, J. Neurochem., 28 (1977) 465-467.
27 T. Kawano, M. Niwa, Y. Fujita, M. Ozaki and K. Mori, Japan. J. Pharmacol., 28 (1978) 168-171.
28 Y. Maruyama and A.E. Takemori, Anal. Biochem., 49 (1972) 240-247.
29 O. Gyllenhaal, L. Johansson and J. Vessman, J. Chromatogr., 190 (1980) 347-357.
30 S. Kawai and Z. Tamura, Chem. Pharm. Bull., 15 (1967) 1493-1497.
31 D.F. Sharman, Br. J. Pharmacol., 36 (1969) 523-534.
32 L.J. Haeffner, J. Magen and O.D. Kowlessar, J. Chromatogr., 118 (1976) 425-428.
33 J.C. Lhuguenot and B.F. Maume, J. Chromatogr. Sci., 12 (1974) 411-418.
34 L.M. Nelson, F.A. Bubb, P.M. Lax, M.W. Weg and M. Sandler, Clin. Chim. Acta, 92 (1979) 235-240.
35 P.S. Doshi and D.J. Edwards, J. Chromatogr., 210 (1981) 505-511.
36 D.F. LeGatt, G.B. Baker and R.T. Coutts, Res. Commun. Chem. Path. Pharmacol., 33 (1981) 61-68.
37 R.T. Coutts, G.B. Baker, D.F. LeGatt, G.J. McIntosh, G. Hopkinson and W.G. Dewhurst, Progr. Neuropsychopharmacol, 5 (1981) 565-568.
38 G.B. Baker, W.A. Cristofoli and R.T. Coutts, manuscript in preparation.
39 M.G. Horning, A.M. Moss, E.A. Boucher and E.C. Horning, Anal. Lett., 1 (1968) 311-321.
40 P. Cancalon and J.D. Klingman, J. Chromatogr. Sci., 10 (1972) 253-256.
41 A.C. Moffat and E.C. Horning, Biochim. Biophys. Acta, 222 (1970) 248-250.
42 C.J.W. Brooks, B.S. Middleditch and G.M. Anthony, Org. Mass. Spectrom., 2 (1969) 1023-1032.
43 G.M. Anthony, C.J.W. Brooks and B.S. Middleditch, J. Pharm. Pharmacol., 22 (197?) 205-213.
44 J. Vessman, A.M. Moss, M.G. Horning and E.C. Horning, Anal. Lett., 2 (1969) 81-91.
45 M.W. Couch and C.M. Williams, Anal. Biochem., 50 (1972) 612-622.
46 F. Cattabeni, S.H. Koslow and E. Costa, Science, 178 (1972) 166-168.
47 F. Bennington, S.T. Christian and R.D. Morin, J. Chromatogr., 106 (1975) 435-439.
48 M. Donike, Chromatographia, 9 (1976) 440-442.
49 J.J. Warsh, A. Chiu, D.D. Godse and D.V. Coscina, Biochem. Med., 18 (1977) 10-20.
50 N. Narasimhachari and K. Leiner, J. Chromatogr. Sci., 15 (1977) 181-184.
51 G.B. Baker, I.L. Martin, R.T. Coutts and A. Benderly, J. Pharmacol. Methods, 3 (1980) 173-179.

52 K. Blau, G.S. King and M. Sandler, Biomed. Mass Spectrom., 4 (1977) 232-236.
53 D.G. Calverley, G.B. Baker, H.R. McKim and W.G. Dewhurst, Can. J. Neurol. Sci., 7 (1980) 237.
54 G.B. Baker, R.T. Coutts and I.L. Martin, Progr. Neurobiol., 17 (1981) 1-24.
55 J. Eyem and L. Bergstedt, in A. Frigerio and N. Castagnoli (Eds.), Advances in Mass Spectrometry in Biochemistry and Medicine, Vol. 1, Spectrum Publications, Inc., New York, 1976, pp. 497-507.
56 M. Donike, R. Gola and L. Jaenicke, J. Chromatogr., 134 (1977) 385-395.
57 B. Holmstedt, W.J.A. Vanden Heuvel, W.L. Gardiner and E.C. Horning, Anal. Biochem., 8 (1964) 151-157.
58 E.C. Horning, M.G. Horning, W.J.A. Vanden Heuvel, K.L. Knox, B. Holmstedt and C.J.W. Brooks, Anal. Chem., 36 (1964) 1546-1549.
59 J.S. Oliver, H. Smith and D.J.Williams, Forensic Sci., 9 (1977) 195-203.
60 E.J. Cone, M.E. Risner and G.L. Neidert, Res. Commun. Chem. Path. Pharmacol., 22 (1978) 211-232.
61 D.E. Coffin, J. Assoc. Off. Anal. Chem., 52 (1969) 1044-1047.
62 N.P. Sen, J. Food Sci., 34 (1969) 22-26.
63 E.R. Kaplan, N. Sapeika and I.M. Moodie, Analyst, 99 (1974) 565-569.
64 S. Yamamoto, S. Wakaboyashi and M. Makita, J. Agric. Food Chem., 28 (1980) 790-793.
65 R.W. Daisley and H.V. Gutka, J. Pharm. Pharmacol., 32 (1980) 77.
66 R.L. Borison, A.D. Mosnaim and H.C. Sabelli, Life Sci., 15 (1974) 1837-1848.
67 J.W. Schweitzer, A.J. Friedhoff and R. Schwartz, Biol. Psychiat., 10 (1975) 277-285.
68 G.P. Reynolds and D.O. Gray, Clin. Chim. Acta, 83 (1978) 33-39.
69 D.J. Edwards and K. Blau, Biochem. J., 132 (1973) 95-100.
70 J.W. Schweitzer and A.J. Friedhoff, in A.D. Mosnaim and M.E. Wolf (Eds.), Noncatecholic phenylethylamines, part 1, Marcel Dekker, 1978, pp. 475-488.
71 W.R. Martin, J.W. Sloan, W.F. Buchwald and S.R. Bridges, Psychopharmacologia (Berl), 37 (1974) 189-198.
72 G.P. Reynolds and D.O. Gray, J. Chromatogr., 145 (1978) 137-140.
73 G.P. Reynolds, M. Sandler, J. Hardy and H. Bradford, J. Neurochem., 34 (1980) 1123-1125.
74 I.L. Martin and G.B. Baker, J. Chromatogr., 123 (1976) 45-50.
75 I.L. Martin and G.B. Baker, Biochem. Pharmacol., 26 (1977) 1513-1516.
76 G.B. Baker, D.G. Calverley, W.G. Dewhurst and I.L. Martin, Br. J. Pharmacol., 67 (1979) 469P.
77 F.D. Chattaway, J. Chem. Soc., (1931) 2495-2496.
78 L.H. Welsh, J. Am. Pharm. Assoc., 44 (1955) 507-514.
79 M. Goldstein, A.J. Friedhoff and C. Simmons, Experientia, 15 (1959) 80-81.
80 M. Hagopian, R.I. Dorfman and M. Gut, Anal. Biochem., 2 (1961) 387-390.
81 C.J.W. Brooks and E. Horning, Anal. Chem., 36 (1964) 1540-1545.
82 R. Laverty and D.F. Sharman, Br. J. Pharmacol., 24 (1965) 538-548.
83 R.T. Coutts, G.B. Baker and S.-F. Liu, Proc. West. Pharmacol. Soc., 23 (1980) 305-309.
84 R.T. Coutts, G.B. Baker and D.G. Calverley, Res. Commun. Chem. Path. Pharmacol., 28 (1980) 177-184.
85 G.B. Baker, D.F. LeGatt and R.T. Coutts, Biochem. Soc. Trans., 8 (1980) 622-623.
86 D.F. LeGatt, G.B. Baker and R.T. Coutts, J. Chromatogr. Biomed. Appl., 225 (1981) 301-308.
87 G.B. Baker, D.F. LeGatt and R.T. Coutts, J. Neurosci. Methods, 5 (1982) 181-188.
88 D.G. Calverley, G.B. Baker, R.T. Coutts and W.G. Dewhurst, Biochem. Pharmacol., 30 (1981) 861-867.
89 R.J. Argauer, Anal. Chem., 40 (1968) 122-124.
90 W.J. Cole, J. Parkhouse and Y.Y. Yousef, J. Chromatogr., 136 (1977) 409-416.
91 J. Singh, W.P. Cochrane and J. Scott, Bull. Environm. Contam. Toxicol., 23 (1979) 470-474.
92 E. Bailey and E.J. Barron, J. Chromatogr. Biomed. Appl., 183 (1980) 25-31.
93 M.C.H. Oon and R. Rodnight, Biochem. Med., 18 (1977) 410-419.
94 J.I. Javaid and J.M. Davis, J. Pharm. Sci., 70 (1981) 813-815.

95 N. Narasimhachari and R.O. Friedel, Clin. Chim. Acta, 110 (1981) 235-243.
96 K. Jacob, C. Falkner and W. Vogt, J. Chromatogr., 167 (1978) 67-75.
97 P.G. Deo and P.H. Howard, J. Offic. Anal. Chem., 61 (1978) 210-213.
98 F.M. Pasutto and R.T. Coutts, unpublished observations.
99 H. Mita, H. Yasueda and T. Shida, J. Chromatogr., 175 (1979) 339-342.
100 H. Mita, H. Yasueda and H. Shida, J. Chromatogr., 181 (1980) 153-159.
101 H. Mita, H. Yasueda and T. Shida, J. Chromatogr., 221 (1980) 1-7.
102 N. Mahy and E. Gelpi, Chromatographia, 11 (1978) 573-577.
103 D.J. Edwards, P.S. Doshi and I. Hanin, Anal. Biochem., 96 (1979) 308-316.
104 P.S. Doshi and D.J. Edwards, J. Chromatogr., 176 (1979) 359-366.
105 N. Mahy and E. Gelpi, J. Chromatogr., 130 (1977) 237-242.
106 H. Navert, J. Chromatogr., 106 (1975) 218-224.
107 H. Navert and A. Wollin, Union Medicale du Canada, 109 (1980) 1507.
108 D.F. LeGatt, G.B. Baker, W.A. Cristofoli and R.T. Coutts, submitted.
109 R.T. Coutts and G.B. Baker, in A. Lajtha (Ed.), Handbook of Neurochemistry,
    Plenum Press, New York (in press).
110 J.F. Tallman, J.M. Saavedra and J. Axelrod, J. Neurochem., 27 (1976) 465-469.

CHAPTER 7

QUANTITATIVE HIGH RESOLUTION MASS SPECTROMETRY OF BIOGENIC AMINES

BRUCE A. DAVIS AND DAVID A. DURDEN

Psychiatric Research Division, University Hospital,
Saskatoon, Saskatchewan S7N 0X0 (Canada)

7.1   INTRODUCTION

7.1.1   Historical background and development

The necessity for the determination of low concentrations of biogenic amines in brain and brain regions, blood plasma, cerebrospinal fluid and small populations of cells has stimulated the search over the past twenty years for analytical methods of improved sensitivity and specificity.   The term biogenic amines includes the trace amines (phenylethylamine, phenylethanolamine, tryptamine, $\underline{m}$- and $\underline{p}$-tyramine, $\underline{m}$- and $\underline{p}$-octopamine and $\underline{m}$- and $\underline{p}$-synephrine), the catecholamines (dopamine, adrenaline, noradrenaline), and their metabolites (3-methoxytyramine, metanephrine, and normetanephrine), the aliphatic amines (piperidine and histamine) and the polyamines (putrescine, cadaverine, spermidine and spermine).

Mass spectrometry in conjunction with chromatographic procedures is one of the most sensitive and specific methods available for the quantitation of biogenic amines.   In this review, the emphasis will be on the analysis of the trace amines by a thin-layer chromatographic-mass spectrometric technique which we have developed, refined and exploited over the last ten years.

The first attempts at quantitating biogenic amines by mass spectrometric measurement of a selected ion were carried out on relatively crude extracts from rat brain [1-3].   It was soon realized, however, that more accurate results could be obtained if the amines were first derivatized and purified by chromatography.   The 1-($\underline{N}$,$\underline{N}$-dimethylamino)naphthalene-5-sulfonyl (dansyl) derivatives proved to be the most suitable with regard to both chromatographic and mass spectrometric properties.   Dansyl chloride was originally introduced as a fluorescent end-group reagent for proteins and peptides [4], and then as a derivatizing reagent for amino acids [5].   This was soon extended to include quantitation by fluorescence measurements [6-14] and confirmation of identity by mass spectrometry [14-19], the latter assisted by the publication of numerous spectra of dansyl amines [20-23].   The partial mass spectrum and structure of bis-dansyl p-tyramine are presented in Figure 7.1.

Since 1973 we have employed a high resolution thin-layer chromatographic-high resolution mass spectrometric method for the determination of a number of biogenic amines in tissues and body fluids as their dansyl derivatives, with deuterium-labelled analogues as internal standards. These studies have included the identification and distribution of the trace amines phenylethylamine (PEA), tryptamine (T), and m- and p-tyramine (mTA and pTA) in rat organs and brain regions [24-28], in rabbit brain [29], in human brain [30], their subcellular distribution in rat brain [31], and their urinary excretion in the rat and human [32-35]. Other amines measured by this method include phenylethanolamine (PEOH) [36], m- and p-synephrine (mSYN and pSYN) [37,38] and m- and p-octopamine (mOA and pOA) [38] in various tissues, benzylamine in rats pre-treated with pargyline [39], tetrahydroisoquinoline and tetrahydro-β-carboline alkaloids [40], γ-aminobutyric acid (GABA) in astrocytes [41], adrenaline (A) in fowl diencephalon (42), and amphetamine and p-hydroxyamphetamine in the rat following amphetamine injection [43-46]. In addition, trace amine levels in brain [46-56] and in urine [57] after the administration of monoamine oxidase inhibitors, anti-psychotics and other drugs have been measured.

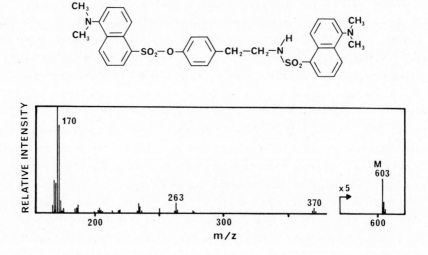

Figure 7.1 The partial mass spectrum and structure of bis-dansyl p-tyramine.

A somewhat similar procedure using low resolution mass spectrometry of the dansyl amines with the internal standard (either a homologue of the compound under investigation or a compound possessing a similar evaporation profile but different mass) added just prior to quantitation in the mass spectrometer has been used by Seiler et al. [58] to estimate tissue levels of putrescine [59-61], serotonin

(5-hydroxytryptamine, 5-HT) and bufotenin [62], GABA [59] and piperidine [63], and by Dolezalova et al. and Stepita-Klauco et al. to measure piperidine [64-67] and cadaverine [68-70] in the brains of snails and mice.

7.1.2   Mass spectrometric quantitative techniques

Mass spectrometric methods for the analysis of biogenic amines have followed two main lines of development: gas chromatography-mass spectrometry (GC-MS) and thin-layer chromatography-mass spectrometry (TLC-MS). Each of these has been applied in a number of ways, depending on the resolution achieved in the chromatography and mass spectrometry. Packed GC columns, which are capable only of low resolution, can be employed with mass spectrometry at low or high resolution (GC-LRMS or GC-HRMS). Capillary or high resolution gas chromatography coupled with low or high resolution mass spectrometry (HRGC-LRMS or HRGC-HRMS) can also be used, although there are few examples in the literature. The higher the resolution of chromatography and mass spectrometry the greater the specificity of the method. Similarly, TLC-MS can be either low or high resolution, depending on the number of TLC separations and the resolution of the mass spectrometer (TLC-LRMS and TLC-HRMS). Our method involves three TLC separations on three different plates in three different solvent systems, which is roughly equivalent to the high resolution obtained with GC capillary columns, and the mass spectrometer is operated at high resolution (7,000-10,000).

In all of the above procedures, the mass spectrometer is adjusted so that one or more ions characteristic of the compound under investigation are monitored continuously as an aliquot of the sample (from the direct probe or outlet of a GC column) enters the ion source. This technique has acquired a number of names, including integrated ion current, mass fragmentography and single, multiple or selected ion detection, monitoring or recording. The International Union of Pure and Applied Chemistry has studied this problem of nomenclature but has made no recommendations [71]. Watson et al. [72] suggest selected ion monitoring (SIM) and it is this term which shall be used in this review. The acronym, SIM, can therefore be added to those given above to complete the description of the procedure. For example, the abbreviation for our method, high resolution thin-layer chromatography-high resolution mass spectrometry with selected ion monitoring, would be HRTLC-HRMS-SIM.

The choice of internal standard and when in the procedure it is added are of crucial importance. Isotopically-labelled (usually deuterium) analogues of the compound to be measured, although expensive to buy or difficult or time-consuming to synthesize, are ideal internal standards for mass spectrometric quantitation. When added to the tissue homogenate or biological fluid, they are carried through all the steps of extraction, derivatization and chromatography in exactly the same way as the endogenous amine, so that losses due to decomposition, adsorption and inefficient extraction are automatically corrected for. All other analytical methods must rely on homologues or other compounds with similar properties for internal standards, with the attendant requirements for calibration curves, recovery estimates and supplementation studies which increase the number of samples to be analyzed.

7.2   PROCEDURE

7.2.1   Chemical and biochemical preparation of the sample for mass spectrometry

The isolation, derivatization and separation of trace amines as currently employed in the preparation of samples for high resolution mass spectrometric analysis was first described by Durden et al. for PEA [24] and the β-hydroxyarylalkylamines [37,38] and by Philips et al. for pTA [26], T[27] and mTA [28]. This procedure, which will be described here only briefly, is outlined schematically in Figure 7.2.

An aliquot containing a known amount of the appropriate dideutero- or tetradeutero-amine internal standard is added to the tissue homogenate or biological fluid before any processing takes place. Two procedures are in use depending on the amount of tissue or fluid to be analyzed. If more than 100 mg of tissue or 0.5 ml of fluid is to be analyzed, the amine fraction is first isolated by ion-exchange chromatography and then dansylated. Otherwise, the homogenate or physiological fluid is derivatized directly by the addition of sodium carbonate and an acetone solution of dansyl chloride. The derivatization solutions are concentrated under a stream of nitrogen and the dansyl amines extracted into benzene. If β-hydroxyamines are to be quantitated, they may be acetylated at this point provided that the separation of m- and p-isomers is not required, otherwise the dansylated isomers are separated by TLC and then acetylated separately. The benzene extract or acetylating mixture is dried under nitrogen and the derivatives are re-dissolved in toluene and transferred to silica gel TLC plates which are developed uni-dimensionally on two or three different plates each in a different solvent system. Following each chromatographic separation, the zone containing the dansyl amine of interest is visualized briefly under ultraviolet light, outlined with a spatula, powdered and extracted. After the final chromatographic separation, the powdered zone is transferred to a micro-extraction tube, and extracted by elution with 25-30 µl of ethyl acetate. The micro-extraction tube is then sealed at both ends with hematocrit sealing clay and stored at -17°C until mass spectrometric analysis.

7.2.2   Mass spectrometric procedure for the quantitation of biogenic amines

The procedure for the mass spectrometric quantitation by SIM of biogenic amines is a modification of earlier procedures developed for other types of compounds [2,73-75]. In our laboratory in Saskatoon, ions are resolved using an AEI MS902S high resolution double-focussing mass spectrometer equipped with a direct insertion probe, operating at 8kV acceleration potential, electron energy of 70 eV and 500 µA emission. The direct probe has been modified so that it is of fixed length and

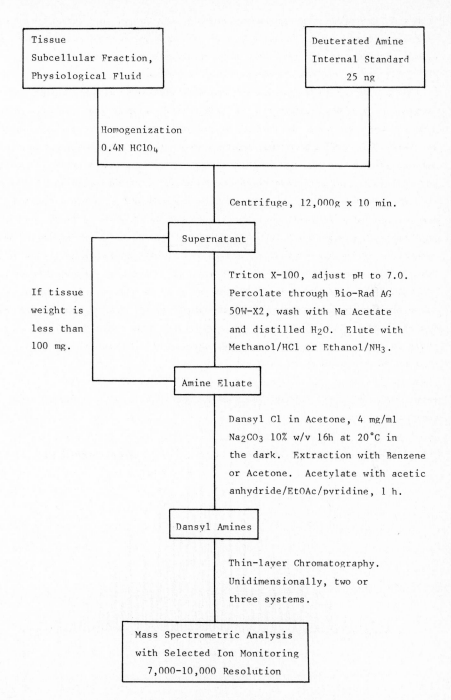

Figure 7.2. Schematic outline of procedure for the HRTLC-HRMS-SIM analysis of trace amines.

such that the probe tip (pyrex or quartz glass) enters the source to within 1-2 mm of the electron beam. Because repeated use of a pyrex or quartz sample holder increases its surface activity with concurrent decomposition of the dansyl compounds, the probe tip is replaced frequently.

The SIM procedure is normally performed with the mass spectrometer tuned to a resolution of 7,000 to 10,000, depending on the substance to be analyzed. The mass spectrometer is first focussed onto an ion characteristic of the compound under investigation (preferably the molecular ion M) and the mass ratio of the peak matching decade box is set so that an ion of known exact mass of a mass reference compound (such as perfluorotri-n-butylamine or perfluorokerosene) is centered on the oscilloscope in the high mass position. Immediately before introduction of the sample, the ratio of the decade box is changed so that the mass of the corresponding ion of the deuterated internal standard occurs in the high mass position. An aliquot of the sample extract (usually 5 µl) is transferred to the probe tip by means of a syringe, the solvent is carefully evaporated and the probe is inserted into the hot ion source. As the sample evaporates the signals from the endogenous amine and deuterated internal standard are recorded alternately. After measuring the areas under the two profiles by planimetry or by computer [36], the amount of endogenous amine can be calculated, correcting for isotopic and chemical impurities in the deuterated standard and natural isotopic contributions of the low to high mass signal. A typical example of the profiles obtained by this procedure can be seen in Figure 7.3 for dansyl PEA and the internal standard dansyl PEA-d4.

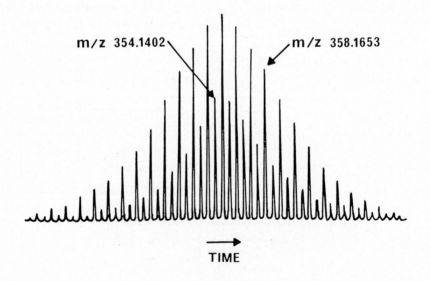

Figure 7.3 Mass spectrometric selected ion monitoring profiles of the molecular ions of dansyl phenylethylamine and dansyl phenylethylamine-d4.

7.3 EVALUATION OF HIGH RESOLUTION MASS SPECTROMETRY VS OTHER ANALYTICAL TECHNIQUES

7.3.1 Specificity

(i) Comparison of values for trace amines measured by different analytical techniques. Specificity is one of the most important criteria by which an analytical technique can be judged. Nevertheless, the literature is rife with examples of analyses which were thought to be specific but which subsequently were proven not to be so. As a rule, as analytical methods become more advanced and sophisticated, reported values for a given compound fall, indicating that earlier methods lacked specificity. In comparing results from different laboratories, one must bear in mind, however, that differences in animal strain, housing conditions, diet, age and post-mortem handling of tissues may affect the tissue levels of some compounds.

TABLE 7.1 Values for phenylethylamine, p-tyramine and tryptamine in whole rat brain as measured by different analytical techniques.

Authors and Reference		Method	PEA (ng/g)	pTA (ng/g)	T (ng/g)
Fischer et al.	[76]	SPF	492		
Durden et al.	[24]	HRTLC-HRMS-SIM	1.8		
Saavedra	[77]	REA	1.5		
Willner et al.	[79]	GC-LRMS-SIM	1.7		
Suzuki & Yagi	[81]	SPF	5.0		
Martin & Baker	[78]	GC-ECD	1.1		
Philips et al.	[30]	HRTLC-HRMS-SIM	2.1		
Karoum et al.	[86]	GC-LRMS-SIM	8.1	4.2	
Edwards et al.	[80]	GC-CI-MS-SIM	1.0		
Philips et al.	[26]	HRTLC-HRMS-SIM		2.0	
Tallman et al.	[84]	REA		12.9	
Duffield et al.	[85]	GC-CI-MS-SIM		2.2	
Saavedra & Axelrod	[87]	REA			22.0
Snodgrass & Horn	[91]	Radio-dansyl			69.0
Philips et al.	[27]	HRTLC-HRMS-SIM			0.5
Philips & Boulton	[49]	HRTLC-HRMS-SIM			0.35
Sloan et al.	[88]	SPF			20.9
Warsh et al.	[89]	GC-LRMS-SIM			not detectable
Artigas & Gelpi	[90]	GC-LRMS-SIM			<0.38

Analyses of the trace amines PEA, pTA and T in rat brain over a period of a few years using different analytical methods illustrate perfectly the importance of specificity (Table 7.1). PEA in rat brain was first measured by Fischer et al. [76] who obtained a value of 492 ng/g by the spectrophotofluorimetric (SPF) method, yet a year later Durden et al. [24] had obtained a value of 1.8 ng/g using HRTLC-HRMS-SIM. Since then other methods have been employed and confirmed the latter result, for example, the radioenzymatic assay (REA) [77], gas chromatography with electron capture detection (GC-ECD) [78], GC-MS-SIM [79] and GC-MS-SIM with chemical ionization (GC-CI-MS-SIM) [80]. Using an improved SPF procedure, Suzuki and Yagi [81] were able to obtain a value of 5.0 ng/g. A similar phenomenon was observed for PEA in rabbit brain, in which early results by Borison et al. [82] (340 ng/g) using gas chromatography with flame ionization detection (GC-FID) and Mosnaim et al. [83] (400 ng/g) using SPF turned out to be much higher than the value obtained by Boulton et al. [29] (0.44 ng/g) using HRTLC-HRMS-SIM. Similar developments can be observed in Table 7.1 for pTA and T. The radioenzymatic assay produced a value of 12.9 ng/g for pTA [84] compared to 2.0 ng/g by HRTLC-HRMS-SIM [26]. GC-CI-MS-SIM gave a value of 2.2 ng/g [85], but another GC-MS procedure yielded a somewhat higher value of 4.2 ng/g [86]. For T the REA gave a value of 22.0 ng/g [87], similar to that reported for a SPF method (20.9 ng/g) [88], but much higher than that obtained by HRTLC-HRMS-SIM (0.50, 0.35 ng/g) [27,49]. One paper [89] describing a GC-LRMS-SIM procedure reported T to be not detectable, even with three pooled brains, and another [90] reported T to be less than 0.38 ng/g, presumably the lower limit of detection. A radio-dansyl assay gave a value of 69 ng/g [91]. PEOH has been reported to be present in whole rat brain at a level of 6.2 ng/g [92] as measured by the REA, 3.8 ng/g by GC-LRMS-SIM [79] and 5.8 ng/g by GC-CI-MS-SIM [80], but by HRTLC-HRMS-SIM [36] a value of only 0.06 ng/g was obtained. m-TA has been measured in tissue only by the HRTLC-HRMS-SIM technique [28] so a comparison with other analytical methods is not possible.

TLC-LRMS-SIM without a true internal standard (that is, a compound similar to the one under investigation is added to the direct probe) has been used to measure the level of piperidine in snail brain. The results of different workers, however, do not agree. Using dansyl derivatives and dansyl pyrrolidine as standard (added to the direct probe not to the homogenate), Dolezalova et al. [65] measured 277 ng/g and Seiler et al. [63], using the corresponding 1-(N,N-di-n-butylamino) napthalene-5-sulfonyl (bansyl) derivatives, found 17 ng/g piperidine in snail brain.

(ii) Evaluation of the specificity of the different analytical techniques.

a. Thin-layer chromatography-mass spectrometry. It is interesting to note that HRTLC-HRMS-SIM is the only method providing the lowest values (and presumably, therefore, the correct ones) for all the above-mentioned amines. The reason for this is undoubtedly the greater specificity of the HRTLC-HRMS-SIM procedure. This greater specificity is due not only to the high resolution of the thin-layer chromatography (three plates each developed unidimensionally in a different solvent system) and the mass spectrometry (resolution of 7,000 to 10,000), but also to the nature of the dansyl derivative. The low mass region of the spectrum of a biological extract is rather complex due to the presence of molecular ions from substances other than the one of interest and fragment ions deriving from higher molecular weight substances. The dansyl derivatives of biogenic amines not only possess high enough molecular weights so as to be relatively free of interfering ions, but also one or more mass-deficient sulfur atoms which facilitate the resolution of the characteristic dansyl amine ions from background ions, many of which are non-mass deficient hydrocarbons.

b. _Radioenzymatic assay_. Since the specificity of the REA is somewhat dependent on the purity and specificity of the enzyme and even more dependent on the selectivity of the isolation procedure, it may be that the high values obtained for T, pTA and PEOH are due to alternative substrates for the methylating enzyme which were not adequately separated before or after methylation. Secondly, the REA as originally utilized was unable to distinguish between positional isomers, for example the o-, m- and p-isomers of octopamine[93]. By dansylation of the methylation product mixture and isolation of the products by two or three TLC separations it has been shown [94] that separation and quantitation of mOA and pOA are possible, and that these values are lower than those of the original REA [93].

c. _Gas chromatography-mass spectrometry_. The GC-MS-SIM procedure usually gives values comparable to those obtained by HRTLC-HRMS-SIM, provided that the choices of derivative, internal standard and ion or ions to be recorded are wisely made. The ideal internal standard, as mentioned above, is the deuterated analogue of the compound under investigation. Many investigators, however, employ homologues or other similar compounds as internal standards, accepting the resulting uncertainty which arises because of possible differences in the extraction, chromatographic and mass spectrometric properties of the standard compared to the amine to be analyzed. A frequently used device to ensure specificity is to monitor two or more ions of the same compound; if both (or all) are present and the relative intensities are in the same ratio as for the authentic compound this is taken as proof of the purity of the substance being analyzed. While it is probably correct to assume that the ratio remains constant for periods of a few hours, it has been observed that the ratio of ion abundances can vary significantly from day to day depending on ion source cleanliness, instrument tuning and changes in source temperature [95]. It is of particular importance to bear this in mind when analyzing biogenic amines as their perfluoroacyl or trimethylsilyl derivatives, the most popular derivatives for GC-MS because of the ability of the corresponding reagents to react readily with the usual functional groups found in the biogenic amines. Unfortunately, the most abundant ions in the spectra of these derivatives are often common to several different amines (and also their acidic and alcoholic metabolites), so that specificity must rely on chromatographic resolution (which is comparatively low for the packed columns which are most often used) and the ratio of ion abundances (which, as mentioned earlier, can be variable due to instrumental factors). Furthermore, these common ions are generally of relatively low mass, which increases the likelihood of interference from background ions. For example, m/z 174 for the trimethylsilyl derivatives of primary amines [96], m/z 58 for dimethyl tertiary amines [97,98] and m/z 140 for the pentafluoropropionyl derivatives of methyl secondary amines [99] are each the base peak in the mass spectrum and all result from cleavage of the α,β-alkyl carbon bond with the charge remaining on the nitrogen fragment which provides no structural information.

The dangers inherent in quantitating biogenic amines by monitoring low mass non-unique ions are illustrated by the results obtained for PEA in rat brain [86] in which the masses m/z 91 (C_7H_7) and 104 (C_8H_8) were monitored. The value obtained for rat brain (8.1 ng/g) [86] is substantially higher than the 1.8 ng/g [24] obtained by HRTLC-HRMS-SIM. The monitoring of low mass, non-unique ions can in some circumstances be acceptable and produce values comparable to those obtained by HRTLC-HRMS-SIM, as, for example, in the measurement of PEA in human urine [100] in which the value obtained (8.0 μg/24h) agrees with that obtained by HRTLC-HRMS-SIM [32,35], even though the masses monitored were m/z 91 and 104. Willner et al [79], using pentafluoropropionyl derivatives and deuterated internal standard obtained a value of 1.7 ng/g for PEA in rat brain using masses m/z 93 for the internal standard and m/z 104 for the endogenous amine. Jacob et al. [101] employed GC-HRMS-SIM in order to differentiate the fragment ions of the N-trifluoro acetyl-O-trimethylsilyl derivatives of catecholamines to be measured from interfering substances having the same nominal mass. A few investigators are using GC-CI MS by which they obtain intense pseudo-molecular ions and therefore an increase in specificity [80,102-106] due to the uniqueness of such ions.

d. Gas chromatography. GC is inherently much less specific than mass spectrometric methods since the specificity is based for the most part on retention time and the resolving power of the column. The specificity can be improved somewhat by a suitable choice of detector and derivative. For example, the electron capture detector with fluorinated derivatives is more specific (and more sensitive) than the flame ionization detector. The relatively low specificity of GC is at least partly responsible for the wide variation in values reported for PEA in human urine and rat brain. In urine, values of 770 [100], 400 [107], 10.3 [108] and 6.8 [109] μg/24 h have been measured by GC, whereas values determined by MS are consistently less than 10 μg/24 h [32,35,86]. Similarly, the high value (400 ng/g) for PEA in rat brain reported by Fischer et al. [76] has been brought down only after extensive cleanup of the sample before GC by Martin and Baker [78] to 1.1 ng/g, which compares favorably with values obtained by HRTLC-HRMS-SIM [24,30]. Obviously, if there is a careful preliminary cleanup of the sample and a suitable choice of column and internal standard, an accurate value can be obtained by GC.

e. High-performance liquid chromatography. The specificity of HPLC rests on the resolution of the column and the unique detection methods available for a limited number of compounds. The catecholamines (and other catechol-type compounds) are usually measured electrochemically (EC), with specificity dependent on these compounds having a low oxidation potential (that is < 0.7 volts). Because of the small amounts present and their high oxidation potentials, the trace amines have not been determined by HPLC-EC. To measure the trace amines by HPLC it would

perhaps be necessary to prepare fluorescent derivatives and quantitate them fluorimetrically. An attempt in this laboratory to do this using the dansyl derivatives
has so far been unsuccessful because of the enormous number of peaks present, the
relatively poor resolving power of HPLC columns and the small amounts of the amines
present [110]. A few types of compounds, such as the indoles, possess a natural
fluorescence and can therefore be detected and quantitated fluorimetrically without
the need for derivatization. When HPLC is compared with MS or the REA, the values
obtained by HPLC are sometimes somewhat higher, particularly in the picogram range,
suggesting an inferior specificity for HPLC [111,112]. For example, Warsh et al.
[111] observed that values for DA measured by HPLC-EC were 9% higher (p < 0.01)
than those obtained by GC-MS.

f. Spectrophotofluorimetry. SPF methods are generally not very specific since
their specificity relies almost entirely on a chemical separation of compounds.
Even with fairly extensive cleanup procedures, contamination by substances forming
similar fluorophors, difficulties in obtaining consistently low blanks, and the
non-linear relationship between concentration and fluorescence intensity (due to
quenching) contribute to uncertainty in the accuracy of the results. Not surprisingly, all the values listed in Table 7.1 measured by SPF are substantially higher
than those obtained by mass spectrometric methods.

It is clear then that, in combination with TLC or GC with capillary columns,
HRMS-SIM of the molecular ion of a high molecular weight amine derivative offers
the greatest specificity of the methods currently available.

7.3.2 Sensitivity

(i) Definition. Considerable caution must be exercised in assessing the relative sensitivities of the various analytical techniques, since the limits of sensitivity reported in the literature are not always based on the same premises. In
this review, the sensitivity will be expressed as the minimum detectable quantity
(MDQ), which we define as the quantity of a substance in a biological sample giving
rise to a signal (or number of counts) twice the size of the blank. The MDQ refers
to the amount of the substance in the original sample, not the amount in the aliquot put into the instrument. The MDQ for a number of biogenic amines measured by
various techniques is summarized in Table 7.2.

(ii) Trace amines. Most trace amine analyses have been carried out using HRTLC-
HRMS-SIM. The MDQ ranges from 100-200 pg [23,113]. The actual amount inserted
into the ion source of the mass spectrometer is usually considerably less than this
since only a fraction is put on the direct probe and recoveries are generally less
than 40%. However, because the samples have been isolated and purified by three
TLC separations, they are very pure so it is possible, when necessary, to put the
entire sample into ion source, whereas in column chromatographic techniques (GC-MS,
GC, HPLC) usually only a small fraction is injected onto the column.

TABLE 7.2. Minimum detectable quantities for some biogenic amines (in picograms).

Method	Trace Amines (PEA, mTA, pTA, OA, SYN)	Catecholamines (A, NA, DA)	Indoleamines (T, 5HT)	Polyamines	Other Amines (Piperidine, Histamine)
HRTLC-MS-SIM	100-200 [23, 113]	3000 [23, 113]	250 (T) [23, 113]	–	300 [63] –
GC-MS-SIM	400 [100]	100 [101,111,119]	500 (T) [89]	100 [134]	2000 [128] 2000 [130]
REA	10-250 [77, 84, 94]	1-5 [116,117]	5-10(5HT)[118] 5000 (T) [87]	–	– 10 [131]
HPLC	–	10-500 [124,126,127]	100 (5HT) [121-123,125]	1000-10,000 [132,133]	100 [123]
GC	20,000 [82] FID 1,000 [78] ECD	– –	2500-5000 [114,120] (5HT) FID	500[136] ECD 20,000[135] FID	3000[129]ECD
Radio-dansyl	–	100 [115]	100 [115] (5HT)	–	–

The MDQ of 100-200 pg is by no means the lowest that can be attained by MS. Each amine will have a much lower basic sensitivity depending only on the mass spectrometric properties of that compound. The higher practical MDQ arises because of losses of the sample during isolation and derivatization and because of contributions to the blank from other substances in the extract and trace impurities present in extracting solvents and chromatographic materials. Contributions to the blank can also arise from isotopic impurities in the internal standard. We have observed, for example, that in using dideutero internal standards the MDQ is somewhat higher (200-500 pg) than when using tetradeutero internal standards (50-150 pg) in which the amount of undeuterated amine present is known to be smaller.

The REA, in which the entire sample is carried through the procedure and counted is capable of MDQ as low as 10 pg for OA [94] and 100 pg for PEA [77]. This method, however, has seldom been applied to the analysis of trace amines.

GC-MS, although potentially a very sensitive and specific technique for measuring trace amines, has also seldom been used for this purpose [89,100]. The procedures commonly employed for the isolation and derivatization of samples for GC-MS result in high recoveries compared to HRTLC-HRMS, but usually only a small aliquot can be injected, otherwise overloading of the column occurs and reduces the resolution. The overall result is a sensitivity similar to that obtained by HRTLC-HRMS. By investing more time and effort in purification of the sample prior to injection into the GC, it may be possible to inject more concentrated solutions of the sample and thus obtain an increase in sensitivity.

GC-ECD is similar to GC-MS in that it is in principle a very sensitive method, but because of the presence of numerous other substances in the biological extract is in practice only moderately sensitive. Martin and Baker [78], for example, apparently can detect about 4 pg of PEA (injected amount) using GC-ECD, but the sample must be so diluted to reduce the concentrations of other compounds that a nanogram or more must be present in the original sample. GC-FID, as one would expect, is even less sensitive, with MDQ of 20 ng for PEA [82] and 5 ng for T [114] being reported, neither of which is sensitive enough to measure these amines in rat brain, even if several are pooled.

The radiodansyl method for T is also not sensitive enough for measuring this amine in most tissues since about 10 ng is the smallest amount detectable [91], although the method is adequate for serotonin and the catecholamines [115].

(iii) Catecholamines and serotonin. HRTLC-HRMS-SIM of the dansyl derivatives of the catecholamines with a MDQ of about 3 ng has not proven to be as sensitive as for the trace amines, mainly because of the very small relative intensity (1% or less) of the molecular ion and the basic conditions required for dansylation (which favour oxidation). In addition, the very high molecular weights of the tris-dansyl catecholamines (nearly 1000 amu) are at or above the upper limit of the mass range in which our mass spectrometer can be operated at maximum sensitivity. Acetylation of the side-chain hydroxy group after dansylation increases slightly the relative intensity of the molecular ion [37].

The REA appears to be the most sensitive method available for quantitating the catecholamines and 5-HT, the most advanced modifications being capable of quantitating as little as 1 pg of A and NA, 5 pg of DA [116,117], and 5-10 pg of 5-HT [118]. GC-HRMS-SIM [101] and GC-LRMS-SIM [111,119] are somewhat less sensitive, having MDQ of about 100 pg. GC-FID has been used to measure 5-HT in brain in amounts of 2.5 ng [120].

In the last five years, HPLC with electrochemical or fluorimetric detection of the catecholamines and indoleamines has become increasingly popular, due to good sensitivity as well as to the low cost per assay, speed of analysis and simplicity of sample preparation. MDQ of 100 pg [111,121-127] can be routinely obtained, and in one procedure [124] in which the entire sample is injected onto the column, an MDQ of 10 pg is claimed.

(iv) Other amines. Seiler [63] has employed the TLC-LRMS-SIM procedure to measure piperidine as its bansyl derivative, in which the entire sample is evaporated on the direct probe and inserted into the ion source. He reports that as little as 300 pg can be measured in this way. Piperidine has also been measured by GC-LRMS-SIM [128] and GC-ECD [129], but in these cases the MDQ is 2 or 3 ng.

Histamine has been measured by GC-LRMS-SIM [130], REA [131] and HPLC with fluorimetric detection of the o-phthalaldehyde derivative [123]. The REA is the most sensitive method with a MDQ of about 10 pg, HPLC can detect 100 pg and GC-LRMS-SIM about 2 ng.

The MDQ for polyamines has not been reported for most methods. Seiler, for example, has used TLC-LRMS-SIM to measure putrescine as its dansyl derivative in rat [60,61] and fish [59] brain in amounts as low as about 300 ng/g, but there is no reason to assume that this is the MDQ. Seiler has also employed the dansyl derivatives for the HPLC fluorimetric quantitation of as little as 1 ng of the polyamines in various tissues [132]. Samejima et al. [133] report that putrescine, cadaverine and spermidine as their fluorescamine derivatives can be detected in quantities of less than 10 ng by HPLC. Smith and Daves [134], working only with standard solutions, used GC-LRMS-SIM and deuterated internal standards to measure down to 100 pg of putrescine, cadaverine and spermidine and 5 ng of spermine, the high values for the latter being due to sample retention on the GC column and peak broadening. About 20 ng is the MDQ for the measurement of polyamines by GC-FID [135], whereas as little as 500 pg can be detected by GC-ECD of the pentafluorobenzoyl derivatives [136].

7.3.3 Convenience and efficiency

In order to achieve the maximum possible specificity and sensitivity all procedures require an extensive purification of samples before instrumental analysis. The HRTLC-HRMS-SIM method involves an extraction, derivatization and TLC purification procedure by which about 60-70 samples can be processed in two days. However, an advantage of this method is that each sample then requires only 5 or 6 minutes of mass spectrometer time and because of its high purity the entire sample can, if necessary, be put on the direct probe and inserted into the instrument. Considering the high cost of purchasing and operating a mass spectrometer such efficient use of the instrument is desirable.

In practice, workup procedures for GC-MS-SIM have been less time-consuming than those for HRTLC-HRMS-SIM, although this can sometimes result in reduced specificity due to the presence of numerous interfering substances not resolved on packed columns and in lower sensitivity due to the requirement of greater dilution of the sample to reduce the concentration of interfering substances. In addition, more mass spectrometer time is required than for HRTLC-HRMS-SIM since one must wait for the compounds of interest to elute, and then wait further while uninteresting compounds are eluted. If an extensive cleanup of the sample is undertaken and the GC conditions changed to reduce elution time and if more than one compound can be determined from a single injection, the mass spectrometer time required per compound analyzed can become competitive with the HRTLC-HRMS-SIM procedure [90].

For GC-ECD or GC-FID, an extensive and efficient cleanup of the sample is of paramount importance since specificity relies mainly on the resolving power of the column and the retention time of the substance. The higher the purity of the sample the less likely is the possibility of an interfering substance co-eluting with the compound of interest. Furthermore, a relatively pure sample permits smaller volumes of solvent to be used and therefore a larger fraction of the sample can be injected.

For maximum specificity, the REA must include chromatographic purification of the sample in addition to selective extraction. Up to 50 samples per day can be analyzed, although the time required to prepare and purify the enzyme is usually in addition to this.

Twenty to thirty samples can be analyzed by HPLC in a day, and for a limited number of types of compounds (catechols and indoles) very little preliminary processing is required. In fact, HPLC lends itself well to automation for the analysis of these compounds [124,137]. It is particularly suited to metabolic studies in which a number of metabolically related compounds in the sample can be determined simultaneously on the basis of a common functionality, as in the cases of indole [138] and catechol compounds [123,138,139].

7.3.4 Other factors

A great strength of the mass spectrometric techniques is that virtually any type of compound with mass less than 2,000-3,000 amu can be analyzed. The REA, on the other hand, is limited to those substances for which suitable enzymes exist and are easily prepared or commercially available. HPLC-EC is strictly limited to those compounds possessing readily oxidizable functional groups (e.g. catechol and 5-hydroxyindole compounds).

Mass spectrometry is becoming increasingly important in the analysis of metabolites formed following the ingestion of stable isotope-labelled precursors [33,140-143]. Because of the hazards associated with them, radioactive compounds are no longer acceptable for use in metabolic studies on humans.

Finally, the question of the costs of purchasing and operating a particular instrument must be considered. With or without a data system, a mass spectrometer is the most expensive instrument of those discussed. Not only is the initial cost high, particularly for a high resolution mass spectrometer, but because of its complexity and size, maintenance and repair costs are high. By comparison, GC, SPF and HPLC are relatively inexpensive to buy, maintain, repair and operate. The REA and radio dansyl procedure are also relatively inexpensive in terms of the initial cost of the liquid scintillation counter, but the cost of the radiochemicals can be exceedingly high if analyses are carried out daily. In addition, special precautions are required in the handling and disposal of the radioactive substances.

7.4 SUMMARY AND PROMISING DEVELOPMENTS

A summary of the advantages and limitations of the various procedures is presented in Table 7.3. Although the choice of method depends to a considerable extent on the background of the investigator, the nature of the problem and the personnel and equipment available, high resolution mass spectrometry is unequalled in terms of specificity and sensitivity. However, a number of approaches can be taken to improve sensitivity and specificity still further. One approach is to use a derivative which produces greater molecular ion or unique fragment ion intensities. The bansyl derivatives exhibit in their mass spectra very large M-43 ions (usually the base peak) as well as significant molecular ions [63,144-146]. Unfortunately the tris-bansyl catecholamines have such high molecular weights that they are out of the range of maximum sensitivity on most mass spectrometers [146]. However, if the derivatization procedure is modified so that only the amino group is bansylated and the phenolic and catecholic hydroxy groups alkylated, the resulting mixed derivative has a more useful molecular weight (400-700 amu) and retains the M-43 ion as base peak. This is reflected, at least for standard solutions, in a

TABLE 7.3 Advantages and limitations of procedures for the analysis of biogenic amines.

Procedure	Advantages	Limitations
HRTLC-HRMS-SIM	Very sensitive for trace amines, highly specific, can be adapted to many amines, efficient use of instrument – can measure 60 samples per day.	Not very sensitive for catecholamines, equipment is expensive, relatively high blanks, tedious sample preparation, tertiary amines cannot be dansylated.
GC-MS-SIM	Sensitive, very specific for high resolution MS, capillary column outlet can be inserted directly into ion source, can be adapted to most amines, can measure more than one compound per injection.	Less efficient use of MS time than above, expensive equipment.
REA	Very sensitive for most biogenic amines, inexpensive equipment, can process up to 50 samples per day.	Not always specific, tedious, requires handling of radioactive substances, expensive reagents.
HPLC	Relatively inexpensive, sensitive, fair specificity, can measure more than one compound per injection, simple sample preparation.	Not always specific.
GC	Inexpensive, can be coupled to a mass spectrometer.	Not well developed for biogenic amines, poor specificity, extensive sample clean-up required.
Radio-dansyl	Fairly sensitive, potential of wide application to amines.	Specificity limited, expensive reagent, requires handling of radioactive substances. Reaction of dansyl chloride is not constant at low amounts.

much greater sensitivity than for dansyl derivatives [146]. Unfortunately, bansyl chloride is not commercially available and is fairly time-consuming to prepare and purify, so that its use on a regular basis is restricted.

An obvious and promising approach is to combine high resolution GC (i.e. capillary columns) with high resolution mass spectrometry. Such a technique has not yet been applied to the analysis of biogenic amines, but has been successfully employed in the determination of the acid metabolites of some biogenic amines [142,143,147-149].

Positive ion CI-MS is a technique which is seeing increased use in the hope of improving both specificity and sensitivity [80,85,103-105]. Although most of the ion current is carried by the molecular ion or protonated molecular ion, this does not necessarily lead to increased sensitivity compared to electron-impact (EI)-MS since CI usually produces a smaller total ion signal due to a smaller ion source aperture. Negative ion CI-MS is more recent, but claims of sensitivity four times [150] greater than EI-MS and thirty times [151] greater than positive ion CI-MS have been reported. Lewy and Markey [152] have been able to detect 1 pg of melatonin in 1 ml of plasma by negative ion CI-MS.

Finally, a new and elegant technique, selected metastable peak monitoring (SMPM), is showing great promise [153-155]. In this method, only selected daughter ions formed from a selected parent ion in the first field-free region of a double focussing mass spectrometer are recorded. This results in a dramatic reduction in the size and number of background ions since most, if not all, other fragments deriving from endogenous contaminants, column bleed or solvent impurities are formed by different mechanisms and are therefore trapped in the electrostatic sector of the mass spectrometer. Although this in itself does not usually result in an absolute increase in sensitivity, the absence of interfering background ions allows the gain of the mass spectrometer to be increased so that in practice there is an improvement in sensitivity. Durden [155] has refined this technique with a relatively inexpensive but sophisticated modification that permits multiple metastable peak monitoring, which is necessary if one is to use as internal standards the deuterium-labelled analogues of the endogenous compounds of interest. The application of this technique to the determination of biogenic amines holds promise of improved specificity and sensitivity.

ACKNOWLEDGEMENTS

We wish to thank our colleagues in the Psychiatric Research Division whose work constitutes a large part of this review, Dr. A.A. Boulton for his advice and critical reading of the manuscript, and Saskatchewan Health and the Medical Research Council of Canada for financial support.

146

REFERENCES

1. A.A. Boulton and J.R. Majer, J. Chromatogr., 48 (1970) 322-327.
2. J.R. Majer and A.A. Boulton, Nature, 225 (1970) 658-660.
3. A.A. Boulton and J.R. Majer, Can. J. Biochem., 49 (1971) 993-998.
4. W.R. Gray and B.S. Hartley, Biochem. J., 89 (1963) 59P.
5. A.A. Boulton and I.E. Bush, Biochem. J., 92 (1964) 11P-12P.
6. N. Seiler and A. Askar, J. Chromatogr., 62 (1971) 121-127.
7. A.A. Boulton, Methods of Biochemical Analysis, 16 (1968) 327-393.
8. E.J. Diliberto and V. DiStefano, Anal. Biochem., 32 (1969) 281-285.
9. N. Seiler and M. Wiechmann, Progr. Thin-Layer Chromatogr. Related Methods, 1 (1970) 95-144.
10. N. Seiler, Research Methods in Neurochemistry, 3 (1975) 410-441.
11. J.H. Fleischer and D.H. Russell, J. Chromatogr., 110 (1975) 335-340.
12. G. Dreyfuss, R. Dvir, A. Harell and R. Chayen, Clin. Chim. Acta, 49 (1973) 65-72.
13. N. Seiler and M. Wiechmann, Z. Anal. Chem., 220 (1966) 109-127.
14. A.A. Boulton and J.R. Majer, Research Methods in Neurochemistry, 1 (1972) 341-356.
15. A.A. Boulton and L. Quan, Can. J. Biochem., 48 (1970) 1287-1291.
16. C.R. Creveling, K. Kondo and J.W. Daly, Clin. Chem., 14 (1968) 302-309.
17. N. Seiler, J. Chromatogr., 63 (1971) 97-112.
18. S. Axelsson, A. Björklund and N. Seiler, Life Sci., 13 (1973) 1411-1419.
19. C.R. Creveling and J.W. Daly, Nature, 216 (1967) 190-191.
20. J. Reisch, H. Alfes, N. Jantos and H. Möllmann, Acta Pharm. Suecica, 5 (1968) 393-397.
21. N. Seiler, H. Schneider and K.-D. Sonnenberg, Z. Anal. Chem., 252 (1970) 127-136.
22. D.I. Chapman, J.R. Chapman and J. Clark, Int. J. Biochem., 3 (1972) 66-72.
23. D.A. Durden, B.A. Davis and A.A. Boulton, Biomed. Mass Spectrom., 1 (1974) 83-95.
24. D.A. Durden, S.R. Philips and A.A. Boulton, Can. J. Biochem., 51 (1973) 995-1002.
25. A.A. Boulton, S.R. Philips and D.A. Durden, J. Chromatogr., 82 (1973) 137-142.
26. S.R. Philips, D.A. Durden and A.A. Boulton, Can. J. Biochem., 52 (1974) 366-373.
27. S.R. Philips, D.A. Durden and A.A. Boulton, Can. J. Biochem., 52 (1974) 447-451.
28. S.R. Philips, B.A. Davis, D.A. Durden and A.A. Boulton, Can. J. Biochem., 53 (1975) 65-69.
29. A.A. Boulton, A.V. Juorio, S.R. Philips and P.H. Wu, Brain Res., 96 (1975) 212-216.
30. S.R. Philips, B. Rozdilski and A.A. Boulton, Biol. Psychiatry, 13 (1978) 51-57.
31. A.A. Boulton and G.B. Baker, J. Neurochem., 25 (1975) 477-481.
32. J.M. Slingsby and A.A. Boulton, J. Chromatogr., 123 (1976) 51-56.
33. A.A. Boulton, L.E. Dyck and D.A. Durden, Life Sci., 15 (1974) 1673-1683.
34. A.A. Boulton and L.E. Dyck, Life Sci., 14 (1974) 2497-2506.
35. N.D. Huebert and A.A. Boulton, J. Chromatogr. Biomed. Appln., 162 (1979) 169-176.
36. D.A. Durden, Research Methods in Neurochemistry, 4 (1978) 205-250.
37. D.A. Durden, A.V. Juorio and B.A. Davis, Anal. Chem., 52 (1980) 1815-1820.
38. D.A. Durden, A.V. Juorio and B.A. Davis, in A.P. deLeenheer, R.R. Roncucci and C. van Peteghem (Eds.), Quantitative Mass Spectrometry in the Life Sciences, Elsevier, Amsterdam, 1978, Vol. 2, pp. 389-397.
39. D.A. Durden, S.R. Philips and A.A. Boulton, Biochem. Pharmacol., 25 (1976) 858-859.
40. D.A. Durden, T.J. Danielson and A.A. Boulton, in A. Frigerio and D.M. Desiderio (Eds.), Advances in Mass Spectrometry in Biochemistry and Medicine, Spectrum, New York, 1976, Vol. 2, pp. 597-605.

41. P.H. Wu, D.A. Durden and L. Hertz, J. Neurochem., 32, (1979) 379-390.
42. A.V. Juorio and D.A. Durden, Can. J. Biochem., 55 (1977) 761-765.
43. T.J. Danielson and A.A. Boulton, Biomed. Mass Spectrom., 1 (1974) 159-162.
44. T.J. Danielson and A.A. Boulton, Eur. J. Pharmacol., 37 (1976) 257-264.
45. T.J. Danielson, B.A. Davis and A.A. Boulton, Can. J. Physiol. Pharmacol., 55 (1977) 439-443.
46. T.J. Danielson, T.B. Wishart, H.A. Robertson and A.A. Boulton, Progr. Neuro-psychopharmacol., 1 (1977) 279-284.
47. T.J. Danielson, T.B. Wishart and A.A. Boulton, Life Sci., 18 (1976) 1237-1244.
48. A.A. Boulton, A.V. Juorio, S.R. Philips and P.H. Wu, Br. J. Pharmacol., 59 (1977) 209-214.
49. S.R. Philips and A.A. Boulton, J. Neurochem., 33 (1979) 159-167.
50. S.R. Philips, G.B. Baker and H.R. McKim, Experientia, 36 (1980) 241-242.
51. A.V. Juorio, Life Sci., 20 (1977) 1663-1668.
52. A.V. Juorio, Brain Res., 126 (1977) 181-184.
53. A.V. Juorio and T.J. Danielson, Eur. J. Pharmacol., 50 (1978) 79-82.
54. A.V. Juorio, Brain Res., 179 (1979) 186-189.
55. A.V. Juorio, Br. J. Pharmacol., 70 (1980) 475-480.
56. A.V. Juorio, B.A. Davis and A.A. Boulton, Res. Commun. Psychol. Psychiat. and Behav., 5 (1980) 255-264.
57. L.E. Dyck and A.A. Boulton, Res. Commun. Chem. Pathol. Pharmacol., 11 (1975) 73-77.
58. N. Seiler and B. Knodgen, Org. Mass Spectrom., 7 (1973) 97-105.
59. N. Seiler and U. Lamberty, J. Neurochem., 20 (1973) 709-717.
60. N. Seiler and U. Lamberty, J. Neurochem., 24 (1975) 5-13.
61. N. Seiler and T. Schmidt-Glenewinkel, J. Neurochem., 24 (1975) 791-795.
62. N. Seiler and K. Bruder, J. Chromatogr., 106 (1975) 159-173.
63. N. Seiler and H.H. Schneider, Biomed. Mass Spectrom., 1 (1974) 381-385.
64. H. Dolezalova, M. Stepita-Klauco and R. Fairweather, Brain Res., 72 (1974) 115-122.
65. H. Dolezalova, E. Giacobini, N. Seiler and H.H. Schneider, Brain Res., 55 (1973) 242-244.
66. H. Dolezalova and M. Stepita-Klauco, Brain Res., 74 (1974) 182-184.
67. M. Stepita-Klauco, H. Dolezalova and R. Fairweather, Science, 183 (1974) 536-537.
68. H. Dolezalova, M. Stepita-Klauco and N. Seiler, Brain Res., 67 (1974) 349-351.
69. H. Dolezalova, M. Stepita-Klauco and R. Fairweather, Brain Res., 77 (1974) 166-168.
70. M. Stepita-Klauco and H. Dolezalova, Nature, 252 (1974) 158-159.
71. International Union of Pure and Applied Chemistry, Physical Chemistry Division, Commission on Molecular Structure and Spectroscopy, Pure and Applied Chem., 50 (1978) 67-73.
72. J.T. Watson, F.C. Falkner and B.J. Sweetman, Biomed. Mass Spectrom., 1 (1974) 156-157.
73. C.C. Sweeley, W.H. Elliott, I. Fries and R. Ryhage, Anal. Chem., 38 (1966) 1549-1553.
74. C.-G. Hammar, B. Holmstedt and R. Ryhage, Anal. Biochem., 25 (1968) 532-548.
75. A.G. Jenkins and J.R. Majer, Talanta, 14 (1967) 777-783.
76. E. Fischer, H. Spatz, B. Heller and H. Reggiani, Experientia, 28 (1972) 307-308.
77. J.M. Saavedra, J. Neurochem., 22 (1974) 211-216.
78. I.L. Martin and G.B. Baker, Biochem. Pharmacol., 26, (1977) 1513-1516.
79. J. Willner, H.F. LeFevre and E. Costa, J. Neurochem., 23 (1974) 857-859.
80. D.J. Edwards, P.S. Doshi and I. Hanin, Anal. Biochem., 96 (1979) 308-316.
81. O. Suzuki and K. Yagi, Anal. Biochem., 75 (1976) 192-200.
82. R.L. Borison, A.D. Mosnaim and H.C. Sabelli, Life Sci., 15 (1974) 1837-1848.
83. A.D. Mosnaim and E.E. Inwang, Anal. Biochem., 54 (1973) 561-577.
84. J.F. Tallman, J.M. Saavedra and J. Axelrod, J. Neurochem., 27 (1976) 465-469.
85. P.H. Duffield, D.F.H. Dougan, D.N. Wade and A.M. Duffield, Biomed. Mass Spectrom., 8 (1981) 170-173.

86. F. Karoum, H. Nasrallah, S. Potkin, L. Chuang, J. Moyer-Schwing, I. Phillips and R.J. Wyatt, J. Neurochem., 33 (1979) 201-212.
87. J.M. Saavedra and J. Axelrod, J. Pharm. Exp. Ther., 182 (1972) 363-369.
88. J.W. Sloan, W.R. Martin, T.H. Clements, W.F. Buchwald and S.R. Bridges, J. Neurochem., 24 (1975) 523-532.
89. J.J. Warsh, D.D. Godse, H.C. Stancer, P.W. Chan and D.V. Coscina, Biochem. Med., 18 (1977) 10-20.
90. F. Artigas and E. Gelpi, Anal. Biochem., 92 (1979) 233-242.
91. S.R. Snodgrass and A.S. Horn, J. Neurochem., 21 (1973) 687-696.
92. J.M. Saavedra and J. Axelrod, Proc. Nat. Acad. Sci., USA, 70 (1973) 769-772.
93. J.M. Saavedra, Anal. Biochem., 59 (1974) 628-633.
94. T.J. Danielson, A.A. Boulton and H.A. Robertson, J. Neurochem., 29 (1977) 1131-1135.
95. B.N. Colby and M.W. McCaman, Biomed. Mass Spectrom., 5 (1978) 215-219.
96. F.P. Abramson, M.W. McCaman and R.E. McCaman, Anal. Biochem., 57 (1974) 482-499.
97. N. Narasimhachari and H.E. Himwich, Biochem. Biophys. Res. Commun., 55 (1973) 1064-1071.
98. R.W. Walker, L.R. Mandel, J.E. Kleinman, J.C. Gillin and R.J. Wyatt, J. Chromatogr. Biomed. Appln., 162 (1979) 539-546.
99. J.M. Midgley, M.W. Couch, J.R. Crowley and C.M. Williams, J. Neurochem., 34 (1980) 1225-1230.
100. H. Anderson and C. Braestrup, Scand. J. Clin. Lab. Invest., 37 (1977) 33-37.
101. K. Jacob, W. Vogt, M. Knedel and G. Schwertfeger, J. Chromatogr. Biomed. Appln., 146 (1978) 221-226.
102. J.C. Lhuguenot and B.F. Maume, Biomed. Mass Spectrom., 7 (1980) 529-532.
103. C.R. Freed, R.J. Weinkam, K.L. Melmon and N. Castagnoli, Anal. Biochem., 78 (1977) 319-322.
104. Y. Hashimoto and H. Miyazaki, J. Chromatogr., 168 (1979) 59-68.
105. Y. Mizuno and T. Ariga, Clin. Chim. Acta, 98 (1979) 217-224.
106. A. Liuzzi, F.H. Foppen, J.M. Saavedra, R. Levi-Montalcini and I.J. Kopin, Brain Res., 133 (1977) 354-357.
107. E. Fischer, H. Spatz, R.S. Fernandez Labriola, E.M. Rodriguez Casanova and N. Spatz, Biol. Psychiatry, 7 (1973) 161-165.
108. J.W. Schweitzer, A.J. Friedhoff and R. Schwartz, Biol. Psychiatry, 10 (1975) 277-285.
109. G.P. Reynolds and D.O. Gray, Clin. Chim. Acta, 70 (1976) 213-217.
110. C.W. Kazakoff, unpublished observations.
111. J.J. Warsh, A. Chiu, P.P. Li and D.D. Godse, J. Chromatogr., 183 (1980) 483-486.
112. D.S. Goldstein, G. Feuerstein, J.L. Izzo, I.J. Kopin and H.R. Keiser, Life Sci., 28 (1981) 467-475.
113. D.A. Durden and A.A. Boulton, in H.L. Kornberg, J.C. Metcalfe, D.H. Northcote, C.I. Pogson and K.F. Tipton (Eds.), Techniques in Metabolic Research, Elsevier, Amsterdam (1979), Vol. B214, pp. 1-25.
114. W.R. Martin, J.W. Sloan, S.T. Christian and T.H. Clements, Psychopharmacologia, 24 (1972) 331-346.
115. M. Recasens, J. Zwiller, G. Mack, J.P. Zanetta and P. Mandel, Anal. Biochem., 82 (1977) 8-17.
116. M. da Prada and G. Zurcher, Life Sci., 19 (1976) 1161-1174.
117. S.R. Philips, Advances in Cellular Neurobiology, 2 (1980) 355-391.
118. M.N. Hussain and M.J. Sole, Anal. Biochem., 111 (1981) 105-110.
119. S.H. Koslow, F. Cattabeni and E. Costa, Science, 176 (1972) 177-180.
120. Y. Maruyama and A.E. Takemori, Biochem. Pharmacol., 20 (1971) 1833-1841.
121. J.J. Warsh, A. Chiu, D.D. Godse and D. Coscina, Brain Res. Bull., 4 (1979) 567-570.
122. D.D. Koch and P.T. Kissinger, Anal. Chem., 52 (1980) 27-29.
123. T.P. Davis, C.W. Gehrke, C.W. Gehrke, Jr., T.D. Cunningham, K.C. Kuo, K.O. Gerhardt, H.D. Johnson and C.H. Williams, Clin. Chem., 24 (1978) 1317-1324.
124. Y. Yui, Y. Itokawa and C. Kawai, Anal. Biochem., 108 (1980) 11-15.
125. S. Sasa and C.L. Blank, Anal. Chem., 49 (1977) 354-359.

126. K.-I. Okamoto, Y. Ishida and K. Asai, J. Chromatogr., 167 (1978) 205-217.
127. P. Hjemdahl, M. Daleskog and T. Kahan, Life Sci., 25 (1979) 131-138.
128. T. Miyata, Y. Okano, K. Murao, K. Takahama and Y. Kasé, Jap. J. Pharmacol., 27 (1977) Suppl. 47p.
129. A.G. Zacchei and L.L. Weidner, Anal. Biochem., 87 (1978) 586-593.
130. H. Mita, H. Yasueda and T. Shida, J. Chromatogr., 181 (1980) 153-159.
131. K.M. Taylor and S.H. Snyder, J. Neurochem., 19 (1972) 1343-1358.
132. N. Seiler, B. Knodgen and F. Eisenheiss, J. Chromatogr. Biomed. Appln., 145 (1978) 29-39.
133. K. Samejima, M. Kawase, S. Sakamoto, M. Okada and Y. Endo, Anal. Biochem., 76 (1976) 392-406.
134. R.G. Smith and G.D. Daves, Biomed. Mass Spectrom., 4 (1977) 146-151.
135. C.W. Gehrke, K.C. Kuo, R.W. Zumwalt and T.P. Waalkes, in D.H. Russell (Ed.), Polyamines in Normal and Neoplastic Growth, Raven Press, New York (1973) p. 343.
136. M. Makita, S. Yamamoto and K. Kono, Clin. Chim. Acta, 61 (1975) 403-405.
137. G. Schwedt, J. Chromatogr. Biomed. Appln., 143 (1977) 463-471.
138. E. Kempf and P. Mandel, Anal. Biochem., 112 (1981) 223-231.
139. I.N. Mefford, M.M. Ward, L. Miles, B. Taylor, M.A. Chesney, D.L. Keegan and J.D. Barchas, Life Sci., 28 (1981) 477-483.
140. C.R. Freed and R.C. Murphy, J. Pharmacol. Exp. Ther., 205 (1978) 702-709.
141. R.A.D. Jones and R.J. Pollitt, J. Pharm. Pharmacol., 28 (1976) 461-462.
142. B.A. Davis and A.A. Boulton, J. Chromatogr. Biomed. Appln., 222 (1981) 161-169.
143. B.A. Davis and A.A. Boulton, Eur. J. Mass Spectrom., 1 (1980) 149-153.
144. N. Seiler, T. Schmidt-Glenewinkel and H.H. Schneider, J. Chromatogr., 84 (1973) 95-107.
145. W.D. Lehmann, H.D. Beckey and H.-R. Schulten, Anal. Chem., 48 (1976) 1572-1575.
146. B.A. Davis, Biomed. Mass Spectrom., 6 (1979) 146-156.
147. B.A. Davis and A.A. Boulton, J. Chromatogr. Biomed. Applns., 222 (1981) 271-275.
148. D.A. Durden and A.A. Boulton, J. Neurochem., 36 (1981) 129-135.
149. D.A. Durden and A.A. Boulton, J. Neurochem., (1981) in press.
150. J.R. Shipe, D.F. Hunt and J. Savory, Clin. Chem., 25 (1979) 1564-1571.
151. S.P. Markey, A.J. Lewy, A.P. Zavadil, J.A. Poppiti and A.W. Hoveling, 25th Annual Conference on Mass Spectrometry and Allied Topics, Washington D.C. (1977) 276-278.
152. A.J. Lewy and S.P. Markey, Science, 201 (1978) 741-743.
153. S.J. Gaskell and D.S. Millington, Biomed. Mass Spectrom., 5 (1978) 557-558.
154. D.J. Harvey, J.T.A. Leuschner and W.D.M. Paton, J. Chromatogr., 202 (1980) 83-92.
155. D.A. Durden, 29th Annual Conference on Mass Spectrometry and Allied Topics, Minneapolis (1981).

Chapter 8

GAS CHROMATOGRAPHY-MASS SPECTROMETRY AND SELECTED ION MONITORING OF BIOGENIC AMINES
AND RELATED METABOLITES

EMILIO GELPI

Analytical Neurochemistry Unit, Instituto de Quimica Bio-Organica, C.S.I.C.
c/ Jorge Girona Salgado, s/n. Barcelona-34 (Spain)

8.1 INTRODUCTION

8.1.1 Historical perspective

It has repeatedly been claimed that the use of combined gas chromatography-mass
spectrometry (GC-MS) represents one of the most powerful and versatile analytical
tools available for the unequivocal identification and quantitation of small amounts
of organic compounds in complex mixtures. Considering the complexity of metabolic
processes as regards the number and variety of endogenous compounds and drug- or
diet-related products that they generate, it is not surprising that the GC-MS
methodology has had a dramatic impact on biological research. This has been
particularly significant in various fields of the neurosciences such as neurochemistry, neuropharmacology and biochemical and clinical psychopharmacology,
where the biogenic amines undisputedly play a major role. In fact, GC-MS has taken
an unquestionable place together with other standard or advanced analytical techniques commonly used for biogenic amine determinations and which are discussed in
other chapters of this volume. However, it must be noted that in biogenic amine
GC-MS analyses the mass spectrometer has almost invariably been reduced to the
level of a highly specialized and sensitive GC detector, operating on the principle
of continuous monitoring of a limited number of preselected ions, all of them
representative of the compound(s) sequentially eluted from the GC column. Thus,
considering the work done in this field, it becomes immediately apparent that the
contributions of the GC-MS techniques are closely interrelated to the development
of the MS technology for selected ion monitoring (SIM). The first biological application of SIM, reported by Sweeley et al. in 1966 (1), was limited to the resolution
of components coeluting from the gas chromatograph. Two years later Hammar et al.
demonstrated that the rapid alternation of the accelerating voltage in the mass
spectrometer, allowing the specific recording of the ion current profiles of certain
preselected mass values, could also be used to characterize and quantitate molecules
and called the technique mass fragmentography (2,3), later more aptly described as
selected ion monitoring (section 8.1.2). Upon reflection, it is very fitting that
this powerful analytical technique, which has become so important in neurobio-
logical research, should have been introduced in practice with the determination
of chlorpromazine, one of the most important neuroactive drugs for the treatment

of mental illness (2) and with the identification of acetylcholine (ACh), one of
the established neurotransmitters of nervous systems, in rat brain (3). In the
first case the new method allowed the identification and quantitation of chlorpromazine and its desmethylated and didesmethylated metabolites in human plasma
and, in the second case, the definite identification of ACh in rat brain by
selective detection of fragment ions common to esters of choline. In 1929, 40
kg of horse spleen were required for the isolation of ACh, and in 1937 it was
isolated from the same amount of ox brain. In contrast, the combined GC-MS
technique allowed the identification of ACh in an homogenate from a single rat
brain and with detection limits of the order of 1 ng/g tissue (3).

After these two initial applications, the recording of selected ion current
profiles became rapidly established as an analytical method for the detection and
quantitation of biological components at subnanogram levels in complex mixtures.
In this regard, the 1974 issue of Analytical Chemistry Bi-Annual Instrumental
Reviews (4) contains a section with 38 references to applications of mass fragmentography as a new analytical technique in neurobiology, most of them dealing
with the determination of various biogenic amines and metabolites in biological
samples. The year 1972 was truly a landmark in the study of biogenic amines by
GC-MS with the first reports on the mass fragmentographic assays of endogenous
noradrenaline (NA) and dopamine (DA) in the picomole range (5), indole-3-acetic
acid (IAA) and 5-hydroxyindole-3-acetic acid (5-HIAA) in human CSF (6,7), 5-hydroxy-
tryptamine (5-HT; serotonin), N-acetylserotonin (NAS), 5-methoxytryptamine (5-MT)
and melatonin (M) in rat pineal (8), and homovanillic acid (HVA) in CSF (9). This
remarkable succession of reports showed that the method could measure picomole or
femtomole concentrations of biogenic amines in small samples of CSF or nervous
tissue and prompted numerous studies on the GC-MS/SIM determination of these
metabolites in plasma, CSF, urine, and brain tissue (see Table 8.1 for references).
Practically all of these determinations utilized SIM because of the diverse
structures of the components of the samples as well as the exceedingly low concentrations of the amines and their related metabolites in body fluids or tissues.
In the case of the catecholamines, the application of this technique has resulted
in the identification and quantitation of DA (5,22) and metabolites such as VPA
(36), VLA (35), DOPAC (18,19,25,35,40), DOPET (35), HVA (9,16,18,19,21,25,32,
39-41), iso-HVA (21,32,39), and MOPET (10,35) as well as NA and A (5,22,43) and
metabolites such as DOMA (36), DOPEG (36), VMA (14,18,19,21,25,30,32,40), MOPEG
(10-12,14,17-20,29,30,40), iso-VMA (39), iso-MOPEG (39), MN (23) and NMN (23).
Comparable studies on indoleamine metabolism has resulted in the identification
and quantitation of 5-HT (8,13,24,28,37), T (24,27) and metabolites such as 5-HIAA
(7,16,24,28,37,40,41), IAA (6,24,26,37,41), 5-MTOL (15,34), 5-MIAA (33,40), 5-MT
(8,38,42), NAS (8) and M (8,31) (see Fig. 8.1).

TABLE 8.1

GC-MS assays of catecholamines, indoleamines and their metabolites

Ref	Cpds	Sample/homogenate	Extraction	Deriv/GC	I.Std.	Sensitiv.
5	NA,DA	b/HCOOH+AA	-	PFP/OV-17	α-MeNA≰DA	0.5 pmol
6	IAA	CSF	HCl/NaCl,EA	MeHFB/XE60	5-MIAA	2ng/ml
7	5-HIAA	CSF	HCl/NaCl,EA	MeHFB/XE60	5-HIAAd$_2$	2ng/ml
8	5-HT,M 5-MT,NAS	Rat pineal/ ZnSO$_4$,Ba(OH)$_2$	-	PFP/OV-17	α-Me5-HT NAcT	10^{-12}- 10^{-13}-mol
9	HVA	CSF	pH2,EA	MeHFB/SE30	HVA-CD$_3$	400pmol/ml
10	MOPEG, MOPET	Urine (F+C)	Na$_2$SO$_4$,ether	PFP/OV-210	TOL	<100 pmol
11	MOPEG	CSF (F+C)	pH5.2,EA	TFA/XE60	MOPEG d$_2$	1ng/ml
12	MOPEG	b/(Acetic+AA)F+C	PCA,EA	PFP/OV-17	VMA	27ng/g
14	MOPEG,VMA VMA	b/40% ethanol CSF,serum-XAD-2	HCl/NaCl,EA ACl/NaCl,EA	PFP MePFP/OV17	VMA-d$_3$	50pmol/ml 10pmol/ml
15	5-HTOL, 5-MTOL	CSF (F+C)	pH8,Diethyl- ether	PFP/SE-54	5Fα-MT	0.1ng/ml 0.3ng/ml
16	HVA,5-HIAA	CSF	pH3,HCOOH,EA	PFP/OV-17	both d$_2$	
17	MOPEG	CSF,plasma,Urine b/40% ethanol(F+C)	EA	TFA/OV-17	MOPEG-d$_2$	30pmol/g
18 & 19	PHPA,PHMA, HVA,DOPAC VMA,PHPE, MOPEG	b/(F+C) Acids:HCl Alcohols:ZnSO$_4$+ Ba(OH)$_2$	Na$_2$SO$_4$+EA	MePFP/SE54 PFP/OV-210	All d.	1.3ng/ml 3ng/ml 12ng/ml
20	MOPEG	Amniotic (F+C)	HCl/NaCl,EA	PFP/OV-210	α-lindane	25ng/ml
22	A,DA	plasma (F+C)	PCA,Al$_2$O$_3$-MeOH	TFA/OV-17	synephr	10 pmol
23	3MT,NMN MN	plasma (F+C)	PCA,Al$_2$O$_3$-GC50 XAD-4	TFA/OV-17	φephrine	10 pmol
24	TP,5-HT,T IAA,5-HIAA	urine,CSF	XAD-2,MeOH	PEP/SE-30 MePFP		54 pg
25	DOPAC,HVA VMA	urine	EA	MePFP/OV17	DHPPA	230ng/ml
26	IAA	b/ZnSO$_4$+AA	pH1,Diethylet.	MePFP/OV17	IAAd$_2$	10ng/g
27	T	b/ ethanol	GC-50,HCl	AC/TFA/OV17	Td$_2$	<0.5ng/g
28	5HT,5HIAA	b/HCOOH+AA	ether,heptane	PFP/SP2100	both d$_2$	10 pmol
29	MOPEG	urine	XAD-2,MeOH	TFA/OV-17	MHPGd$_2$	0.7μg/ml
30	MOPEG,VMA	Serum+HClO$_4$ (F+C)	pH8-9,IPrAc≠ HCl/NaCl,EA	AcTFA/OV17 MePFP	both d$_2$	1-2ng/ml
31	M	plasma	Chloroform	TMS/OV-17	hexanoyl-T	20pg/ml
32	VMA,HVA iso-HVA	plasma,LCR urine,XAD-4	HCl/NaCl,EA	TFA-HFIP OV-1	HMPE	2ng/ml
33	5-MIAA	urine,pH4	ether,Dowex	TMS/OV-17	5MIAAd$_2$	
34	5-HTOL	LCR+cyst+serum	pH8,NaCl,EA	PFP/OV-17	5HTOLd$_4$	0.15ng/ml

Ref	Cpds	Sample/homogenate	Extraction	Deriv/GC	I.Std.	Sensitiv.
35	DOPAC,VLA MOPET, DOPET	urine (F+C)	HCl,NaCl,EA pH6.2,EA	TMS/OV-1 OV-225	all d_3-d_7	0.5ng/g
36	VPA,DOPEG DOMA	urine	NaCl,pH6,EA	MePFP/OV-1	all d_2-d_3	
37	TP,5HT,T IAA,5HIAA	b/HCl/KCl.NaHSO$_3$	XAD-2,MeOH	PFP/OV17 MePFP	all d	
38	5-MT	b/HClO$_4$+AA+EDTA	NaOH/NaClCACl$_3$	PFP/SP2100	5-MTd$_4$	3-4 pmol
39	iHVA,iVMA, iMOPEG	urine	same as 35	MeTFA/35	all d_3-d_5	
40	DOPAC,VMA MOPEG, HVA 5HIAA,MIAA	CSF+AA (F+C)	HCOOH,EA	PFP/OV-17	all d_2-d_5	1-6 pmol
41	HVA,DOPAC IAA	urine+Ba(OH)$_2$	DEAE Sephadex	TMS/OV-225	all d_2	

Ref, reference number; Deriv/GC, type of derivative and GC column used; I. Std, internal standard; Sensitiv., sensitivity reported; b., brain tissue; AA., ascorbic acid; PFP, pentafluoropropionyl derivatives; F+C, free + conjugate fractions assayed; MeHFB, methyl heptafluorobutyryl derivatives; TFA, trifluoroacetylated derivatives; NAcT, N-Acetyltryptamine; PCA, perchloroacetic acid; EA, ethyl acetate; All d, all of the internal stds were deuterated; synephr., synephrine; φephrine, phenylephrine cyst., cysteine.
The table provides a non-comprehensive survey of most of the representative work in this field with indication of the compounds determined, the type of samples and homogenization media plus preservatives used, the extraction procedure, the type of chemical derivatives and GC columns used as well as the internal standards and sensitivity levels reported.

In fact some of these metabolites were identified for the first time in different biological samples mainly as a consequence of the remarkable sensitivity and specificity of the GC-MS/SIM method. Typical examples are VMA in CSF and brain (14), 5-HTOL and 5-MTOL in CSF (15), and IAA in brain extracts (26,37). Histamine (HA) (44) and polyamines (45) have also been assayed by the same GC-MS procedures. The number of papers on the GC-MS identification and quantitation of the acidic or alcoholic metabolites of catecholamines and indoleamines far surpasses the number of reports on the parent amines. This reflects the higher concentrations and wider distribution of metabolites in body fluids compared to the more restricted availability in tissue samples and lower concentrations of these amines in plasma or CSF. There is also a current interest in the analysis of the so-called trace amine or microamines (46,47) and their acidic metabolites. Various reports are available in the literature on the GC-MS identification and quantitation of these compounds (18,19,48,55). The high resolution MS determination of some of these amines is discussed in the chapter by Davis and Durden in this volume.

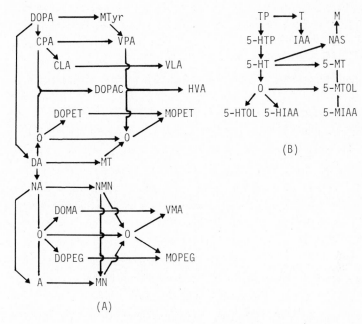

Fig. 8.1 (A) The metabolism of catecholamines. Abbreviations: DOPA, 3,4-dihydroxy-phenylalanine; MTyr, 3-methoxytyrosine; CPA, 3,4-dihydroxyphenylpyruvic (catechol-pyruvic) acid; VPA, 3-methoxy-4-hydroxyphenylpyruvic (vanilpyruvic) acid; CLA, 3,4-dihydroxyphenyllactic (catechollactic) acid; VLA, 3-methoxy-4-hydroxyphenyllactic (vanillactic) acid; DOPAC, 3,4-dihydroxyphenylacetic acid; HVA, 3-methoxy-4-hydroxy-phenylacetic (homovanillic) acid; DOPET, 3,4-dihydroxyphenylethanol; MOPET, 3-methoxy-4-hydroxyphenylethanol; DA, dopamine; MT, 3-methoxytyramine; NA, noradrena-line; NMN, normetanephrine; DOMA, 3,4-dihydroxymandelic acid; VMA, 3-methoxy-4-hydroxymandelic (vanillylmandelic) acid; DOPEG, 3,4-dihydroxyphenylethyleneglycol, MOPEG, 3-methoxy-4-hydroxyphenylethyleneglycol; A, adrenaline; MN, metanephrine. (B) The metabolism of indolealkylamines. Abbreviations: TP, tryptophan; 5-HTP, 5-hydroxytryptophan; 5-HT, 5-hydroxytryptamine, serotonin; T, tryptamine; M, mel-atonin; IAA, indole-3-acetic acid; NAS, N-acetyl-5-hydroxytryptamine; 5-MT, 5-methoxytryptamine; 5-MTOL, 5-methoxytryptophol; 5-HTOL, 5-hydroxytryptophol; 5-HIAA, 5-hydroxyindole-3-acetic acid; 5-MIAA, 5-methoxyindole-3-acetic acid (O: inter-mediary aldehydes).

8.1.2 Conventional GC-MS and selected ion monitoring (SIM)

The first successful coupling of two complementary analytical instruments, the gas chromatograph and the mass spectrometer, took place during the late 1950's. Both instruments had been employed independently until it was realized that the GC could be a very convenient sample inlet system for the mass spectrometer. The latter, acting as a very sensitive and specific GC detector, was capable of providing the identity of most of the components of complex samples, provided they were effectively resolved by the gas chromatograph. A number of detailed reviews of combined GC-MS and its applications are available (56 and references therein).

Gas chromatography is a well established quantitative technique in all scientific fields where separation power and efficiency are required, whereas mass spectrometry is a technique especially suited for the identification of unknown compounds and for the detection of exceedingly small amounts of known components of metabolic pathways. Accordingly, some of the principal biomedical uses of the combined GC-MS instrumentation are: a) the identification of new structures or metabolites (e.g. refs. 14,15,26,37,52); b) the verification of identifications made by other less specific methods such as enzymatic or fluorimetric assays (see section 8.5); c) the diagnosis of clinical pathologies by metabolic profiling, identification of single metabolites or pattern recognition (see section 8.1.3); d) the qualitative detection and quantitative determination of known compounds (e.g. Table 8.1); and, e) tracer studies with stable isotopes (section 8.4.3).

To date, more than 100 new metabolites have been identified by mass spectrometry. This requires amounts of the order of 10-100 ng of the unknown for an interpretable mass spectrum, which usually is not a problem in the study of inborn errors of metabolism, such as phenylketonuria, where normal metabolites and toxic by-products may accumulate in the body (53). However, once the mass fragmentation pattern of a given compound has been established it can be readily detected or quantitated at substantially lower levels by focusing the mass spectrometer on a few pre-selected masses characteristic of the compound rather than scanning through the entire mass range. When operated in this mode, the mass spectrometer can be considered as a highly specific and sensitive GC detector. The technique, as indicated above, variously known as mass fragmentography, multiple ion detection (MID) or selected ion monitoring (SIM), has been described in detail in the literature (57,58). Briefly, the remarkable specificity of the GC-MS/SIM technique stems from the interplay of three factors that when added together increase the reliability of the quantitative detection of known compounds or metabolites in complex biological samples. These factors are: a) the specific retention time of the compounds, a characteristic parameter related to the nature of the solute and the stationary phase in GC; b) the ion current profiles generated at the predicted GC retention time by recording the abundance of preselected ion fragments representative of the compound of interest. These are specifically chosen from the most prominent or characteristic ions of the full mass spectrum of the compound or a chemical derivative of the same. Proper derivative selection is important in this respect as it can enhance the detectability of a given compound by producing more characteristic or abundant ions (section 8.2.2); and, c) the abundance ratio of the preselected ions. The ion current profiles of the separated component of a mixture must be coincident in relative abundances with the same ions in the complete mass spectrum of an authentic sample of the compound analyzed by GC-MS.

Bearing this in mind, it can be predicted that the coelution of one or more compounds with the compound of interest usually will not affect the detection and quantitation of the latter as long as ions characteristic of the other coeluting substances do not coincide in mass with the values preselected as characteristic of the compound in question. Thus, quantitative analysis by SIM benefits from this unique capability to discriminate between unresolved compounds because it allows the use as internal standards of stable isotope-labelled analogues of the same molecular composition and retention time as the sample. This effectively compensates for losses due to irreversible absorption in the GC system. Accordingly, as discussed below, most SIM determinations of biogenic amines have been carried out using the deuterated homologues as internal standards (Table 8.1). An alternative approach when no deuterated homologue is available would be the use of a related compound with different GC retention time but generating either an ion of the same mass (usually a closely related compound with a similar mass spectrometric pattern) or ions shifted by a few amu in the mass scale. This was the method used in the earliest determinations of catecholamines (5) and indolealkylamines (8). In all of these cases α-methyl homologues were used as internal standards.

There are many applications where sensitivity is important to the point where it can make the difference between success or failure in a given assay and this is especially critical in biogenic amine assays in tissue or blood samples. In this case, concentrations of the order of a few nanograms per gram or milliliter of samples are typical, and considering the small volumes injected in the GC together with the practical limitations of the usual preconcentration steps, this dictates a requirement for absolute detectabilities of the order of a few picograms or less. In instrumental analysis by combined GC-MS these sensitivity levels are only possible by SIM since in this case the mass spectrometer spends all of the measurement time on a few ions so that the output from the electron multiplier can be integrated over a longer time interval, with the resulting improvement of the signal-to-noise ratio. Furthermore, in magnetic sector instruments the slit widths can be increased with the corresponding sensitivity gain because resolution is not as critical as in conventional scanning of the entire mass spectrum. Thus, one can obtain by SIM an average increase in sensitivity of the order of 1000 fold relative to conventional GC-MS.

8.1.3 Metabolic profiling assays vs. single component determinations

Combined GC-MS can be used for the diagnosis of clinical conditions (59) either by profiling techniques leading to the recognition of normal and abnormal metabolic patterns or by identification of single metabolites (e.g. HVA in urine for diagnosis of certain classes of tumors). However, in the latter case much of the scope inherent in GC-MS methodology for rapid and accurate multicomponent analysis is not being utilized when one considers the wealth of information which can be obtained through metabolic profiling studies. The possibilities of GC methods for multi-

component analysis led directly to the concept of "metabolic profile" (60) analyses
This term refers to the simultaneous detection and identification of all or most o
the members of a group or class of interrelated compounds (e.g. the components of
a given metabolic pathway). In this manner the changes in the GC ratios of various
key constituents in a "normal" profile, characteristic of a given body fluid or
tissue, can be readily connected in many cases to metabolic changes due to diseases
or specific inborn errors of metabolism (59 and refs. therein). For example, in
addition to expanding our knowledge of metabolism in known diseases or inborn
errors, GC-MS profiling techniques have allowed the discovery of more than twenty
new defects (61). Likewise, the appearance of previously unidentified compounds
in a GC profile from a biological extract has facilitated the identification of new
metabolites through the study and interpretation of their corresponding mass
spectra (62).

In the field of biogenic amines there are various reports in the literature
which describe the simultaneous determination of different components of the
catecholamine pathways (21,25,36,37,40,63,64). However, only a few directly addres
the topic of biogenic amine metabolic profiling (65), although it has recently been
shown that GC-MS/SIM allows the detailed study of TP metabolism through both the
serotoninergic and tryptaminergic pathways (24,37).

8.2 EXPERIMENTAL APPROACHES
8.2.1 Sample preparation for GC-MS work

In general analytical systems demand specific sample work-up schemes adequately
tailored to their performance characteristics. These schemes, including sample
collection and extraction, compound separation and detection, are inversely related
in terms of their experimental complexity, to the specificity of the detector used
Lack of specificity in the detection of a given compound is translated into long
and generally tedious sample clean-up processes, usually leading to large sample
losses at the various steps of the procedure with the consequent loss of sensitivit

Combined GC-MS, in the total ion mode or in the SIM mode, is unique amongst
analytical techniques in the degree of specificity that it brings to the determination of selected compounds in very complex biological mixtures such as body fluids
and tissue extracts. However, even in this case proper sample preparation is
essential for a successful and reproducible assay. In this context, biogenic amine
analyses are not exempt from careful attention to the necessary pretreatment of
biological extracts, although this can be significantly simplified when using GC-MS
SIM relative to other less specific and sensitive techniques. The potential of
GC-MS along these lines was soon realized by Costa's group at the NIMH in Washingto
who measured various endogenous amines and metabolites in the picomole range in
different rat brain structures by selected ion monitoring (5,8,66,67) without any
sample pretreatment other than tissue homogenization and direct chemical derivatiz

ation (section 8.2.2) of the homogenates for GC-MS analysis. In these early applications of the new GC-MS methodology to biogenic amine assays, it was demonstrated
that it could be applied to the study of catecholamines in very discrete brain
structures weighing less than 100 µg (5) and with sufficient instrumental sensitivity
to analyze for NA and DA in a volume of homogenate equivalent to 15 µg of tissue
(66). However, the enthusiasm generated by these detectability levels must be
tempered by the consideration of the concentration of the amines in the brain
structures assayed. For example, although it is true that Koslow et al. (5) were
able to use less than 100 µg of tissue from the Locus coeruleus of rat brain for
the SIM assay of NA, the amount reported was 20 ± 1.6 picomoles per sample, which
is equivalent to a concentration of 33 µg/g vs only 54 ± 2 picomoles or 0.3 µg/g
in striatum, for whose determination a much larger amount (27 mg) of tissue had to
be homogenized. Nevertheless, the sensitivity of the technique is high relative
to other conventional assay procedures such as spectrophotofluorometry (66). A
condensed summary of different sensitivity values is included in Table 8.1.

The very high specificity and sensitivity of GC-MS techniques allows the direct
assay of biogenic amines in nuclei punched out from brain cryostat sections (400 µm
thick) with stainless steel tubing (0.8 - 1.2 mm, i.d.) under stereomicroscope
observation (13). The process of sample preparation for conventional scanning
GC-MS or selected ion monitoring of biogenic amines involves the steps of sample
collection, homogenization in case of tissues, extraction of the amines or their
metabolites into a suitable organic phase, occasional hydrolysis of conjugates
and chemical derivatization of the free amines or metabolites to render them amenable to GC separation in the vapor phase.

(i) Sample collection. The samples may be in the form of body fluid specimens
(urine, blood, CSF) or tissues. In the case of urine, samples are collected in
plastic containers and kept at 4° or -20°C during the collection period. Usually
24 hr specimens are acidified to pH 1-3 with HCl. The urine samples thus treated
can be stored at -20° to -80°C until analyzed (32,35). Creatinine content is
usually determined by any of various modifications of the Jaffe reaction.

Blood is usually collected by venous puncture or occasionally by indwelling
catheters. The assay of biogenic amines by GC-MS does not impose any specific
constraints so that plasma or serum samples are prepared according to standardized
procedures, maintaining the specimens at a temperature of 4° or -20°C if the
analytical determination is not carried out immediately after collection. For
example, venous blood has been collected into lithium heparin tubes, centrifuged
and the plasma stored at -20°C (31). Also, blood can be mixed with heparin, centrifuged and the plasma stored at -20°C in 6M HCl (32).

Lumbar CSF for biogenic amine assays is usually protected with the addition
of ascorbic acid (1 mg/ml) and stored at -20°C or -70°C until analyzed (14-16, 31,
32,34,40). Occasionally 6M HCl or cysteine HCl have also been added to the CSF

samples before storing them at -20°C (32,34). The lumbar punctures are performed after an overnight fast and the aliquots taken for analysis should not contain any blood.

For the studies of body fluid metabolites of biogenic amines or of the amines themselves it is recommended that the subjects be maintained on a low amine and no-alcohol diet for at least a week before sample collection.

(ii) <u>Homogenization and extraction</u>. Tissue samples for biogenic amine assays have been variously homogenized in acetone:formic acid, 40% ethanol, 0.1 or 1N HCl, 0.1M $ZnSO_4$ with $Ba(OH)_2$, formic acid or $HClO_4$ (see Table 8.1). Deuterated internal standards are usually added at this step. Ascorbic acid or $NaHSO_3$ are recommended as antioxidants to prevent structural changes of the amines or metabolites during the extraction procedures (5,26,28,37,38). The homogenates are next centrifuged at 2000 to 100000 g for periods of 10 to 45 minutes, depending on the optimization of the different homogenization procedures by various authors.

The supernatants are usually prepared for an efficient extraction of the compounds of interest by saturation with NaCl and the extraction, after pH adjustment with HCl for the acidic or alcoholic metabolites or NaOH for the basic amines, is carried out into ethyl acetate, diethylether or chloroform as indicated in Table 8. The extracts are then taken to dryness and reconstituted in a suitable solvent, depending on the derivatization procedure to be used, as discussed below.

(iii) <u>Hydrolysis of conjugates</u>. The determination of the acidic and alcoholic metabolites of the catecholamines and indoleamines isolated from the samples according to the usual extraction procedures gives a quantitative account of their free forms. However, as most of these metabolites can be conjugated with glucuroni or sulfuric acid, the determination of total amounts of metabolites must take into account both the free and conjugated moieties. For that purpose, the samples have to be hydrolyzed with the proper enzyme preparations. This releases the metabolite from their conjugates allowing their GC-MS/SIM determination in the same manner as described for the unconjugated fractions. For example, in connection with GC-MS assays, various authors have reported on the determination of the conjugated forms of different metabolites such as the conjugated tryptophols in CSF (15) or conjugated MOPEG in human urine, CSF, plasma or rat brain and liver (10,12,14,18-20, 29,30,40). In these cases the sulfate ion can be precipitated with $CaCl_2$ and the supernatants incubated for 16-20 hours at 37°C with a mixture of sulfatase and glucuronidase. A wider range of conjugated alcoholic and acidic metabolites was assayed by Karoum <u>et al</u>. (18,19) after incubation of rat brain samples with glusula for 20 hours at 40°C (see Table 8.1) and more recently by Faull <u>et al</u>. (40), who incubated 1 ml aliquots of CSF with arylsulfatase H1 (mixture of arylsulfatase and β-glucuronidase, Sigma) for 1 hour at 37°C.

An interesting alternative to the use of sulfatases has been reported by Murray <u>et al</u>. (29). These authors isolated MOPEG and its sulfate conjugate from urine passed through XAD-2. This fraction was further resolved on a column of Sephadex LH-20 into an eluate containing MOPEG and another containing its conjugated forms. The sulfate ester was then displaced by reaction with trifluoroacetic anhydride (TFAA), thus accomplishing in a single step the hydrolysis of the conjugate and the derivatization of the liberated MOPEG. The MOPEG glucuronide is stable to TFAA so that the method allows a specific determination of the sulfate conjugate.

8.2.2 Specific derivatizations

The need for the chemical derivatization of biogenic amines and their acidic or alcoholic metabolites is a direct consequence of the relatively low volatility and high polarity of these compounds which render them unfit for GC analysis. Accordingly, the different acid, alcohol or amino groups have to be blocked by reaction with suitable derivatization reagents. From the point of view of the mass spectrometer itself, the derivatization can also serve the useful purpose of improving several fold the detectability of a given compound by providing a more specific or directed fragmentation pattern. For example, certain derivatives can enhance the SIM sensitivity level by concentrating the fragmentation on a few ions at relatively higher m/z values and thus free of interferences from the rest of the compounds that constitute the biological matrix (10,15,16,20,24,26,27,34,38, 40,52,63,65,68). On the other hand, in the case of polyfunctional compounds such as the biogenic amines and their metabolites it is important to avoid those procedures yielding multiple derivatives from a single starting compound (68-69). Otherwise, this may counteract the inherent multicomponent assay capabilities of GC-MS in metabolic profiling studies by introducing additional interferences generated by the derivatization procedures so that the profiles become complicated by a number of unnecessary extra components. These may overlap and mask many components of interest. Furthermore, the derivative selected must be sufficiently stable to limited storage and often cumbersome assay procedures, and totally inert to possible interactions in the gas phase within the GC system (69).

A relatively wide variety of derivatization procedures have been described for biogenic amine GC-MS determinations. The usual procedures are those involving N- and O-acylation or silylation of amino, alcohol and phenolic groups (70 and references therein). Several other methods, based for instance on the formation of permethyl, nitrophenyl, isothiocyanate and dansyl as well as mixed acyl-silyl derivatives, are also described in a comprehensive reference publication (70). More recently Davis (71) has reported the use of N,N-dialkylamino-naphthalene-sulfonyl chloride reagents with C_1 to C_5 alkyl groups for the MS of biogenic amines.

Nevertheless, a glance at Table 8.1 shows that the acylated derivatives are the preferred choice, the pentafluoropropionyl (PFP) group being the most commonly used. These derivatives were shown to be stable for at least three days in ethyl

acetate (72) and, while initially designed for the electron capture detector in
GC, they were later shown to be very good in GC-MS profiling studies (65).

   In one of the first applications of GC-MS/SIM to biogenic amines, Koslow
et al. claimed that the PFP derivatives were stable over 24 hours (5) and Takahashi
et al. reported that PFP derivatives of 5-HTOL were stable for 2 weeks if kept
refrigerated. These same authors also indicated that the trifluoroacetyl (TFA)
derivatives gave a response much lower than that of PFP derivatives (34). In a
comparison of PFP, TFA and TMS derivatives in assays of tryptophol, Curtius et al.
(15) came to the conclusion that the first of these derivatives was more appropriat
within the constraint of the limited mass range usually covered in the SIM mode
by a magnetic sector MS. A detailed account of the GC and MS performance of the
PFP derivatives of catecholamines, indolealkylamines and their metabolites has
been published (65). As a result of this study, it was subsequently feasible
to define compromise experimental conditions needed for the simultaneous derivatization and SIM assay of TP, 5-HT, T, 5-HIAA and IAA (24,37) in single body fluid or
tissue samples. The advantages of using pentafluoropropanol in PFPA were discussed
by Fri et al. (16). In this way, both the alkylation of acidic groups and acylatio
of amino and/or alcoholic groups is carried out in a single step. On the other
hand, Warsh et al. (27) used a double acylation procedure to extract previously
acetylated tryptamine into diethylether at basic pH. The residue thus obtained was
then reacted with TFAA. The disadvantages of trimethylsilylated derivatives in SIM
determinations have been discussed by Zagalak et al. (73). Basically, the problems
lie in the major unspecific ion fragments generated at m/z 218 for all α-amino
acids and at m/z 174 for biogenic amines. However, this can also be an advantage
for functional group mass spectrometry, as Abramson et al. (63) have shown. Anothe
important fact to consider is that the signals arising from the natural isotopes of
silicon would interfere with a quantitation when deuterated standards are used.
These authors also comment on the problem of multiple peak formation and lack of
reproducibility of mixed acyl-TMS derivatives (73). The kind of problems and
pitfalls associated with the use of silylated derivatives for biogenic amine
determinations have been addressed in detail recently by Martinez and Gelpi (69).
This study has shown how multiple derivative formation and derivative stability
are difficult to control when serotonin and tryptamine are reacted with TMS.
These studies have indicated that indolealkylamines are best determined by GC-MS
in the form of PFP derivatives. In general, the amines or their metabolites are
fully acylated though occasionally it may be advantageous to use the partially
acylated molecules. In some cases the partially acylated amine or metabolite can
be obtained in higher yields and a more favorable fragmentation pattern in the
MS may be produced (34,69).

8.2.3   Some practical considerations and neglected aspects

The concentrations of endogenous biogenic amines in body tissues or fluids are
in general very low, sometimes exceedingly low, as in the case of A and NA circulating in plasma (22,43) or tryptamine in mammalian brain (27,37,49). Furthermore, these substances as well as the corresponding chemical derivatives used for
GC-MS analysis are rather labile and prone to decomposition or absorption during
collection, extraction or analysis. This calls for great care and special attention
to experimental details during the processing of these compounds through a given
analytical scheme. In consequence, the analytical reproducibility of the GC-MS
procedures for biogenic amine assays is dependent on a number of factors, many of
them closely interrelated. If any of these factors is overlooked, unreliable data
may be the result. The list is indeed long and space considerations prevent a
detailed treatment, which nevertheless would not really be possible, because not
enough is known about the physico-chemical bases underlying many of the puzzling
experimental observations sometimes made in the laboratory. For example, occasionally compounds that previously were detected almost routinely for unexplained
reasons seem to vanish from the extraction and analytical schemes. What might
possibly have gone wrong? Several factors, including those which follow, need to
be considered in this regard.

i) The stability of compounds during storage prior to analysis. It is known
that catecholamines or their metabolites can be readily lost from biological samples
if not properly protected from oxidative breakdown processes. For example, it has
been reported that DOPAC in amounts of 24 and 48 picomoles completely disappears
from artificial CSF solutions incubated for 2 hrs at 37°C prior to extraction and
derivatization (40). This can be prevented by addition of ascorbic acid to a
concentration of 1-2 mM in the aqueous phase. A similar situation applies to
5-hydroxyindole compounds in tissue or urine samples. In this regard if samples
have to be stored for a number of weeks before analysis, it would be advisable to
verify the possible rate of sample loss vs. time by sequential analysis at different
time intervals of aliquots stored under identical conditions. In this sense,
Takahashi et al. (30) reported that MOPEG and VMA are stable at -20°C for periods
of up to 8 weeks and Bertilsson et al. (7) reported less than 5% loss of 5-HIAA
kept at -15°C for two months. However, although other authors also give data of
this sort, the stability of the samples is often overlooked. To avoid unwanted
losses, the samples can be protected against oxidative decomposition by the addition
of antioxidants and preservatives immediately after collection or prior to the
start of the extraction and analytical procedures. In GC/MS biogenic amine assays
the most popular preservative is ascorbic acid although sodium metabisulfite,
thioglycolic acid and cysteine have also been used (see Table 8.1). The use of
pooled human serum as a preservative to avoid the oxidative decomposition of 5-HTOL
has also been reported (34).

ii) <u>The catalytic effects of residual impurities in the solvents or reagents used</u>. A specified upper limit of oxidant residues of 0.00025% in a 15 ml volume of the solvent used to extract a given compound would amount to a total of 30 μg coming in contact with the compound (74). This would be more than enough to induce the oxidative breakdown of the few nanograms of endogenous compounds that one usually attempts to recover. This topic has recently been discussed in some detail (74) and would also apply to possible traces of the peroxides in detergents absorbed or retained on the walls of glassware. Faull <u>et al</u>. (40) report that, to avoid variations in the GC-MS analysis of DOPAC and 5-HIAA in replicate samples, the ethyl acetate used for the extraction has to be freshly redistilled.

iii) <u>Contamination of enzymes used in the hydrolysis of conjugates of biogenic amines and incomplete hydrolysis</u>. It is known that the arylsulfatases and β-glucuronidases used to free conjugated metabolites may contain detectable amounts of interfering compounds (12,15) and this has to be taken into account by running enzyme blanks, especially since purification of commercial enzyme preparations may result in losses of activity (15). When attempting to determine the free and total content of biogenic amines or their metabolites by enzymatic digestion of the conjugated moieties there is always the possibility of incurring variations due to insufficient hydrolysis. This has been suggested as a reason for the discrepancies in reported values of certain metabolites of biogenic amines (30). However, any required increase in the time of the hydrolytic process can affect the stability of the desired metabolite, as indicated above for DOPAC. On the other hand, insufficient hydrolysis can also be a result of the addition of preservatives such as bisulphite to the incubation media (12).

iv) <u>Variability of extraction efficiency</u>. As already discussed, the process of selective transfer of a given substance from a tissue or fluid matrix to a suitable organic or aqueous phase requires the use of relatively large volumes of solvents. These solvents are not always sufficiently pure to insure their absolute chemical inertness toward the compound(s) extracted. A similar situation arises with the media used to homogenize tissue samples since the homogenizing agents themselves may introduce a relatively high amount of oxidating impurities, occasionally leading to almost quantitative losses of the metabolites present in the nanogram range (74). It is also noteworthy that in some instances lower extraction efficiencies are associated with higher reproducibilities, provided that this can be compensated for by the sensitivity of the detection system. This is the case in GC/MS/SIM work where sometimes it may be more practical to accept a 30% or 40% recovery (29,37), duly compensated for by the high sensitivity of SIM, than to attempt to improve it by more elaborate experimental procedures which may result in a higher overall variability due to excessive sample processing.

v) <u>Carrier effects in the extraction derivatization and clean-up processes</u>. It has been claimed that the recoveries of certain biogenic amines and metabolites can be substantially improved by coextraction of a previously added large excess of structurally related molecules. This would be in line with the extension of the effectiveness of the so-called "carrier effect" in chromatographic systems (75) to the processes of extraction, derivatization and sample clean-up. Although carriers were initially introduced in the early days of chromatography as adsorption-reducing compounds to minimize losses of sample by decomposition and irreversible binding on column, the fact is that interactions with surface active sites can take place during the whole analytical system; from the syringe or receptacle used for sample collection to the GC-MS instrument (69). For instance, Faull et al. (40) reported that the recovery of 5-HIAA (500 picomoles) in ethyl acetate increases from a mere 5% to a 37% in the presence of 100 μg of 5-MIAA. Likewise, Bertilsson et al. (7) reported an increased recovery of 5-HIAA in the presence of 4-HIAA (2 μg), although these authors suggested that this could be due to the antioxidant properties of 4-HIAA. A clear experimental illustration of the enhancement of analytical sensitivity using deuterated analogues of polyamines has been reported by Smith and Dales (45). These authors showed that the GC-MS/SIM detection of a one picomole sample of trifluoroacetylated putrescine was not possible without the addition of 100 picomoles of the deuterated analogue. In contrast, Millard et al. (76) could not show a carrier effect for octopamine in the extraction process or GC elution. In this laboratory 5-MeOT was initially used with good results in the concurrent assay of various TP metabolites (37). However, later on, certain irreproducibilities and experimental difficulties were traced to the large excess of carrier and accordingly its use has been discontinued. It has been argued that in the case of labelled analogue carriers, the large excess of these compounds may introduce in the analytical system an amount of unlabelled impurity, sufficient to produce a carrier effect by itself (75). It may also be sufficient to cause unwanted oxidations and decomposition of the relatively much smaller amounts of the analytes of interest. Also, one has to consider the possible interferences in the GC-MS assay of derivatized carrier by-products which, the parent carrier molecule being an analogue of the substance analyzed, would be structurally related and thus potentially apt to interfere with certain selected ions.

   vi) Derivative reproducibility and stability. The variability in the GC-MS assay of biogenic amines introduced by a lack of sufficient reproducibility in the derivatization process, if not compensated by the use of a deuterated analogue as internal standard, can result in significant sample losses. In our experience, the optimization of reaction conditions for certain amines and acidic or alcoholic metabolites using a given batch of solvents and reagents is no absolute guarantee that the reaction yield will be exactly reproducible with other batches of solvents or reagents from the same or different sources. Also, the use of closely related derivatization reagents such as PFPA and HFBA does not necessarily give the same results. In our hands, the use of HFBA in the derivatization of indolealkylamines and related metabolites would not be recommended since the resulting HFB products are much less stable compared to the PFP analogues.

vii) Other important effects, like the type of reaction vessels, losses on evaporation and column and system deactivation are discussed and documented in some detail in a recent report from this laboratory (74).

8.2.4 Specificity and sensitivity

In all instrumental spectrometric analytical systems, including mass spectrometry, higher specificity translates into higher resolution, which in principle is opposed to the operational requirements of high sensitivity. However, the operation of a GC-MS combination in the selected ion monitoring mode provides unique analytical capabilities since it combines an extremely high specificity with surprising sensitivities, often reaching down to the femtomole level (5,45,63,66). As indicated above, the unique specificity of the SIM technique is based on a) the information derived from the specific retention time of a given substance, b) the response obtained monitoring the appearance of one or more preselected and characteristic ions within the expected retention time windows, and c) the relative abundances of the recorded ion current profiles.

On the other hand, the number of ions striking the electron multiplier per second is not hampered in SIM by resolution requirements as in conventional scanning mass spectrometry; consequently, the multiplier is entirely dedicated to acquiring the ion current profiles of a few selected masses. Thus, level of sensitivity can be 100-1000 times higher than in conventional GC-MS (57-59). The combination of these two properties, specificity and sensitivity, in the field of biogenic amine assays, which in many ways has been associated with rather unspecific analytical procedures, led to the downward revision of previously reported levels of certain amines by several orders of magnitude. This is the case, for instance, for β-phenylethylamine (46,48,49) and tryptamine (46,49,27,37) in rat brain. It has also allowed the absolute identification of previously unreported metabolites in biological materials, such as 5-HTOL and 5-MTOL (15), VMA (14), melatonin (31), IAA (26,37), o- and m-hydroxymandelic acids (OHMA and MHMA) (52) and 5-MT (8,38), although in the last case this identification has been recently questioned as a possible artefact produced from endogenous melatonin (42).

The high specificity of this technique has enabled the use of deuterium labelled internal standards of the same molecular composition and GC retention time (9,14, 17,18,26-29,30,33,35-38,40), thus bringing a higher reliability and precision to the overall procedure. This capability has been extensively exploited in the field of biogenic amines as shown in Table 8.1. The table also includes some representative values of the sensitivity levels achieved by different authors. A detailed evaluation of the data published reveals that there are no unified criteria for

reporting the precision of GC-MS procedures in a form suitable for meaningful
comparison amongst different laboratories and methods. Furthermore, it is well
known that the precision in the instrumental assay of a given sample varies significantly depending on the working range selected on the calibration curve. For
example, Bertilsson et al. (7) reported standard deviation (SD) values of
7% in the 8-20 ng/ml range and 1-2% above this range. However, in many cases a
value is given without due mention of the concentration level to which it applies.
Also, occasionally there is no indication whether the SD given corresponds to
standard solutions or to the compounds extracted from a biological matrix. The
various factors influencing sensitivity and precision in SIM work have been
covered in some detail in the literature (57,58).

An important aspect of the high specificity of the MS measurements is the reduction it allows in sample-clean up and purification schemes, with the resulting
gain in recoveries and thus overall sensitivities. For example, the specificity
of the technique has allowed the direct assay of catecholamines and indoleamines
in rat brain homogenates without extraction or further manipulations (5,8,13) and,
although this has been considered as a limitation of the usefulness of the method
(28), the possibility of determining a given metabolite in the presence of the
many other components of a complex biological matrix is precisely one of the strong
points of the technique. This has recently been demonstrated by this laboratory
with the development of a SIM procedure for the concurrent determination of most
of the more significant components of the tryptophan metabolic pathway in single
rat brain samples (37), as discussed below.

8.2.5 General equipment

Leaving aside all of the standard ancillary laboratory equipment necessary to
prepare the sample for injection into GC-MS unit, the essential instrumentation
for this type of application would be a GC-MS equipped with a multiple ion detection system capable of monitoring simultaneously a minimum of four preselected
ions. This would provide for a minimum concurrent detectability of two different
amines or metabolites and their corresponding deuterated analogues as internal
standards, or possibly more substances if they produce common ions to focus upon
(24,63).

While almost any mass spectrometer system can be used or modified for GC-MS work
in the selected ion mode (SIM), most of the work done in this field to date has
been carried out with LKB magnetic sector (Models 9000 or 2091) or Finnigan quadrupole systems (Models 3000, 3200 or 4000), connected to gas chromatographs through
jet type molecular separators of one (Finnigan units) or two (LKB units) stages (56).
Other magnetic sector or quadrupole instruments also used in biogenic amine
analyses are the Varian Mat 311A connected to a Varian Aerograph 1400GC, the
Micromass 16F coupled to a Carlo Erba GI 450, the Hitachi GC-MS Model 52 and the

Hewlett-Packard 5985. Most of these systems are also equipped with data acquisitio
and reduction systems. Usually the molecular separator is kept at a temperature of
250-290°C and the ionizing energy varied between 20eV and 80eV depending on the
fragmentation pattern of the compounds analyzed. Interestingly, as discussed in
the next section, most of the work has been carried out on standard GC packed
columns and silicon phases (Table 8.1).

8.3   SPECIAL TECHNIQUES

8.3.1   Capillary GC-MS

Glass capillary columns offer a unique degree of versatility to permit the
separation of closely related substances in multicomponent samples. However,
although this type of gas chromatographic column has been known for more than 20
years (62) and was used successfully, for instance, in hydrocarbon and fatty acid
analyses from biological and geochemical materials, the last few years have witnes-
sed an upsurge in the use of these columns in the biomedical field (59,62). The
wider application of capillary columns has been enhanced by the commercial introduction of reasonably priced thermostable glass open tubular capillary columns
that provide highly inert systems to assay labile substances as well as sufficient
efficiency and resolution to separate most components. This has been especially
useful in metabolic profiling studies and diagnosis of human diseases (59,62).

Nevertheless, aside from our own work (77,78) and other isolated reports (79),
there does not seem to be much enthusiasm for the use of capillary columns for the
GC-MS determination of biogenic amines in biological samples. This is likely due
to a combination of two factors; on the one hand, these columns demand much more
care and attention to operating parameters than standard packed columns, and on
the other hand the use of such a specific and sensitive detector as the mass spectro
meter in the SIM mode may partly cancel one of the main advantages of capillary
columns by allowing the detection and quantitation of GC peaks even if not resolved
by the packed columns. In other words, by increasing the specificity of detection
as in GC-MS/SIM one may dispense in many cases with the need for greater GC
efficiency. Perhaps one of the main drawbacks of capillary columns for the GC-
MS determinations of biogenic amines stems from the inherent low sample capacities
of these columns which requires the use of sample splitting injection systems so
that only a minor part of the total volume injected actually enters the column
(77). Furthermore, aside from the quantitative loss caused by the injection
splitters, which may be a limiting step in attempting to analyze the very small
amounts of endogenous biogenic amines, the variability inherent to injection
splitting prevents the obtaining of reproducible quantitative data (79).
Recently, various attempts have been made to bypass the splitter (80,81). However, it must be kept in mind that the dynamic operation of a combined GC-MS system
does pose some restrictions to the type of injection system that can be used.

Also, in capillary GC-MS operation, part of the sample can be lost in the GC-MS interface, depending on the carrier gas flow rate and the maximum allowable ion source pressure. For example, a wall coated open tubular column operated at an optimum carrier gas flow rate of 2 ml/min may be incompatible with an ion source connected to a vacuum system capable of removing not more than 0.75 ml/min. Accordingly, open split systems are commonly used at the GC-MS interface (77), although the outlet of the capillary column can also be connected to a molecular separator, provided the necessary make-up gas is added for optimum performance of the separator which works best at an inlet flow of 20-30 ml/min.

Perhaps as a reflection of these practical drawbacks, the application of capillary GC-MS to the quantitative assay of biogenic amines is limited to two recent communications (77,78) on a simple method that allows the injection of biological extracts of biogenic amines in the solid state without any modification of the GC-MS system. Briefly, the method involves the use of a totally splitless solid sample injector and interface connection. The mass spectrometer was equipped with a four-channel multiple ion monitoring system designed and built in this laboratory (82) and was operated in the direct connection mode at a pressure range of $1 - 5 \times 10^{-5}$ mm Hg with a column flow rate of up to 3 ml/min. The glass capillary columns (20m x 0.3mm i.d.) were coated with OV-17 or SE-30. Samples were injected with a solids injection syringe equipped with a special plunger fitted with a spiral section where the liquid sample is deposited and the solvent evaporated prior to injection. The dried samples were volatilized in less than 2 seconds.

For comparative purposes the resolution factors of the peaks of methylated 5-HIAA $(PFP)_2$ and IAA (PFP) in OV-17 capillary and packed columns are 1.92 and 0.88, respectively, whereas T $(PFP)_2$ and 5-HT $(PFP)_3$, which coelute on these packed columns, can be baseline resolved on the capillary columns (78). It has already been suggested (62) that in practice the higher resolving power of capillary columns can be very useful in coping with complex biological extracts because the number of components that can be resolved increases very significantly. On the other hand, from a practical point of view, it may compensate to use a less efficient packed column in connection with a very specific detector such as the MS in the SIM mode, whereby many unresolved compounds can be assayed in a very selective manner with the exclusion of unwanted signals. This is what most investigators have been doing in the field of biogenic amines (Table 8.1). However, when peaks from unidentified substances generate the same fragments or different fragments of identical mass as those selected for SIM this calls for improved GC separation (18).

In the capillary GC-MS system described by Sunol and Gelpi (77,78), injection variability was compensated for by the use of internal standards labelled with deuterium. The system has been applied to the study of indole metabolism in extracts of brain tissue or human urine with results equivalent to those previously obtained with standard packed columns (78).

8.3.2 New ionization techniques

(i) Positive chemical ionization. In 1974 Miyazaki et al.(83) described the first experiments in combined GC-chemical ionization mass spectrometry (CIMS) of biogenic amines in CSF. They demonstrated that the mild ionization conditions characteristic of this technique prevented the extensive fragmentation of the compounds that enter the ion source of the CI mass spectrometer, thus improving the detectability of protonated molecular ions, contrary to what happens in electron impact (EI) mass spectrometry. According to these authors (83) CIMS was ten times more sensitive than EI, allowing the detection of 1 pg of NA. In CI the ionization of the GC effluent occurs through a process of interaction between ionized reagent gas and the molecule under investigation in which the energy transferred to the molecule is much less than in direct EI, so that there is much less fragmentation. For instance, the molecular ion of the PFP derivative of DA almost completely fragments to lower mass ions under electron impact, whereas the CI spectra obtained with methane, isobutane or ammonia as reagent gases are dominated by the protonated $[M+H]^+$ or $[M+NH_4]^+$ quasimolecular ions of m/z 592 and 609, respectively.

The use of methane or isobutane as reagent gases usually produces $[M+H]^+$ or $[(M+H) - (PFPOH)]^+$ ions as base peaks in the spectra of indoleamines and catecholamine PFP derivatives (83), whereas with ammonia the base peaks correspond to the $[M+NH_4]^+$ ions. In general, the lesser fragmentation of the molecule results in higher sensitivities in SIM work. It is important to remark in this context that, in addition to the sensitivity gain, the diminished molecular fragmentation in GC-CIMS provides another degree of specificity relative to conventional GC-MS work. Sometimes the specificity inherent to a given molecular structure is considerably reduced by its extensive fragmentation under electron impact to rather unspecific ions. In other words, although one may achieve a very high degree of specificity by monitoring a GC effluent in search of a preselected ion, part of this specificity may be lost in the fragmentation of the molecule to relatively unspecific ions. Thus, in the case of GC-MS analysis of choline esters by EI, the best ion for SIM detection is the base peak at m/z = 58 (3). However, this is also a major fragment characteristic of many other tertiary amines. This fragment accounts for approximately 65% of the total ionization of choline analogue and their demethylated esters (84) and if focussed upon would not distinguish between them; so these compounds have to be identified on the basis of their chromatographic retention times. This constitutes just an example where a common ion, while advantageous for functional group monitoring (24,63), cannot provide sufficient selectivity and specificity per se, requiring the use of retention times for identification purposes. One way to recover specificity in this particular case, which can serve as an illustrative example of many other similar situations, is with GC-CIMS. For instance, the CI-methane mass spectrum of dimethylaminoethyl acetate shows a base peak at m/z = 72 with only about 5% relative abundance at m/z = 58 and almost 75% at the [M+H]$^+$ quasimolecular ion at m/z = 132 (84). The corresponding molecular ion at m/z = 131 by EI/MS is barely discernible at 1.3% abundance (3). Apart from the higher specificity achievable in this way since all other analogues will be differentiated by their respective [M+H]$^+$ ions, the important fact to consider in SIM is that the m/z = 132 ion represents almost 35% of the total ion current. This can be increased to almost 80% of the total ioniz-ation by using isobutane instead of methane as a reagent gas (84).

This advantage has not been overlooked by other groups who have also applied the CI techniques to the GC-MS measurements of biogenic amines and their metabolites (85-91). For instance, Edwards et al. (90) recently described a method for the simultaneous analysis of PFP derivatives of phenylglycols and phenylethanols in human urine by GC-CIMS. The method allowed the first identification of p-hydroxy-phenylglycol in biological samples, which, as the authors suggest, may provide a means for the determination of the turnover of octopamine in vivo. This group has also described the application of GC-MS in the CI/SIM mode of operation to the analysis of phenylethylamine, phenylethanolamine, m- and p-tyramine and octopamine as their corresponding N-DNP, O-TMS derivatives (89). This work was carried out using methane or isobutane as reagent gases and, in contrast to the low abundance of the molecular ions in the EI spectra, the CI spectra contained relatively abundant [M+H]$^+$ ions, except for octopamine which presented an [(M+H) - (TMSOH)]$^{\cdot+}$ base peak. By monitoring the ions corresponding to the [M+H]$^+$ ions, a simultaneous SIM profile of these compounds in whole rat brain samples could be obtained, establishing a concentration of phenylethylamine of 1 ng/g in untreated animals (89).

Likewise, the simultaneous determination of endogenous NA and DBH activity in rat adrenal tissue with ^{2}H-DA as a substrate has also been carried out by a combined GC-CIMS/SIM procedure (88). The same procedure has also been used for the assay of PFP derivatives of catecholamines (NA, A, DA) in small pieces punched out from different regions of human autopsy brain specimens (87). As little as 5 mg of tissue were sufficient for the quantitative assays. A GC-CIMS/SIM method has also been used for the analysis of DA and NA in adrenal medullary cell cultures (91). The CI/SIM allows detections at the nanogram level in a higher mass range than in the EI mode so that this results in a substantial gain in overall specificity. Furthermore, the inherent specificity of the CI mode is such that it allows the simultaneous determination of the PFP derivatives of DA and α-methyldopamine and β-ethyl ether PFP derivatives of NA and α-methylnoradrenaline from rat brain samples, without any previous gas chromatographic separation. The samples contain-ing the mixture of these amines were directly introduced and evaporated in the

ion source of an MS-9 mass spectrometer, modified for chemical ionization (85), via
a solid direct insertion probe. The protonated molecular ions of all amines appear
in the mass range 657 to 592 amu. In a similar manner, the levels of tryptophol,
5-HTOL and IAA in the urine of patients with carcinoid tumours have been determined
by CIMS techniques (86).

    ii) <u>Negative chemical ionization</u>. In 1976 Hunt <u>et al</u>. (92) showed that electron
capturing compounds could be detected with a 10-100 fold increase in sensitivity
by the new technique of negative ion chemical ionization (NICI) mass spectrometry,
which can provide useful complementary information relative to positive CIMS. The
first biological application of this new technique belongs to the field of biogenic
amines since it allowed the routine measurement of melatonin in human plasma at the
level of 1 picogram per milliliter (93). This level of sensitivity is in line with
the theoretically postulated ratio of negative and positive sample ion currents
when a mass spectrometer operated under CI conditions affords a mixture of positive
reagent ions and a population of electrons with near thermal energies (94). These
low energy electrons can induce ionization by a process of resonance electron
capture (ECNI), (95). The analytical potential of this methodology was subsequently
demonstrated by Hunt <u>et al</u>. (94) for several derivatives of phenols and amines
under GC-MS conditions. The authors presented the tracings obtained from 25 femto-
grams or 53 atomoles of a suitable derivative of DA recorded by GC-ECNI/CIMS/SIM
(94). This impressive sensitivity results from the fact that in the NICI mode,
detection and quantitation of every molecule capable of forming a molecular anion
on collision with a thermal electron can be carried out at levels 2 orders of
magnitude lower than what is possible with positive CI or EI methodology.

8.3.3 GC-High resolution MS/SIM

    In the selected ion monitoring mode the specificity depends on the recording
of the ion current profiles of fragment ions bearing enough specific structural
information to fully characterize a given compound or set of compounds. The
specificity of the ions thus monitored can be further increased in the positive
or negative CI modes as discussed above. Furthermore, the high inherent specificity
of these MS techniques can be remarkably enhanced by a consideration of the
specific GC retention time windows at which these ions appear in GC-MS operation.
Such windows can even be defined with much more accuracy by using high efficiency
capillary columns (section 8.3.1). However, all of these factors combined do not
necessarily mean the ultimate in specificity has been attained. A few authors
(96,97) have described the recording of preselected ion current profiles by
increasing the resolving power of the mass spectrometer. The potential of this
MS mode of operation has been demonstrated for the detection of steroids and
urinary acids (96) and more recently it has been used for the simultaneous
determination of A and NA from human plasma (97). In this work, the N-TFA-N-TMS

derivatives of both catecholamines were detected at an MS resolving power of 5000 by monitoring the benzylic fragment at m/z 355.1568. The detection limit was 2 pg. As the authors point out, this MS resolution was needed to avoid interferences from other fragment ions of the same nominal mass 355. For example, an ion of this mass is characteristic of bleeding from the silicon phases used in GC.

8.4 SELECTED APPLICATIONS

Different applications of GC-MS in its various operating modes have already been described and discussed in earlier sections of this chapter and some of them are summarized in Table 8.1. Consequently, this section will deal with a few topics not covered in sufficient detail and which deserve a closer look from the point of view of instrumental and practical reasons. These are the trace amines, the polyamines and histamine and the application of GC-MS to metabolic studies of biogenic amines or related products with stable metabolic tracers.

8.4.1 Trace amines

The trace amines, also referred to as the microamines (46) constitute a group of phenyl or indolealkylamines that includes β-phenylethylamine (PEA), and the ortho-, meta- and para-isomers of tyramine (TA), octopamine (OA) and synephrine (SN), all related to phenylalanine and tyrosine (see Fig. 8.2) as well as tryptamine (T), a metabolite of tryptophan.

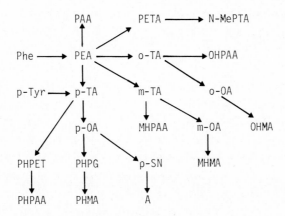

Fig. 8.2. The metabolism of trace amines. Abbreviations: PAA, phenylacetic acid; PETA, phenylethanolamine; N-MePETA, N-methyl PETA; Phe, phenylalanine; PEA, phenylethylamine; o-TA, ortho-tyramine; OHPAA, ortho-hydroxyphenylacetic acid; o-, m-, p-TA, ortho-, meta-, para-tyramine; o-, m-, p-OA, ortho-, meta-, para-octopamine; MHPAA, meta-hydroxyphenylacetic acid; OHMA, ortho-hydroxymandelic acid; PHPET, para-hydroxyphenylethanol; PHPG, para-hydroxyphenylglycol; p-SN, para-synephrine; MHMA, meta-hydroxymandelic acid; PHPAA, para-hydroxyphenylacetic acid; PHMA, para-hydroxymandelic acid; A, adrenaline.

Some of these substances are known to be pharmacologically active or to interfere with aminergic neurotransmission, facilitating or modulating the effect of established neurotransmitters (46-49). As the name implies, their concentrations in brain tissue are about 1000 times lower relative to the concentrations of other classical amines such as DA, NA and 5-HT. For example, in rat and rabbit brain the concentration of the last three amines range from 400-500 ng/g of fresh tissue whereas those of PEA, m- and p-TA and T range from 0.05 to 12.9 ng/g (46,49).

The existence of these amines in such low endogenous levels in mammalian brain, coupled to the lack of sufficiently sensitive and specific assay procedures, probably explains why many early determinations established falsely elevated values. For instance, it has been said that due to the lack of sensitive and specific procedures, practically all of the quantitative data collected for the tyramines before about 1970 were ambiguous (47). In this sense, while mass spectrometry can readily differentiate each isomer of TA, the radioenzymatic assays provide combined values. Considering the literature on the identification and quantitation of these amines, it could be stated that from 1970 onwards, the MS procedures have played a unique role in terms of unequivocal and reliable determinations, having contributed to the discovery of significant inconsistencies and analytical inaccuracies. For example, using a GC method it was reported that PEA was present in rabbit brain in a concentration of the order of 340.9 ± 45.8 ng/g of wet brain while more specific MS techniques have shown that this value is erroneously high, the true value being 0.44 ± 0.06 ng/g (49). A survey of the existing literature shows that the MS/SIM methodology has contributed significantly to the study of trace amine metabolism with the identification and quantitation of PHPA, PHMA (18,19), PEA, m- and p-TA (46,51), o- and m-OA (98), OHPAA, MHPAA, PHPAA (50), OHMA, MHMA and PHMA (52), OHMA (53), MHPAA, PHPAA and PAA (54), PEA and PETA (48) and MHMA (55).

Regarding T, the application of mass spectrometric procedures has been decisive due to the exceedingly low levels of this amine and its acidic metabolite, IAA, in mammalian brain (see Fig. 8.1). For example, it has been reported that T is present in rabbit and rat brain at concentrations of only 0.05 ng/g and 0.5 ng/g, respectively (46). On the other hand, IAA had not been detected until Warsh et al. (26) and Artigas and Gelpi (37) recently reported it in rat brain at levels of the order of 10-13 ng/g. The existing GC-MS methodology is summarized in Fig. 8.3. The GC-MS/SIM methodology not only provides the necessary sensitivity and specificity but it also allows a simultaneous determination (metabolic profiling) of the serotoninergic and tryptaminergic pathways, along with their common precursor, TP (37), in single rat brain samples. Perusal of the values given in Fig. 8.3 demonstrates another unique advantage of GC-MS, namely its remarkable dynamic range. It should be noted that the detection limits reported for the acylated derivatives of T and IAA (24,37) represent 543 and 300 femtomoles, respectively.

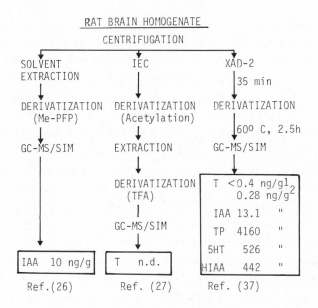

Fig. 8.3. Summarized scheme indicating the reported methods for T and IAA GC-MS assays. Abbreviations: IEC, ion-exchange chromatography; r.t., room temperature; n.d., not detected; 1, value obtained on packed columns; 2, value obtained on capillary columns (78).

8.4.2 Polyamines and histamine

As recently as 1971 it was claimed in a paper on the excretion of diamines in human urine that the main difficulty in the study of these substances was the lack of methods for the separation and quantitation of di- and polyamines (99). At about the same time it was demonstrated that the urinary excretion of polyamines is abnormal in patients with cancer (45 and refs. therein), and this stimulated a search for suitable analytical procedures. Compared to other biogenic amines, the GC-MS techniques have not been extensively used for polyamine determinations. The number of reports to date is rather limited in spite of the fact that spermine, spermidine, putrescine and cadaverine can be measured by GC-MS/SIM to a level of 1 picomole (45). These amines can be N-permethylated (100), trifluoroacetylated or trimethylsilylated (45) for GC-MS work.

The situation concerning histamine is somewhat similar, with many different methods reported for the determination of this amine but few reports on its measurement by GC-MS/SIM in a biological sample (44). This study used the N^a-heptafluorobutyryl-N^T-ethoxycarbonyl derivative and reports a 50 pg detection limit at a signal to noise ratio of 5:1. Other adequate derivatives potentially useful for GC-MS have been reported by this laboratory (101), though no biological applications were carried out.

8.4.3 Studies of metabolic pathways with stable isotope tracers

The mass spectrometer is the best instrument for the measurement of stable isotopes and thus is uniquely suited for the detection and quantitation of metabolites endogenously labelled with stable isotopes by prior loading with a labelled precursor. This allows in vivo metabolic studies which otherwise require the use of potentially hazardous radioactive isotope labelled precursors. In this regard, the GC affords the required separation of the metabolites of interest from the rest of the components in the sample while the MS in the SIM mode can detect and quantitate both the labelled and unlabelled species (these are not resolved by the GC) of a given metabolite. Their relative amounts give an indication of the degree of incorporation of the stable isotope, which in turn is an index of the metabolic or enzymatic activity. Along these lines, the group of Curtius at the Children's Hospital in Zurich has recently applied the GC-MS/SIM technology to the investigation of phenylalanine-tyrosine-dopa and tryptophan metabolism in vivo. For this purpose healthy subjects and patients with different diseases were loaded with L-phenylalanine-d_5 (102), L-tyrosine-d_2 (64) and tryptophan-d_5 (103) and the resulting deuterated-metabolites, in plasma or urine samples, were analyzed by SIM of the corresponding N-, O-, TFA methyl esters. For example, in a study of the activity of phenylalanine-4-hydroxylase it was determined that healthy controls converted about 10% of the deuterated phenylalanine into tyrosine, whereas patients with phenylketonuria produced only 1% or less (102). In a similar manner, two other groups simultaneously described the labelling of acetylcholine in mouse brain by feeding (104) or injecting (105) the animals with deuterium labelled choline (Ch). Labelled and unlabelled choline and ACh were measured by GC-MS/SIM as previously described (3,84). This work showed that plasma Ch is partly supplied from the diet and partly from endogenous sources, that it rapidly equilibrates with a pool of 30-40 nmol/g of Ch in brain, responsible for ACh biosynthesis (104) and that there could be two functional pools with different turnover rates (105).

Deuterium labelled L-Dopa has been used as an analogue tracer for the study of DA and NA turnover in rat hypothalamus (106). The hypothalamic replacement of DA and NA by DA-d_3 and NA-d_3 after L-Dopa-d_3 infusions was determined by GC-MS/SIM of their respective PFP derivatives. The results showed that total DA is proportional to the dose of L-Dopa, while NA is produced at a constant rate. A detailed account of the use of SIM techniques in the study of DA metabolism has been given by Sedwal et al. (107).

8.5 COMPARISON WITH OTHER ANALYTICAL PROCEDURES

In general there is a lack of reports like the one recently published by Sjöquist and Johansson (108) dealing with the direct comparison of GC-MS determinations with other analytical procedures. These authors carried out a controlled comparison between fluorometric and SIM methods for HVA and 5-HIAA determination in human CSF. Their results showed that HVA and 5-HIAA values by SIM were higher than those obtained by fluorometry, the difference being clearly significant in the first instance but not in the second. The statistical correlation for the HVA values obtained by the two methods was high (0.90) whereas it was poor for 5-HIAA (0.55); however, in the latter case the mean values obtained by both methods were similar. The authors claim that this effect could be due to the fact that 5-HIAA is sensitive to autoxidation. The reason for the lower HVA values by fluorometry could be the presence of unreacted HVA (the fluorescence of HVA depends on the coupling of two HVA molecules to form a fluorescent dimer) and formation of non-fluorescent oxidation products. This shows clearly the importance of using methods sufficiently selective and specific to discriminate against possible interferences. Lack of specificity and sensitivity relative to that achievable in GC-MS is the most frequently cited characteristic of some of the previously reported analytical procedures, including a) the trihydroxyindole and ethylenediamine condensation methods used for the estimation of plasma catecholamines (22), b) the colorimetric, fluorometric and GC methods reported for the quantitation of the 3-0-methylated catecholamines, MT, NMN and MN (23), c) the bioassay, radioimmunoassay and fluorometric procedures used for quantitation of a number of amines, and d) chromatographic methods combined with colorimetric reactions for the identification or quantitation of 3-0-methylated CA metabolites in body fluids and tissue which have been considered responsible for the observed discrepancies in the values reported by different authors (39). Along these lines Cattabeni et al. (8) reported differences in the GC-MS/SIM values for M and NAS relative to those previously obtained by the o-phthalaldehyde fluorescence measurements. This is not to say that these other methods are of no value in biogenic amine assays since, in many cases, the results obtained (for instance by spectrofluorometry) are coincident to those obtained by GC-MS methods (8,14,66). However, one has to be perfectly aware of the limitations and the capabilities of the technique chosen for a given problem. The more complex the sample, the higher the specificity required for an accurate and unambiguous determination. In this sense, it is acknowledged that the specific derivatization of the compound, and the recognition of characteristic ions and the GC retention time provide three dimensions of accuracy that make the assay very specific (66). The importance of both specificity and sensitivity is demonstrated by the widely divergent values obtained for the amine T in rat brain homogenates, where a radioisotopic enzymatic method gave a value of 22 ng/g, a TLC-radioassay procedure 70 ng/g while no detection could be achieved by a fluorometric or a GC-ECD method (see Artigas and Gelpi (37) and Suñol and Gelpi (78) for a detailed discussion and references). As mentioned in section 8.4.1, the concentration of T in these samples is not higher than 0.5 ng/g, which represents 140 times less than the TLC-radioassay value.

It has also been said that GC-ECD and GC-MS/SIM essentially complement each other as the former is free of some of the disadvantages of GC-MS, such as its technical difficulties and high cost, although GC-MS/SIM is 100 times more sensitive (10). Furthermore, even in cases where both techniques may give results essentially similar in terms of sensitivity, the higher selectivity of SIM provides an effective increase in sensitivity when dealing with complex biological samples (89). Also, in this regard it should be kept in mind that a further enhancement of selectivity and sensitivity is possible through the use of GC-CIMS/SIM (see section 8.3.2). However, this increases the overall cost and complexity of the instrumentation.

8.6   SUMMARY

A survey of the literature shows that in the past ten years, combined GC-MS techniques have contributed to a remarkable extent to the detailed study of biogenic amine metabolism and disturbances in metabolism. In this regard, it is interesting that practically all studies have relied on the use of selected ion monitoring (SIM). In this way the sensitivity and specificity that usually characterize a combined GC-MS system are further enhanced to the point where certain amines or their metabolites can be directly assayed even in the presence of many other components of the biological matrix. In fact, the discriminating power of the GC-MS/S technique is such that the coelution of the compound of interest with two or more substances usually does not represent a problem for its identification and quantitation. This is a clear advantage over other techniques requiring the use of elabor sample clean-up and concentration procedures to remove potential interferences. Th the SIM technique has provided the means to detect biogenic amines and their metabolites in the pico- or femtomole range with a remarkably high level of selectivity, and this has resulted in the identification for the first time of several metabolit of biogenic amines in biological samples. Furthermore, the selectivity and specificity of the technique can be increased by the use of the new sample ionization variants known as positive and negative chemical ionization mass spectrometry. Thi in many cases provides ions more specific for a given compound, thus increasing its detectability. However, since both the GC and the MS work in the vapour phase, the samples have to be properly derivatized to improve their volatility. In this regar various derivatization schemes have been described in the literature. Those most commonly used are based on the perfluoroacylation of N- and O-groups. It is also worth noting that chemical derivatization of biogenic amines for GC-MS can improve the specificity of the assay by directing the fragmentation pattern of the molecule to the production of very characteristic and unique ion fragments. Also, the reliability of these assays is further enhanced by the possibility of using stable isotope labelled analogues as internal standards. Perhaps a good reflection of the overall specificity and sensitivity of GC/MS/SIM is the fact that virtually all of the determinations of biogenic amines in complex extracts from biological sources have been carried out on low resolving power standard packed GC columns. However, providing that proper precautions are taken it is to be hoped that in the future capillary columns will contribute more to this area of research as they have done in many other fields. In conclusion, GC-MS techniques are of particular value in the field of biogenic amines. They are extremely useful for the identification of new metabolites, the validation of results obtained by other less specific methods, the study and diagnosis of clinical pathologies by metabolic profiling of the amines and their metabolites, the quantitative assay of known compounds and also for tracer or metabolic studies with stable isotope labelled precursors.

REFERENCES

1   C.C. Sweeley, W.H. Elliot, I. Fries and R. Ryhage, Anal. Chem., 38 (1966) 1549-1553.
2   C.G. Hammar, B. Holmstedt and R. Ryhage, Anal. Biochem., 25 (1968) 532-548.
3   C.G. Hammar, I. Hanin, B. Holmstedt, R.J. Kitz, D.J. Jenden and B. Karlen, Nature, 220 (1968) 915-917.
4   A.L. Burlingame, R.E. Cox and P.J. Derrick, Anal. Chem., 46 (1974) 248R-287R.
5   S.H. Koslow, F. Cattabeni and E. Costa, Science, 176 (1972) 177-180.
6   L. Bertilsson and L. Palmer, Science, 177 (1972) 74-76.
7   L. Bertilsson, A.J. Atkinson, Jr., J.R. Althaus, A. Harfast, J-.E. Lindgren and B. Holmstedt, Anal. Chem., 44 (1972) 1434-1438.
8   F. Cattabeni, S.H. Koslow and E. Costa, Science, 178 (1972) 166-168.
9   B. Sjöquist and E. Änggard, Anal. Chem., 44 (1972) 2297-2301.
10 F. Karoum, H. Lefevre, L.B. Bigelow and E. Costa, Clin. Chim. Acta, 43 (1973) 127-137.
11 L. Bertilsson, J. Chromatogr., 87 (1973) 147-153.
12 C. Braestrup, Anal. Biochem., 55 (1973) 420-431.
13 S.H. Koslow, G. Racagni and E. Costa, Neuropharmacol., 13 (1974) 1123-1130.
14 B. Sjöquist, J. Neurochem., 24 (1975) 199-201.
15 H. Ch. Curtius, M. Wolfensberger, U. Redweik, W. Leimbacher, R.A. Maibach and W. Isler, J. Chromatogr., 112 (1975) 523-533.
16 C.G. Fri, F.A. Wiesel and G. Sedvall, Life Sci., 14 (1975) 2469-2480.
17 B. Sjoquist, B. Lindstrom and E. Anggard, J. Chromatogr., 105 (1975) 309-316.
18 F. Karoum, J.C. Gillin, R.J. Wyatt and E. Costa, Biomed. Mass Spectrom., 2 (1975) 183-189.
19 F. Karoum, J.C. Gillin and R.J. Wyatt, J. Neurochem., 25 (1975) 653-658.
20 F. Zambotti, K. Blau, G.S. King, S. Campbell and M. Sandler, Clin. Chim. Acta, 61 (1975) 247-256.
21 N. Narasimhachari, K. Leiner and C. Brown, Clin. Chim. Acta, 62 (1975) 245-253.
22 M. Wang, K. Imai, M. Yoshioka and Z. Tamura, Clin. Chim. Acta, 63 (1975) 13-19.
23 M. Wang, M. Yoshioka, K. Imai and Z. Tamura, Clin. Chim. Acta, 63 (1975) 21-27.
24 J. Segura, F. Artigas, E. Martinez and E. Gelpí, Biomed. Mass Spectrom., 3 (1976) 91-96.
25 E. Peralta and E. Gelpí, Clin. Chim. Acta, 73 (1976) 13-18.
26 J.J. Warsh, P.W. Chan, D.D. Godse, D.V. Coscina and H.C. Stancer, J. Neurochem., 29 (1977) 955-958.
27 J.J. Warsh, D.D. Godse, H.C. Stancer, P.W. Chan and D.V. Coscina, Biochem. Med., 18 (1977) 10-20.
28 O. Beck, F.A. Wiesel and G. Sedvall, J. Chromatogr., 134 (1977) 407-414.
29 S. Murray, T.A. Baillie and D.S. Davies, J. Chromatogr. Biomed. Appl., 143 (1977) 541-555.
30 S. Takahashi, D.D. Godse, J.J. Warsh and H.C. Stancer, Clin. Chim. Acta, 81 (1977) 183-192.
31 B.W. Wilson, W. Snedden, R.E. Silman, I. Smith and P. Mullen, Anal. Biochem.,

81 (1977) 283-291.

32 S. Takahashi, M. Yoshioka, S. Yoshiue, and Z. Tamura, J. Chromatogr. Biomed. Appl., 145 (1978) 1-9.

33 J.A. Hoskins and R.J. Pollit, J. Chromatogr. Biomed. Appl., 145 (1978) 285-289.

34 S. Takahashi, D.D. Godse, A. Naqui, J.J. Warsh and H.C. Stancer, Clin. Chim. Acta, 84 (1978) 55-62.

35 F.A.J. Muskiet, D.C. Fremouw-Ottevangers, J. van der Meulen, B.G. Wolthers and J.A. de Vries, Clin. Chem., 24 (1978) 122-127.

36 F.A.J. Muskiet, D.C. Fremouw-Ottevangers, G.T. Nagel, B.G. Wolthers and J.A. de Vries, Clin. Chem., 24 (1978) 2001-2008.

37 F. Artigas and E. Gelpí, Anal. Biochem., 92 (1979) 233-242.

38 O. Beck and T.R. Bosin, Biomed. Mass Spectrom., 6 (1979) 19-22.

39 F.A.J. Muskiet, D.C. Fremouw-Ottevangers, G.T. Wagel and B.G. Wolthers, Clin. Chem., 25 (1979) 1708-1713.

40 K.F. Faull, P.J. Anderson, J.D. Barchas and P.A. Berger, J. Chromatogr. Biomed. Appl., 163 (1979) 337-349.

41 E.F. Domino, B.N. Mathews and S.K. Tait, Biomed. Mass Spectrom., 6 (1979) 331-334.

42 N. Narasimhachari, E. Kempster and M. Anbar, Biomed. Mass Spectrom., 7 (1980) 231-235.

43 J. Yoshida, K. Yoshino, T. Matsunaga, S. Higa, T. Suzuki, A. Hayashi and Y. Yamamura, Biomed. Mass Spectrom., 7 (1980) 396-398.

44 H. Mita, H. Yasueda and T. Shida, J. Chromatogr. Biomed. Appl., 181 (1980) 153-159.

45 R.G. Smith and G.D. Dales, Jr., Biomed. Mass Spectrom., 4 (1977) 146-151.

46 A.A. Boulton, in E. Costa, E. Giacobini and R. Paoletti (Eds.), Adv. Biochem. Psychopharmacol., Vol. 15, Raven Press, New York, 1977, pp. 57-67.

47 A.A. Boulton, Life Sci., 23 (1978) 659-672.

48 J. Willner, H.F. LeFevre and E. Costa, J. Neurochem., 23 (1974) 857-859.

49 A.A. Boulton, A.V. Juorio, S.R. Philips and P.H. Wu, Brain Res., 96 (1975) 212-216.

50 N. Narasimhachari, U. Prakash, E. Heigeson and J.M. Davis, J. Chromatogr. Sci., 16 (1978) 263-267.

51 N.D. Heubert and A.A. Boulton, J. Chromatogr. Biomed. Appl., 162 (1979) 169-176.

52 J.M. Midgley, M.W. Couch, J.R. Crowley and C.M. Williams, Biomed. Mass Spectrom 6 (1979) 485-490.

53 J.R. Crowley, J.M. Midgley, M.W. Couch, A. Garnica and C.M. Williams, Biomed. Mass Spectrom., 7 (1980) 349-353.

54 B.A. Davis and A.A. Boulton, J. Chromatogr. Biomed. Appl., 222 (1981) 161-169.

55 B.A. Davis and A.A. Boulton, J. Chromatogr. Biomed. Appl., 222 (1981) 271-275.

56 M.C. Ten Noever de Brauw, J. Chromatogr., 165 (1979) 207-233.

57 F.C. Falkner, B.J. Sweetman and J. Throck Watson, Appl. Spectrosc. Reviews, 10 (1975) 51-116.

58 C.C. Sweeley, S.C. Gates, R.H. Thompson, J. Harten, N. Dendramis and J.F. Holla in A.P. de Leenheer and R.R. Roucucci (Eds.), Quantitative Mass Spectrometry in Life Sciences, Elsevier, Amsterdam, 1977, pp. 29-48.

59 H.M. Liebich, J. Chromatogr. Biomed. Appl., 146 (1978) 185-196.

60 E.C. Horning and M.G. Horning, in A. Zlatkis (Ed.), Chromatography Symposium, University of Houston, Houston, 1970, pp. 226-243.

61 S.I. Goodman, P. Helland, O. Stokke, A. Flatmark and E. Jellum., J. Chromatogr. 142 (1977) 497.

62 E. Jellum, P. Størseth, J. Alexander, P. Helland, O. Stokke and E. Teig, J. Chromatogr., 126 (1976) 487-493.

63 F.P. Abramson, M.W. McCaman and R.E. McCaman, Anal. Biochem., 51 (1974) 482-499.

64 H. Ch. Curtius, M. Mettler and L. Ettlinger, J. Chromatogr., 126 (1976) 569-580.

65 E. Gelpí, E. Peralta and J. Segura, J. Chromatogr. Sci., 12 (1974) 701-709.

66 F. Cattabeni, S.H. Koslow and E. Costa, in E. Costa and G.L. Gessa (Eds.), Adv. Biochem. Psychopharmacol., Vol. 6, Raven Press, New York, 1972, pp. 37-59.

67 S.H. Koslow and A.R. Green, in E. Costa and B. Holmstedt (Eds.), Adv. Biochem.
    Psychopharmacol., Vol. 7, Raven Press, New York, 1973, pp. 33-43.
68 E. Martinez and E. Gelpi, J. Chromatogr., 167 (1978) 77-90.
69 E. Martinez and E. Gelpi, J. Chromatogr., 186 (1979) 619-636.
70 D.R. Knapp in Handbook of Analytical Derivatization Reactions, John Wiley,
    New York (1979), pp. 65-145.
71 B.A. Davis, Biomed. Mass Spectrom., 6 (1979) 149-156.
72 E. Änggard and G. Sedvall, Anal. Chem., 41 (1969) 1250-1256.
73 M.J. Zagalak, H. Ch. Curtius, W. Leimbacher and U. Redweik, J. Chromatogr.,
    142 (1977) 523-531.
74 F. Artigas, E. Martinez and E. Gelpi, J. Chromatogr. Sci., (1981) in press.
75 R. Self, Biomed. Mass Spectrom., 6 (1979) 315-316.
76 B.J. Millard, P.A. Tippett, M.W. Couch and C.M. Williams, Biomed. Mass Spectrom.,
    4 (1977) 381-384.
77 C. Suñol and E. Gelpi, J. Chromatogr., 142 (1977) 559-574.
78 C. Suñol and E. Gelpi in H. Jaeger (Ed.), Glass Capillary Gas Liquid Chromato-
    graphy: Clinical and Pharmacological Analyses, Marcel Dekker, (1981) in press.
79 B.W. Wilson, W. Snedden, J. Neurochem., 33 (1979) 939-941.
80 K. Grob and K. Grob, Jr., J. Chromatogr., 151 (1978) 311-320.
81 K. Grob and K. Grob, Jr., J. High Resol. Chromatogr. C.C., 1 (1978) 57-64.
82 F. Artigas, E. Gelpi, M. Prudencio, J.A. Alonso and J. Baillart, Anal. Chem.,
    49 (1977) 543-549.
83 H. Miyazaki, Y. Hashimoto, M. Iwanaga and T. Kubodera, J. Chromatogr., 99
    (1974) 575-586.
84 I. Hanin and R.F. Skinner, Anal. Biochem., 66 (1975) 568-583.
85 C.R. Freed, R.J. Weinkam, K.L. Melmon and N. Castagnoli, Anal. Biochem., 78
    (1977) 319-332.
86 A.J. Greenberg and R. Ketcham, J. Pharm. Sci., 67 (1978) 478-480.
87 Y. Mizuno and T. Ariga, Clin. Chim. Acta, 98 (1979) 217-224.
88 Y. Hashimoto and H. Miyazaki, J. Chromatogr., 168 (1979) 59-68.
89 D.J. Edwards, P.S. Doshi and I. Hanin, Anal. Biochem., 96 (1979) 308-316.
90 D.J. Edwards, R. Rizk and J. Neil, J. Chromatogr. Biomed. Appl., 164 (1979)
    407-416.
91 J.C. Lhuguenot and B.F. Maume, Biomed. Mass Spectrom., 7 (1980) 529-532.
92 D.F. Hunt, G.C. Stafford, Jr., F.W. Crow and J.W. Russell, Anal. Chem., 48
    (1976) 2098-2105.
93 A.J. Lewy and S.P. Markey, Science, 201 (1978) 741-743.
94 D.F. Hunt and F.W. Crow, Anal. Chem., 50 (1978) 1781-1784.
95 E.C. Horning, D.I. Caroll, I. Dzidic, S.N. Lin, R.N. Stillwell and J.-P. Thenot,
    J. Chromatogr., 142 (1977) 481-495.
96 A.L. Burlingame, B.J. Kimble, F.C. Walls, R.V. McPherson and R.W. Olsen in
    N.R. Daly (Ed.), Advances in Mass Spectrometry, Vol. 7B, Heyden, London,
    (1978) pp. 873-877.
97 K. Jacob, W. Vogt, M. Knedel and G. Schwertfeger, J. Chromatogr. Biomed. Appl.,
    146 (1978) 221-226.
98 C.M. Williams and M.W. Couch, Life Sci., 22 (1978) 2113.
99 H.J. Bremer, E. Kohne and W. Endres, Clin. Chim. Acta, 32 (1971) 407-418.
100 A.G. Giumanini, G. Chiavari and F.L. Scarponi, Anal. Chem., 48 (1976) 484-489.
101 N. Mahy and E. Gelpi, Chromatographia, 11 (1978) 573-577.
102 H. Ch. Curtius, J.A. Völlmin and K. Baerlocher, Clin. Chim. Acta, 37 (1972) 277.
103 H. Wegman, H. Ch. Curtius, R. Gitzelmann and A. Otten, Helv. paed. Acta, 34
    (1979) 497.
104 I. Hanin and J. Schuberth, J. Neurochem., 23 (1974) 819-824.
105 D.J. Jenden, L. Choi, R.W. Silverman, J.A. Steinborn, M. Roch and R.A. Booth,
    Life Sci., 14 (1974) 55-63.
106 C.R. Freed and R.C. Murphy, J. Pharmacol. Exp. Ther., 205 (1978) 702-709.
107 G. Sedvall, L. Bjerkenstedt, C.-G. Swahn, F.-A. Wiesel and B. Wode-Helgodt, in
    E. Costa and G.L. Gessa (Eds.), Adv. Biochem. Psychopharmacol., Vol. 16, Raven
    Press, New York, 1977, pp. 343-348.
108 B. Sjöquist and B. Johansson, J. Neurochem., 31 (1978) 621-625.

Chapter 9

ANALYSIS OF BIOGENIC AMINES USING RADIOENZYMATIC PROCEDURES

IAN L. MARTIN

MRC Neurochemical Pharmacology Unit, Medical Research Council Centre, Medical
School, Hills Road, Cambridge, CB2 2QH (England)

9.1 INTRODUCTION

The development of radioenzymatic methods for the quantitation of the biogenic
amines in the late 1960s marked the opening of a new era in the analysis of these
substances. The radioenzymatic procedures developed over the succeeding years
surpassed in both sensitivity and specificity the spectrofluorimetric methods
which they rapidly began to replace, and today methods of this type remain amongst
the most sensitive for the quantification of certain amines of biological interest.
Their popularity developed largely from the ease with which they could be carried
out, their sensitivity, and limited requirement for capital equipment.

These methods rely on the use of methyl transferase enzyme preparations to catalyse the transfer of a radiolabelled methyl group to the compound of interest.
The compound, so modified, is purified and subsequently quantitated by measurement
of the radioactivity associated with the derivatives produced.

In principle, the method is extremely simple and, provided there is an efficient
enzyme-catalyzed derivatization, the sensitivity attainable will be dependent only
on the specific activity of the radioactive donor and on the ability to separate
the derivatives of interest from other radioactive products of the reaction. In
practice, however, a number of factors have limited the sensitivity of assays of
this type and in order to achieve both the sensitivity and specificity now available considerable efforts have had to be made.

It is not the purpose of this chapter to provide comprehensive details of the
assays, but rather to review the approaches used and the technical developments
made which have led to the assays available today.

9.2 THE CATECHOLAMINES
9.2.1 Assays based on phenylethanolamine-N-methyl transferase (PNMT)

The observation made by Axelrod in 1962 (1) that the ^{14}C-methyl group of ^{14}C-
methyl-S-adenosyl-L-methionine (^{14}C-SAM) could be efficiently transferred to the
primary amino function of a number of phenylethanolamine derivatives in the presence of the enzyme PNMT (Fig. 9.1) subsequently allowed Saelens et al. (2) to
develop a sensitive assay procedure for the estimation of noradrenaline (NA) in
tissue. Tissue samples were homogenized in perchloric acid containing inhibitors

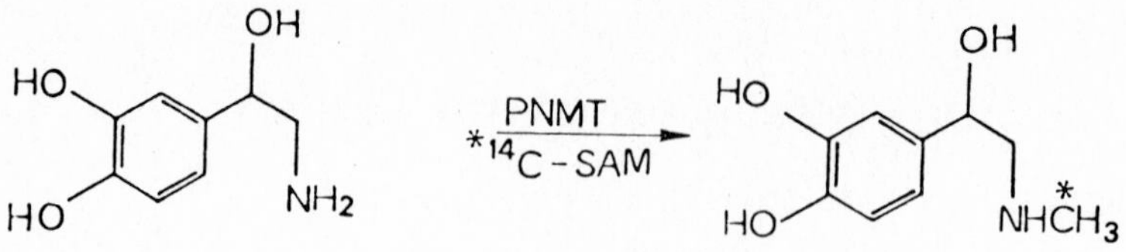

Fig. 9.1. N-methylation in the radioenzymatic assay for noradrenaline.

for both monoamine oxidase (MAO) and catechol-O-methyl transferase (COMT) together
with EDTA. Subsequent to protein precipitation, neutralization with magnesium
carbonate and centrifugation, an aliquot of the clear supernatant was incubated
with a PNMT preparation obtained from rabbit adrenal glands and ^{14}C-SAM. The
reaction was allowed to proceed for 1 hour at 37° and was terminated by placing
the samples on ice. The adrenaline (A) formed was separated from the incubation
mixture by paper chromatography and quantified by liquid scintillation counting
(LSC). The method had a sensitivity of about 1 ng and was specific for NA, with
no interference from normal tissue concentrations of dopamine (DA), metanephrine
(MN), normetanephrine (NMN) or A.

Some three years later Iversen and Jarrot (3) modified the procedure of Saelens
et al., improving the limit of sensitivity of the assay for NA to about 100 pg.
This increase in sensitivity was achieved largely by the replacement of ^{14}C-SAM
in the incubation medium with the higher specific activity ^{3}H-SAM. The stabilizing
agent mercaptoethanol was also introduced into the incubation medium and these
authors used tissue homogenates rather than tissue extracts as, in their hands, it
proved difficult to accurately adjust the pH of the sample extract with magnesium
carbonate as suggested by Saelens et al. (2). Inhibitors for MAO and COMT were
included in the incubation medium and internal standards of known amounts of NA
were added to duplicate samples, allowing the determination of overall recoveries
and thus the calculation of absolute values of the amine. The assay had the required specificity, since other naturally occurring substrates for the enzyme
yielded radioactive products which were resolved at the chromatographic stage.
An important feature of this assay procedure was the observation that enzymic
methylation proceeded to completion when standard samples were used, but some inhibition of the enzymic activity occurred in tissue homogenates; this was compensated for by inclusion of internal standards in the procedure.

Henry et al. (4) in 1975 introduced a further modification to the procedure
which resulted in the limits of sensitivity being reduced to 25 pg. In this method
perchloric acid homogenates were used as such or after concentration by a preliminary alumina adsorption stage. In addition, the incubation medium contained di-

thiothreitol as a stabilizing agent for the enzyme. Several attempts were made
to purify the product in an effort to increase sensitivity. These consisted of
alumina adsorption of the radioactive product A, precipitation of the excess ^{3}H-SAM
with phosphotungstic acid and finally the ion pair extraction of the A with di-(2-
ethylhexyl)phosphoric acid.

9.2.2 Assays based on COMT

The limited substrate specificity of the enzyme PNMT, while conferring a certain
degree of specificity on the assay for NA, precluded its general usefulness for the
assay of other catecholamines. As a result, the use of catechol O-methyltransferase
(COMT) as an enzyme preparation to transfer a radioactive methyl group to the 3-
phenolic substituent of the catechols has become popular.

The first method to employ COMT for the quantitation of the catecholamines was
introduced by Engelman et al. in 1968 (5). These authors used ^{3}H-NA as an internal
standard in each of the plasma samples to be assayed. Subsequent to pH adjustment
to 6, samples were percolated through a cation exchange resin column as a preliminary purification procedure for the catecholamine fraction. The eluate containing
the amines was then lyophilized, and the residue reconstituted in an incubation
mixture containing ^{14}C-SAM and a COMT enzyme preparation obtained from rat liver.
The radioactive products were isolated by cation exchange resin chromatography and
subsequently oxidized with sodium periodate to produce vanillin which, after further
purification by solvent extraction, was quantified by LSC. The method provided a
measure of the total catecholamine (NA and A) content of the original sample as
the normetanephrine (NMN) and metanephrine (MN) produced in the methylation reaction were not separated by the cation exchange resin chromatography stage and
both products were oxidized to vanillin by the periodate (Fig. 9.2). The procedure

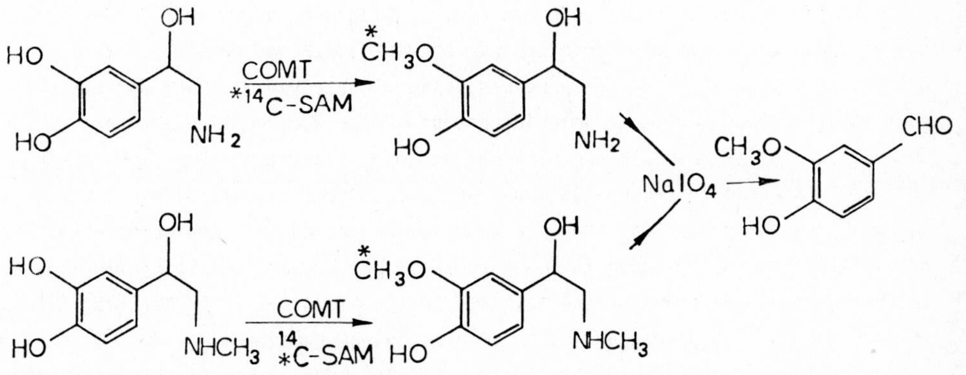

Fig. 9.2. Radioenzymatic determination of catecholamines using COMT, ^{14}C-SAM and
sodium periodate.

had a sensitivity of about 200 pg for NA, and of possible interfering compounds investigated, only isoproterenol and alpha-methyl-NA produced significant interference.

A similar assay was published by Nikodijevic et al. in 1969 (6). These authors determined the overall recovery of the amine in the assay not by the use of ^{3}H-NA, but by the addition of internal standards of cold NA to duplicate samples. Less rigorous attempts were made to purify the resultant products, these being obtained simply by solvent extraction procedures prior to LSC. Despite the much simpler procedure used by these authors, and the fact that the total recovery of NA through the whole procedure was only 10-20%, the sensitivity recorded was about 300 pg. The specificity of the assay was similar to that of the assay mentioned in the previous paragraph, with interference from the same compounds.

Shortly afterwards, Engelman and Portnoy (7) modified their original procedure by introducing a thin layer chromatographic separation of the radioactive products of the enzymic methylation (MN and NMN) prior to their individual oxidation to vanillin and purification as described previously. This modification produced little change in the sensitivity of the assay but allowed the separate quantitation of NA and A; the interference by isoproterenol and alpha-methyl-NA was limited to NA, the A assay being spared this complication. Contamination by equimolar concentrations of DA was restricted in both cases to 1%, and cross contamination between the amines NA and A was less than 0.5%. Again the total recovery of tracer amounts of ^{3}H-NA and A was very low: 10-11%.

Significant improvements were made to this assay procedure by Passon and Peuler (8), with the sensitivity being increased by a factor of 10. The major contribution to the increase in sensitivity resulted from the replacement of ^{14}C-SAM in the incubation medium with higher specific activity ^{3}H-SAM; this necessitated internal standardization with cold amines in duplicate samples rather than the additic of tritiated tracer amines to each sample. Subsequent to the enzymic methylation, the products were extracted into organic solvent which was evaporated to dryness prior to TLC separation of the products, oxidation to vanillin and further solvent purification stages. The enzymic incubation medium was modified by the inclusion of reduced glutathione in place of ascorbic acid, since the latter, while adequately protecting the amines from oxidation, was able to competitively inhibit the methylation of the catechols (9). Reduced glutathione served the same purpose but did not suffer from this disadvantage. Using this procedure, cross contamination between the amines NA and A was reported to be less than 0.5% by Cryer (10) but interference from both isoproterenol and alpha-methyl-NA remained.

Hortnagl et al. (11) were able to increase the sensitivity of the assay procedure of Passon and Peuler by a factor of 4 by the use of a higher specific activity ^{3}H-SAM, a reduction in the molarity of the SAM used in the incubation medium and by improving the solvent extraction efficiency of the methylated products by a factor of 2 simply by saturation of the aqueous phase with sodium chloride.

Coyle and Henry (12) in 1973 extended the assay of Engelman and Portnoy to the quantitation of dopamine (DA). This was accomplished by the extraction of the 3-methoxytyramine (3-MT) (Fig. 9.3) formed in the methylation reaction from DA subsequent to the oxidation of the beta-hydroxylated products to vanillin. This allowed the quantification of DA in the original sample with a sensitivity of 100 pg. However, a simpler assay procedure for DA was introduced by Cuello et al. (13); perchloric acid extracts of tissue, together with duplicate samples containing internal standards, were incubated under similar conditions to those previously used by Nikodijevic et al. (6), although it was found unnecessary to use the highly purified COMT preparation suggested by these latter authors. The radioactive products were separated by paper chromatography and quantitated by LSC after elution of the products from the appropriate areas of the chromatogram. The limits of sensitivity of the assay were found to be 1-3 ng although a footnote to the paper suggested that the sensitivity could be increased by a factor of 10 by the introduction of a simple solvent purification stage prior to the chromatography. The method was used to considerable effect by van der Gugten et al. (14) and Versteeg et al. (15) who obtained sensitivities of about 10 pg for all three amines (NA, A and DA) using essentially this procedure.

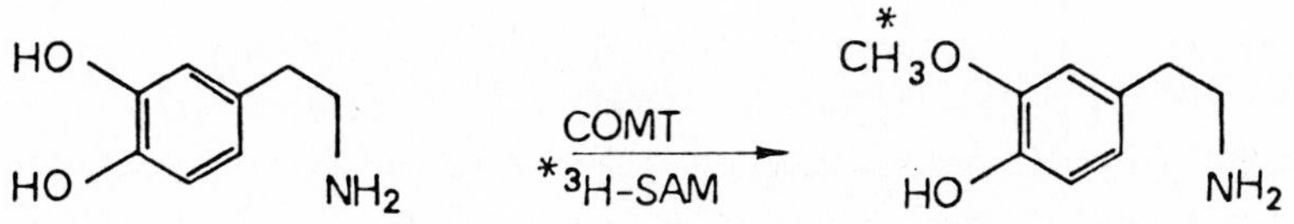

Fig. 9.3. Determination of dopamine using COMT and ^{3}H-SAM.

The recovery of the phenolic amine products from the incubation medium can be improved considerably by their acetylation prior to chromatographic separation. It was found by Chattaway (16) that acetylation of amino and phenolic groups could be carried out with facility in aqueous solution under slightly basic conditions, and this observation was exploited by Fry et al. (17) to modify the assay procedure of Cuello et al. (13). Interestingly, these authors also found that acetylation of the amino function improved the stability of the radioactive products during the subsequent paper chromatographic stage. By replacement of the paper chromatographic separation by TLC, Martin et al. (18) were able to separate the radioactive acetylated products from the incubation medium with a considerable saving in time; approximately 40 samples could be completely assayed for the catecholamines within a single working day. The method gave sensitivities of 40-100 pg for NA, 20-40 pg for DA and 10-30 pg for A, while cross examination between the samples was less than 2% with the exception of overlap of DA into A which amounted to nearly 4%. A similar approach has been adopted by McCaman et al. (19), however, these authors used an enzyme-catalyzed acetylation to quantitate DA with a sensitivity of 15 pg.

An alternative derivatization procedure has been adopted by Philips (20) in which dansylation of the NMN and MN was carried out subsequent to the O-methylation. The derivatives were then recovered by extraction into ethyl acetate and resolved by two successive TLC separations, which allowed the quantitation of both amines with a sensitivity of 2-3 pg.

The methods that have been discussed so far resulted in considerable technical improvements in the assay procedures with the result that they have become relative simple to use; they produce the required sensitivity for most purposes, and provide suitable specificity (as suggested by a comparison of the data with results obtaine using gas chromatography-mass spectrometry). However, with many of the procedures it was impossible to achieve the required sensitivity for the quantitation of the catecholamines in blood plasma samples where concentrations are extremely low. Considerable efforts were therefore made to elucidate, in detail, the optimum conditions for each stage of the assays in the hope that the required sensitivities could be achieved.

Consideration was given to the efficiency of the enzymic conversion of the catecholamines to the appropriate methylated product. It had been noted by both Coyle and Henry (12) and Hortnagl et al. (11) that the enzyme COMT suffered inhibition from substances that were present in samples obtained from biological tissue, thus necessitating the use of internal standardization procedures which have become commonplace. Weinshilboum and Raymond (21) pointed out in 1976 that calcium ions were able to cause non-competitive inhibition of the enzyme COMT with respect to both the substrate and the methyl donor SAM. In a detailed study of partially purified COMT preparations, they suggested that maximal enzyme activity was produced at 1mM $MgCl_2$ and that 0.5mM Ca^{2+} caused about 50% inhibition under these conditions; they also found that dithiothreitol increased enzyme activity considerably.

De Champlain et al. (22), in their attempts to determine catecholamine concentrations in plasma or serum using the procedure of Coyle and Henry, found that considerable inhibition of enzyme activity occurred. This was thought to be due to the relatively high concentration of Ca^{2+} in the samples. In order to circumvent this problem they added EGTA to chelate this cation and added $MgCl_2$. This resulted in restoration of enzymic activity to 95% of the level that was obtainable in the absence of plasma. Quantitation of 25 pg of the catecholamines could be achieved, allowing estimations to be carried out in as little as 300 µl of serum. No differentiation was obtained between NA and A, and DA caused some 4.8% cross contamination with the NA and A fraction, though the method was considerably simple to perform than that of Engelman and Portnoy (7). In order to measure both NA and A, Weise and Kopin (23) used an assay procedure based on COMT to estimate the concentrations of NA and A, and in a duplicate sample determined the concentration of NA alone with PNMT; the concentration of A was then simply found by the difference. The method was rather tedious although it allowed the separate quantification of all three catecholamines with sensitivities of 20 pg for NA, 17 pg for A and 130 pg for DA in 1 ml of plasma.

Ben-Jonathan and Porter (24) by addition of EGTA and $MgCl_2$ to the reaction mixture and the use of ^{14}C labelled standards, added subsequent to the incubation as O-methylated products, were able to obtain 10-30 pg sensitivities for each of the amines, but there was 1-2% cross contamination between the compounds of interest, even after two-dimensional TLC. Peuler and Johnson (25), by the addition of EGTA to the original blood serum sample, the use of internal standardization with non-radioactive catecholamines and an extensive purification of the methylated products gained sensitivities of 1 pg for both NA and A and 6 pg for DA, thus allowing the quantification of all three amines in very small samples of biological fluids.

In order to circumvent the difficulties found by the presence of endogenous inhibitors in certain tissue and plasma samples, Gauchy et al. (26) introduced a preliminary purification of the catecholamines by alumina adsorption. The amines were eluted from the alumina with oxalic acid. It was found that both oxalate and aluminium ions caused some inhibition of COMT activity, but by judicious selection of the experimental conditions it proved possible to quantify as little as 17 pg of NA and 24 pg of DA. However, the method did not permit the differentiation of NA and A. This procedure did introduce the use of sodium tetraphenylborate (NaTPB) as a means of improving considerably the extraction efficiency of 3-MT formed in the reaction, and this reagent has subsequently been employed by others. Da Prada and Zurcher (27) used NaTPB in diethyl ether to extract the O-methylated amines from the aqueous reaction mixture. The amines were then back extracted into HCl and subsequently separated by TLC, then the beta-hydroxylated amines were separately converted to vanillin prior to LSC. This allowed the estimation of both NA and A with a sensitivity of 1 pg and DA with a sensitivity of 5 pg.

A more sophisticated approach for the separation of the O-methylated reaction products has been adopted by Endert (28). NMN and MN formed in the COMT-based assay procedure were separated by high performance liquid chromatography using a reverse phase column with a total chromatography time of only 3 min. The appropriate fractions were collected; the derivatives were oxidized to vanillin, recovered by solvent extraction and quantified by LSC. The method produced no greater sensitivity than the method of Da Prada and Zurcher (27) and suffered from the disadvantage that HPLC equipment was required.

One aspect of these assay procedures not so far considered is that of the sample blank. In order to achieve accurate and reproducible quantitation of the compound of interest it is essential that analytical values for the sample blank should be both low and reproducible. Many different approaches have been used, from attempts to selectively decompose the catecholamines in the tissue sample prior to assay to incubation in the absence of tissue or incubation of the blank samples at 0° instead of 37°. An alternative approach, adopted by Saller and Zigmond (29), was to cause complete inhibition of the COMT enzyme activity by addition of high concentrations of Ca^{2+} to the blank samples. Blank and assay samples were then treated in an identical manner and the contribution of radioactivity from the decomposition of radioactive SAM could be estimated accurately. This, together with the subsequent precipitation of unreacted ^{3}H-SAM with phosphotungstic acid and a detailed study of TLC conditions for the purification of the products, allowed sensitivities of 1 pg for NA and A and 5-10 pg for DA to be achieved.

Radioenzymatic methods have been modified considerably since the initial work of Saelens et al. (2). The introduction of high specific activity ^{3}H-SAM allowed a large increase in sensitivity, while considerable attention to the detail of the assay procedure to ensure maximal enzyme activity, minimal tissue blanks and careful chromatographic separation of the radioactive products has resulted in the development of extremely sensitive assay techniques. Developments still continue. Methods published recently, however, have concentrated on the marriage of techniques previously reported, with the result that the radioenzymatic assays for the catecholamines have become much simpler to use (e.g. Cheng and Wooten (30)) although no greater sensitivity has been achieved over methods such as those of Da Prada and Zurcher (27) and Saller and Zigmond (29). Nevertheless, radioenzymatic procedures remain amongst the most sensitive assay techniques for the catecholamines. The study of Hjemdahl et al. (31) makes a useful comparison between radioenzymatic and HPLC/electrochemical detection methods for the quantification of these compounds, the two procedures showing excellent agreement.

9.3 PRECURSORS AND METABOLITES OF THE CATECHOLAMINES

Space does not permit a detailed review of the methods available for the quantitation of the metabolites of the catecholamines, most of which are simple extensions of methods previously developed for the parent amines. The methods are, however, of considerable importance since the ability to quantitate a number of catabolites has allowed estimation of the effect of various drug treatments on the metabolic turnover of the catecholamines.

Precursors or metabolites possessing the catechol moiety can be converted to their corresponding O-methylated derivatives using COMT while those which retain an amino function can be derivatized with PNMT.

9.3.1 Catecholamine conjugates

There is evidence that the catecholamines are present in blood plasma as con-conjugates (32-34), but the possible importance of these substances remains unclear. Johnson et al. (35) found that inclusion of a commercial sulphatase preparation in the enzyme incubation mix used for the estimation of the catecholamines according to the method of Peuler and Johnson (25) did not interfere with the assay and catalysed the release of the catecholamines from conjugates; this permitted measurement of these conjugates in blood plasma.

9.3.2 3,4-Dihydroxyphenylalanine (DOPA)

DOPA accumulation after inhibition of DOPA decarboxylase activity is thought to be a valuable indicator of DA synthesis. Using the catecholamine assay of Palkovits et al. (37), Hefti and Lichstensteiger (36) were able to quantify DOPA as 3-0-methyl-DOPA in the CNS of animals previously treated with the DOPA decarboxylase inhibitor 3-hydroxybenzylhydrazine. The purification procedure for the 3-0-methyl-DOPA produced in the methyl transferase reaction consisted of cation exchange chromatography followed by adsorption of the substance onto partially deactivated charcoal prior to the removal of ^{3}H-SAM decomposition products by anion exchange resins and quantification by LSC. The method was specific and suffered no significant interference from other catecholamine metabolites at concentrations thought to occur in CNS tissue, but the sensitivity (50-200 pg) was insufficient to allow measurement of normal DOPA concentrations.

An alternative approach was used by Rossor et al. (38). By an extension of the method of Da Prada and Zurcher (27), they were able to measure DOPA and DA concentrations in blood plasma samples taken from Parkinsonian patients receiving DOPA therapy. The method was based on the observation that the COMT enzyme preparation obtained from rat liver possessed considerable DOPA decarboxylase activity which resulted in the conversion of DOPA contained in the assay sample to DA. Peuler and Johnson (25) included in their incubation medium benzylhydroxylamine, a DOPA decarboxylase inhibitor, to prevent this conversion taking place. By assaying blood samples in the presence and absence of this inhibitor, Rossor et al. (38) were able to quantify DOPA simply from the difference of the two values obtained. The method was sensitive (20 pg/ml) and possessed the required specificity for this particular application.

9.3.3 3,4-Dihydroxyphenylacetic acid (DOPAC)

Comoy and Bohoun (39) developed an assay for the quantitation of DOPAC in urine based on the use of COMT and ^{14}C-SAM, but the sensitivity of the procedure (10 ng) made it suitable for use in CNS tissue. An improved assay based on the method of Nikodijevic et al. (6), with a sensitivity limit of 50 pg, was developed by Argiolas and Fadda (40). Subsequent to conversion of DOPAC to homovanillic acid, purification of the product was carried out by Sephadex G10 column chromatography (after Westerink and Korf (41)). This was followed by extraction from the acidified column eluate with ethyl acetate. The method achieved an excellent recovery of DOPAC (80-85%), the only major interference resulting from the addition to the original sample of 3,4-dihydroxymandelic acid (DOMA), an acid which is normally present in very low concentrations in the CNS. Essentially the same method was used for the estimation of DOPAC in the rat mesolimbic dopaminergic pathway by Beart and Gundlach (42), and these workers were able to demonstrate the specificity of the method for DOPAC in these particular brain regions.

Fekete et al. (43) estimated DOPAC, together with the catecholamines, by differential solvent extraction of the acidic and basic products from the enzymic incubation medium. After purification by paper chromatography, sensitivities of 30-100 pg could be achieved.

9.3.4 Normetanephrine (NMN)

Vlachakis and DeQuatro (44) developed a radioenzymatic assay for NMN in urine based on enzymic methylation with PNMT using ^{3}H-SAM as the methyl donor. The method was later modified by Vlachakis et al. (45) for use in brain tissue; the ^{3}H-MN produced was purified by TLC prior to quantification by LSC.

Kobayashi (46) reported a procedure based on the same principle for the estimation of NMN in blood plasma. This method, however, included a preliminary purification of the NMN using a cation exchange resin prior to the enzymic methylation. The product was purified by TLC and exhibited a sensitivity limit of 30 pg/ml blood plasma.

Many methods used for the quantitation of the catecholamines have been modified and utilized for the assays of various catecholamine metabolites. Essentially all that is required is that the compound of interest is a substrate for the enzyme used and that a suitable purification procedure for the product is obtained. While a number of such methods have been developed, interest in the quantitation of the catecholamine metabolites has developed to a lesser degree than that for the parent amines.

9.4 5-HYDROXYTRYPTAMINE (SEROTONIN, 5-HT)

The first radioenzymatic procedure to be developed for 5-HT was published in 1973 by Saavedra et al. (47). The direct methylation of 5-HT cannot be accomplished conveniently, so the amine is initially converted enzymatically to N-acetyl-5-HT. This compound possesses a higher affinity for the methyl transferase enzyme which accomplishes the conversion to ^{3}H-melatonin. The use of this coupled enzyme technique confers a considerable degree of specificity to the analysis and allows tissue blanks to be generated by the omission of the appropriate enzyme or cofactors involved in the initial acetylation stage. The assay is based on the following reactions (Fig. 9.4).

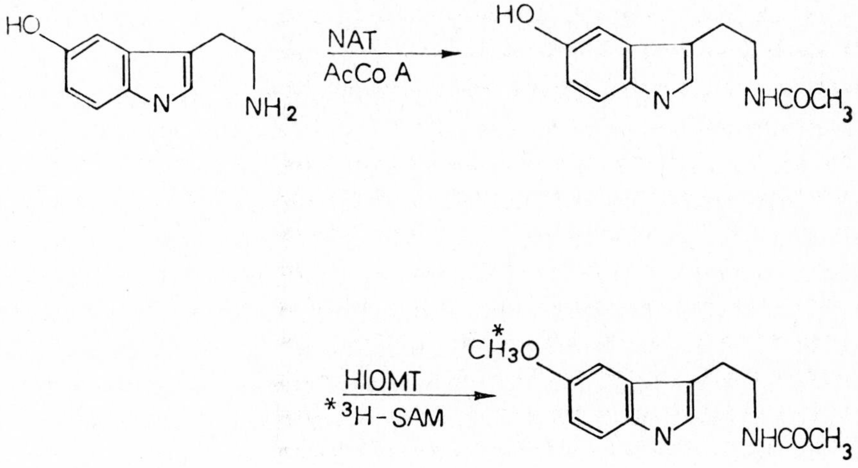

Fig. 9.4. Enzymatic reactions in the analysis of 5-hydroxytryptamine.

The N-acetyltransferase (NAT) enzyme preparation was obtained from rat liver according to the method of Weissbach et al. (48), while the hydroxyindole-O-methyl transferase (HIOMT) was purified from bovine pineals by the method of Axelrod and Weissbach (49). Tissue samples were homogenized in 0.1N HCl and aliquots of the supernatant were incubated with the NAT preparation in phosphate buffer at pH 7.9 in the presence of acetyl CoA at 37° for 30 min. Under these conditions 60-70% of the 5-HT was acetylated; addition of increased amounts of the enzyme resulted in no greater formation of the N-acetyl-5-HT though it did produce increased blank values. Subsequently HIOMT was added, together with ^{3}H-SAM, and the reaction was stopped after a further 10 min incubation by the addition of borate buffer, pH 10. The ^{3}H-melatonin formed was then extracted into toluene. The organic phase was evaporated to dryness at 80° in order to remove volatile radioactive contaminants, and the remaining radioactivity was quantitated by LSC. Tissue blank values were obtained by omitting acetyl CoA in the initial acetylation reaction. The limit of sensitivity of the assay was 50 pg. The specificity of the assay, as determined by TLC analysis of the toluene extracts obtained subsequent to the above analysis in brain, pineal and platelets, indicated that over 90% of the radioactivity in the extract co-chromatographed with melatonin. The use of reference compounds indicated that interference could occur from contamination by N-acetyl-5-HT, 5-hydroxytryptophol or bufotenin, although under normal conditions the low concentrations of these substances in CNS tissue would indicate that such interference would be insignificant. This method has formed the basis of all further developments in radioenzymatic assay procedures for 5-HT.

Saavedra et al. (50) included a decarboxylase inhibitor in assays carried out
in hypothalamic tissue in order to prevent the conversion of 5-hydroxytryptophan
to 5-HT by decarboxylase contamination known to be present in the NAT preparation
obtained from rat liver. However, they found no significant effect of the inclusio
of this inhibitor, and concluded that the concentration of the 5-hydroxytryptophan
in hypothalamic tissue was very low.

In order to enable the quantitation of 5-HT in CSF, Boireau et al. (51) modified
the above procedure. It was found that salts present in this particular biological
fluid interfered with the enzymic conversion of 5-HT to melatonin. A preliminary
purification of the 5-HT in the sample was therefore carried out on a Sephadex G 10
column; this not only removed the salts but allowed the separation of the 5-HT from
5-hydroxytryptophan and N-acetyl-5-HT, two possible interfering substances, and
resulted in the recovery of 95% of the 5-HT by elution of the column with formic
acid. The acid was subsequently evaporated to dryness. The remainder of the
procedure followed essentially the method of Saavedra, although a better enzymic
conversion was achieved by increasing the amount of HIOMT used at the methylation
stage. A final TLC separation allowed Boireau et al. (51) to achieve a sensitivity
of 10-20 pg in up to 3 ml of CSF.

The initial acetylation of 5-HT achieved by Saavedra with NAT was only 60-70%,
and as a result Hammel et al. (52) carried out the acetylation chemically rather
than enzymatically. Subsequent to the chemical acetylation, the ^{3}H-melatonin was
produced enzymatically with HIOMT. The resultant sensitivity of the assay was abou
35 pg.

9.5 HISTAMINE (HA)

Prior to the development of a radioenzymatic method for the estimation of HA by
Snyder et al. in 1966 (53), this amine was measured by either bioassay or fluorimetric procedures, many of which suffered from a lack of specificity. The enzymati
procedure relies upon the ability of histamine-N-methyl transferase, first characte
ized by Brown et al. (54), to transfer the ^{14}C-methyl group from ^{14}C-SAM to HA as
shown in Fig. 9.5.

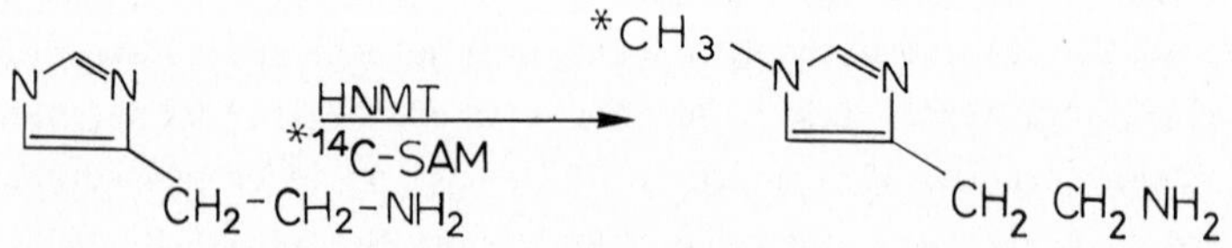

Fig. 9.5. Radioenzymatic determination of histamine.

Supernatants from tissue homogenates prepared in pH 7.9 phosphate buffer were incubated with ^{14}C-SAM and a histamine-N-methyl transferase preparation for 60 min at 37° in the presence of tracer amounts of ^{3}H-HA. The enzymic methylation was stopped by basification of the incubation mixture and the ^{14}C-N-methyl-HA was extracted into chloroform. The organic layer, after washing with 1N sodium hydroxide, was evaporated to dryness. Dual label LSC allowed the calculation of the original HA concentration. Paper chromatography indicated that the only radioactivity in the chloroform extract co-chromatographed with N-methyl-HA. The sensitivity of the assay procedure was limited to 2 ng. It was, however, essential that each sample had its own internal standard of ^{3}H-HA since the conversion of HA to its N-methyl derivative appeared to be both low and extremely variable, a peculiarity of the assay which was at that time unexplained.

Further studies carried out by Beavan et al. (55) were able to elucidate a number of the difficulties experienced with the initial assay procedure of Snyder et al. (53), with the result that marked improvements in the estimation of this amine were made. Beavan et al. used essentially the same technique as Snyder et al., but they found that by increasing the incubation time to 90 min, virtually complete methylation of the HA occurred. The enzymic reaction was stopped with perchloric acid rather than base, and this appeared to give lower and more reproducible assay blanks (which in this case were obtained by omission of the tissue extract). However, probably the most significant finding was the observation that unless large amounts of carrier N-methyl-HA (25 μg) were added, subsequent to the enzymic conversion, considerable and variable losses of ^{14}C-N-methyl-HA occurred; these were probably due to its adsorption on the glassware used in the assay. Under these conditions Beavan et al. were able to increase the limit of sensitivity of the assay to 100 pg.

Taylor and Snyder (56) improved the limit of sensitivity of the original assay procedure to 20 pg by reducing incubation volumes and increasing the concentrations of ^{14}C-SAM used, thus increasing the efficiency of the methylation without any serious deleterious effects on the assay blanks. However, in this assay procedure plasticware was used at the incubation stage, and this undoubtedly resolved many of the adsorption problems previously experienced. These workers concluded that it was no longer necessary to use internal standardization for each individual sample since reproducible methylation and extraction of the product was obtained in this modified procedure. It was therefore possible to eliminate the use of ^{3}H-HA and replace the ^{14}C-SAM with the much higher specific activity ^{3}H-SAM, accounting in large part for the increased sensitivity of this procedure.

The assay method introduced by Beavan et al. (55) was used by Brownstein et al. (57) to determine the HA content of individual hypothalamic nuclei in the rat. It was pointed out by these workers that a more efficient recovery of tissue HA was obtained if homogenization was carried out under strongly alkaline conditions. One

of the earlier criticisms of such a harsh extraction procedure was that the histidi
would be decarboxylated to HA; this appeared to be without foundation. The procedu
used by Brownstein et al. (57) appeared to be specific for tissue HA and exhibited
a limit of sensitivity of 10-20 pg.

It is clear that the radioenzymatic assay for HA has proved more convenient and
more sensitive than the previously popular fluorimetric method of Shore et al. (58)
The latter method suffered from considerable interference in CNS tissue preparation
The values obtained by Snyder et al. (53) for whole brain HA (60 ng/g) were considerably lower than those obtained using the fluorimetric procedure (310 ng/g),
although similar discrepancies were not apparent in other organs of the rat such
as skin, heart and spleen.

9.6   THE TRACE AMINES

Phenylethylamine (PEA), phenylethanolamine (PEOH), the tyramines (TAs) and
octopamines (OAs) and tryptamine (T) fall into a category known variously as the
trace amines, micro-amines or the non-catecholic arylalkylamines. The concentratio
of the trace amines found in mammalian tissues are extremely low compared to those
of the catecholamines and 5-HT. Radioenzymatic methods used to quantitate these
compounds are now considered.

9.6.1   2-Phenylethylamine (PEA) and tyramine (TA)

Saavedra (59) used a radioenzymatic assay procedure for PEA based on a coupled
enzyme procedure in which PEA was first converted to PEOH using a dopamine-beta-
hydroxylase preparation and the product was subsequently N-methylated in the
presence of ^{3}H-SAM and PNMT to produce N-methyl-PEOH (Fig. 9.6).

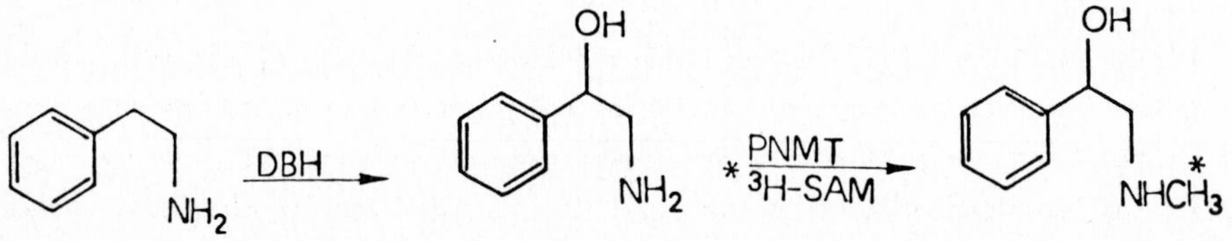

Fig. 9.6. Sequence of reactions in radioenzymatic assay for 2-phenylethylamine.

Tissue was first homogenized in HCl containing an MAO inhibitor, and after centrifugation, an aliquot of the supernatant was adjusted to pH 11 and extracted with
toluene. The PEA was then back extracted into a small volume of HCl, thus providin
an initial purification together with a concentration of the amine. The sample was
then incubated with a dopamine-beta-hydroxylase preparation at pH 5.5, resulting
in conversion of 40-50% of the PEA to PEOH under the conditions used. The pH of
the incubation medium was adjusted to 8.6 and the conversion to the product N-
methylPEOH was accomplished in the presence of PNMT and ^{3}H-SAM. Extraction of

the product was carried out with toluene/isoamyl alcohol which was subsequently evaporated to dryness, with the loss of volatile contaminants. The assay exhibited a sensitivity of 100-200 pg. TLC indicated that in excess of 95% of the radio-activity recovered was isographic with authentic N-methyl-PEOH. The levels of PEA in whole rat brain determined with this procedure (1.5 ng/g) were in good agreement with mass spectrometric methods (60,61). Tissue blanks were obtained by elimination of dopamine-beta-hydroxylase from the initial enzymic conversion reaction; this indeed was essential as it allowed determination of the endogenous concentration of PEOH, an amine which would otherwise lead to an elevated estimation of the con-centration of PEA.

A similar procedure has been used for the estimation of p-TA by Tallman et al. (62). TA is first converted to octopamine with dopamine-beta-hydroxylase and subsequently N-methylated to form synephrine (Fig. 9.7).

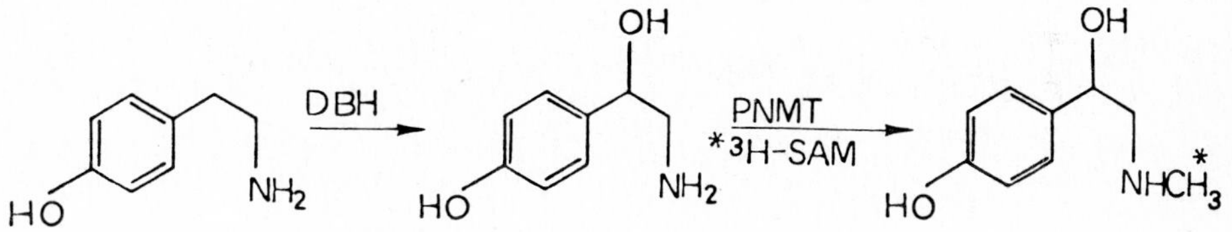

Fig. 9.7. Radioenzymatic assay for p-tyramine.

The pre-extraction and concentration was carried out with methyl acetate as the organic solvent. The efficiency of the initial beta-hydroxylation was found to be 90%. The overall assay sensitivity was reduced to a limit of 1 ng. It was sug-gested that this decreased sensitivity compared to the PEA assay was due to the poor recovery of the methylated product synephrine with the toluene/isoamyl alcohol extraction solvent. Because of the more polar nature of synephrine compared to N-methylphenylethanolamine, the isoamyl alcohol content in the extraction solvent was increased in order to facilitate recovery of the product; this in turn resulted in higher blank values being obtained. Reliance in such an approach simply on the use of more or less polar solvents for the final recovery of the product in-evitably results in a lack of specificity. While TLC analysis of the extract in-dicated that in excess of 90% of the radioactivity co-chromatographed with authentic synephrine, the high tissue blanks resulted in the relatively low sensitivity of this assay. This assay of rat whole brain values of p-TA gives results which are somewhat higher than those reported using mass spectrometry (see 63) or electron-capture gas chromatographic (64) techniques, and does not provide separation of the m- and p-isomers of TA.

9.6.2 Phenylethanolamine (PEOH) and octopamine (OA)

A method based on the conversion of PEOH and OA to their appropriate N-methyl derivatives in the presence of PNMT and ^{3}H-SAM was used by Harmar and Horn (65) to quantitate these amines (Fig. 9.8).

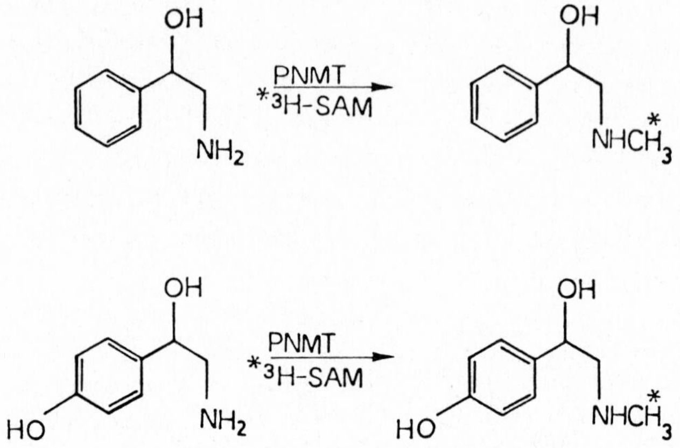

Fig. 9.8. N-methylation in the radioenzymatic analysis of phenylethanolamine and octopamine.

Tissues were homogenized in Tris buffer, pH 8.6, containing an MAO inhibitor, and the homogenates were subsequently boiled for 5 min at 100° in order to pre-cipitate protein and destroy some unknown interfering substances. The supernatants after centrifugation, were incubated with PNMT and ^{3}H-SAM at 37° and the reaction was stopped after 40 min by the addition of borate buffer, pH 11. The reaction products were extracted into ethyl acetate, which was then evaporated to dryness prior to the purification of the products by TLC. The method exhibited a sensi-tivity of 100 pg but this was reported to be insufficient to allow the quantitation of normal levels of PEOH in brain tissue. Harmar and Horn found it necessary to use TLC to purify the raction products because, in their hands, the solvent ex-traction procedures of Saavedra (66) lacked specificity.

Subsequently Danielson et al. (67) modified the above procedure by the conver-sion of the N-methylated products to their dansyl derivatives after their solvent extraction with ethyl acetate. The appropriate dansyl derivatives were then sub-jected to three sequential TLC separations in different solvent systems prior to their quantification by LSC. The method had a limit of sensitivity of 10 pg and allowed the separation of m- and p-OA. The former isomer could not be detected in mammalian brain tissue.

9.6.3 Tryptamine (T)

Saavedra and Axelrod have employed a radioenzymatic procedure for the analysis of T in a number of tissues, including brain (68). The method involved extracting T with a mixture of toluene and isoamyl alcohol (97:3) and its N-methylation using ^{14}C-SAM and partially purified N-methyltransferase prepared from rabbit lung. The radiolabelled product was separated from interfering substances by differential solvent extraction and was shown to co-chromatograph on TLC, in a number of solvent systems, with authentic T standards carried through the same procedure. The value obtained for T in brain (22 ng/g) was in good agreement with that reported using a fluorometric procedure (69), but was considerably higher than levels found using mass spectrometric (70,71,72) or electron-capture gas chromatographic (73) techniques.

9.7 CONCLUSIONS

The introduction of the radioenzymatic assay technique for the biogenic amines and its subsequent extension to the estimation of a number of metabolites of these substances gave considerable impetus to investigations into the functional relevance of these compounds. These methods displayed advantages in sensitivity and, in some cases specificity, over the procedures previously used to quantitate these compounds. The limited requirement for capital equipment ensured the rapid and widespread acceptance of the methodology. In addition, most derivatizations and analyses are done at relatively low temperatures, so that thermolability, a factor which may arise in gas chromatographic and mass spectrometric techniques, does not usually pose a problem in radioenzymatic work.

In principle the approach has few disadvantages, but in practice a number have become apparent. The enzyme preparations required to accomplish the methyl transfer reactions have been commonly used in an impure form; this is of no particular disadvantage if the limitations which it imposes are realized. Many contain additional enzymes; e.g. decarboxylase activity is frequently present in preparations obtained from rat liver and care must be taken in their use to ensure that decarboxylation of some substance in the tissue extract will not falsely elevate the estimation of the compound of interest. This problem is relatively easy to circumvent by the addition of the appropriate enzyme inhibitors. Considerable emphasis has been placed on the specificity endowed upon the analysis by the very specificity of the enzyme preparations used (e.g. COMT). It should be made clear, however, that in order to achieve the sensitivity and specificity obtained in a number of the assays developed (e.g. Da Prada and Zurcher (27) and Saller and Zigmond (29)) considerable care must be taken in the separation and purification of the radioactive products. In this regard it must also be stressed that an accurate estimate of the tissue blank value must be obtained, and in order to make accurate estimates of low sample concentrations (and therefore low levels of radioactivity associated with it), a low blank value is required. It is largely due to the efficient reduction of blank value radioactivity that such high sensitivities have been achieve

A further problem applicable to the quantitation of low amounts of substances is the ability of certain materials to adsorb to glass surfaces. This problem, which must be considered in all assay methods, was exemplified in the first radio-enzymatic method developed for the estimation of HA (53), in which low and variable recovery of the product was encountered. The problem was circumvented by the addition of relatively large amounts of carrier material. In general, radio-enzymatic procedures permit the analysis of a large number of samples in a rel-atively short period of time. However, these assays have in some cases been criticized for their lack of specificity. Another problem which is usually a minor one but should be considered by the potential user of a radioenzymatic method is the disposal of radioactive materials used in the procedure.

Radioenzymatic methods continue to be reported, which attests to their con-tinuing popularity; their sensitivity and the ease with which they can be accom-plished ensure their place in the continued investigations of the functional role of the biogenic amines.

REFERENCES

1   J. Axelrod, J. Biol. Chem., 237 (1962) 1657-1660.
2   J.K. Saelens, M.S. Schoen and G.B. Kovacsics, Biochem. Pharmacol., 16 (1967) 1043-1049.
3   L.L. Iversen and B. Jarrot, Biochem. Pharmacol., 19 (1970) 1841-1843.
4   D.P. Henry, B.J. Starman, D.G. Johnson and R.H. Williams, Life Sci., 16 (1975) 375-384.
5   K. Engelman, B. Portnoy and W. Lovenberg, Amer. J. Med. Sci., 255 (1968) 259-268.
6   B. Nikodijevic, J. Daly and C.R. Creveling, Biochem. Pharmacol., 18 (1969) 1577-1584.
7   K. Engelman and B. Portnoy, Circ. Res., 26 (1970) 53-57.
8   P.G. Passon and J.D. Peuler, Anal. Biochem., 51 (1973) 618-631.
9   B. Blashke and G. Hertting, Biochem. Pharmacol., 20 (1971) 1363-1370.
10 P.E. Cryer, J.V. Santiago and S. Shah, J. Clin. Endocrinol. Metab., 39 (1974) 1025-1029.
11 H. Hortnagl, C.R. Benedict, D.G. Grahame-Smith and B. McGrath, Brit. J. Clin. Pharmacol., 4 (1977) 553-558.
12 J.T. Coyle and D. Henry, J. Neurochem., 21 (1973) 61-67.
13 A.C. Cuello, R. Hiley and L.L. Iversen, J. Neurochem., 21 (1973) 1337-1340.
14 J. Van der Gugten, M. Palkovits, H.L.J.M. Wijnen and D.H.G. Versteeg, Brain Res., 107 (1976) 171-175.
15 D.H.G. Versteeg, J. Van der Gugten, W. Dejong and M. Palkovits, Brain Res., 113 (1976) 563-574.
16 F.D. Chattaway, J. Chem. Soc., (1931) 2495.
17 J.P. Fry, C.R. House and D.F. Sharman, Brit. J. Pharmacol., 51 (1974) 116-117P.
18 I.L. Martin, G.B. Baker and S.M. Fleetwood-Walker, Biochem. Pharmacol., 27 (1978) 1519-1520.
19 M.W. McCaman, J.K. Ono and R.E. McCaman, J. Neurochem., 32 (1979) 1111-1113.
20 S.R. Philips, Can. J. Neurol. Sci., 7 (1980) 230.
21 R.M. Weinshilboum and F.A. Raymond, Biochem. Pharmacol., 25 (1976) 573-579.
22 J. de Champlain, L. Farley, D. Cousineau and M.-R. Van Ameringen, Circ. Res., 38 (1976) 109-114.
23 V.K. Weise and I.J. Kopin, Life Sci., 19 (1976) 1673-1686.

24 N. Ben-Jonathan and J.C. Porter, Endocrinology, 98 (1976) 1497-1507.
25 J.D. Peuler and G.A. Johnson, Life Sci., 21 (1977) 625-636.
26 C. Gauchy, J.P. Tassin, J. Glowinski and A. Cheramy, J. Neurochem., 26 (1976) 471-480.
27 M. Da Prada and G. Zurcher, Life Sci., 19 (1976) 1161-1174.
28 E. Endert, Clin. Chim. Acta, 96 (1979) 233-239.
29 C.F. Saller and M.J. Zigmond, Life Sci., 23 (1978) 1117-1130.
30 C.H. Cheng and G.F. Wooten, J. Pharmacol. Methods, 3 (1980) 333-344.
31 P. Hjemdahl, M. Daleskog and T. Kahan, Life Sci., 25 (1979) 131-138.
32 J. Haggendal, Acta Physiol. Scand., 59 (1963) 242-254.
33 N.T. Buu and J. Kuchel, J. Lab. Clin. Med., 90 (1977) 680-685.
34 T. Unger, N.T. Buu and O. Kuchel, Am. J. Physiol., 235 (1978) F542-F547.
35 G.A. Johnson, C.A. Baker and R.T. Smith, Life Sci., 26 (1980) 1591-1598.
36 F. Hefti and W. Lichtensteiger, J. Neurochem., 27 (1976) 647-649.
37 M. Palkovits, M. Brownstein, J.M. Saavedra and J. Axelrod, Brain Res., 77 (1974) 137-149.
38 M.N. Rossor, J. Watkins, M.J. Brown, J.L. Reid and C.T. Dollery, J. Neurol. Sci., 46 (1980) 385-392.
39 E. Comoy and C. Bohoun, Clin. Chim. Acta, 36 (1972) 207-212.
40 A. Argiolas and F. Fadda, Experientia, 34 (1978) 739-741.
41 B.H.C. Westerink and J. Korf, Eur. J. Pharmacol., 38 (1976) 281-291.
42 P.M. Beart and A.L. Gundlach, Brit. J. Pharmacol., 69 (1980) 241-247.
43 M.I.K. Fekete, B. Kanyicska and J.P. Herman, Life Sci., 23 (1978) 1549-1556.
44 N.D. Vlachakis and V. DeQuatro, Biochem. Med., 20 (1978) 107-114.
45 N.D. Vlachakis, N. Alexander and R.F. Maronde, Life Sci., 26 (1980) 97-102.
46 K. Kobayashi, V. DeQuatro, R. Kolloch and L. Miano, Life Sci., 26 (1980) 567-573.
47 J.M. Saavedra, M. Brownstein and J. Axelrod, J. Pharmacol. exp. Ther., 186 (1973) 508-515.
48 H. Weissbach, B.G. Redfield and J. Axelrod, Biochim. Biophys. Acta, 54 (1961) 190-192.
49 J. Axelrod and H. Weissbach, J. Biol. Chem., 236 (1961) 211-213.
50 J.M. Saavedra, M. Palkovits, M.J. Brownstein and J. Axelrod, Brain Res., 77 (1974) 157-165.
51 A. Boireau, J.P. Ternaux, S. Bourgoin, F. Hery, J. Glowinski and M. Hamon, J. Neurochem., 26 (1976) 201-204.
52 I. Hammel, Y. Naot, E. Ben-David and H. Ginsburg, Anal. Biochem., 90 (1978) 840-843.
53 S. Snyder, R.J. Baldessarini and J. Axelrod, J. Pharmacol. exp. Ther., 153 (1966) 544-549.
54 D.D. Brown, R. Tomchick and J. Axelrod, J. Biol. Chem., 234 (1959) 2948-2950.
55 M.A. Beavan, S. Jacobsen and Z. Horakova, Clin. Chim. Acta, 37 (1972) 91-103.
56 K.M. Taylor and S.H. Snyder, J. Neurochem., 19 (1972) 1343-1358.
57 M.J. Brownstein, J.M. Saavedra, M. Palkovits and J. Axelrod, Brain Res., 77 (1974) 151-156.
58 P.A. Shore, A. Burkhalter and V.H. Cohn, J. Pharmacol. exp. Ther., 127 (1959) 182-186.
59 J.M. Saavedra, J. Neurochem., 22 (1974) 211-216.
60 D.A. Durden, S.R. Philips and A.A. Boulton, Can. J. Biochem., 51 (1973) 995.
61 J. Willner, H.F. LeFevre and E. Costa, J. Neurochem., 23 (1974) 857-859.
62 J.F. Tallman, J.M. Saavedra and J. Axelrod, J. Neurochem., 27 (1976) 465-469.
63 P.H. Wu, G.B. Baker and R.W. Henwood, in A.D. Mosnaim and M.E. Wolf (Eds.), Noncatecholic Phenylethylamines: Part 2 Phenylethanolamine, Tyramines and Octopamine, Marcel Dekker Inc., 1980, pp. 307-339.
64 G.B. Baker, D.F. LeGatt and R.T. Coutts, J. Neurosci. Methods, 5 (1982) 181-188.
65 A.J. Harmar and A.S. Horn, J. Neurochem., 26 (1976) 987-993.
66 J.M. Saavedra, Anal. Biochem., 59 (1974) 628-633.
67 T.J. Danielson, A.A. Boulton and H.A. Robertson, J. Neurochem., 29 (1977) 1131-1135.
68 J.M. Saavedra and J. Axelrod, J. Pharmacol. exp. Ther., 182 (1972) 363-369.
69 J.W. Sloan, W.R. Martin, T.H. Clements, W.F. Buchwald and S.R. Bridges, J.

202

Neurochem., 24 (1975) 523-532.

70 S.R. Philips, D.A. Durden and A.A. Boulton, Can. J. Biochem., 52 (1974) 366-373.

71 J.J. Warsh, D.D. Godse, H.C. Stancer, P.W. Chan and D.V. Coscina, Biochem. Med., 18 (1977) 10-20.

72 F. Artigas and E. Gelpi, Anal. Biochem., 92 (1979) 233-242.

73 D.G. Calverley, G.B. Baker, H.R. McKim and W.G. Dewhurst, Can. J. Neurol. Sci., 7 (1980) 237.

Chapter 10

DETERMINATION OF BIOGENIC AMINES AND THEIR METABOLITES BY HIGH-PERFORMANCE LIQUID CHROMATOGRAPHY

J.J. WARSH, A.S. CHIU AND D.D. GODSE
Clarke Institute of Psychiatry, University of Toronto, Toronto, Ontario, Canada.

10.1 INTRODUCTION

Understanding the role of biogenic amines in central nervous system (CNS) function and in neuropsychiatric disorders has been contingent upon the development of highly sensitive and accurate assays of these substances and their metabolites in biological samples. The earliest physico-chemical methods for measurement of biogenic amine compounds (the parent amines, precursors and metabolites; biogenic amine compounds; BACs) involved their isolation and purification from complex biological matrices by liquid-liquid partitioning or adsorption on ion-exchange columns, followed by quantitative fluorescence detection [1-4]. Although these methods were very useful in the investigation of CNS biogenic amine metabolism and function, they were limited especially with respect to practical working assay sensitivity, which rarely exceeded 25 ng per sample. This prevented the reliable measurement of BACs in small brain regions or in small (< 1 ml) amounts of cerebrospinal fluid (CSF) or blood.

In the early 1970's several new highly sensitive analytical techniques were introduced for the determination of BACs. Radioenzymatic procedures which employed methylating enzymes and ^{3}H-S-adenosyl-methionine permitted determination of methyl-accepting BACs at low picogram levels. Gas chromatography with electron capture detection also provided a highly sensitive technique for the determination of some BACs, particularly when used in conjunction with capillary gas chromatography (see Baker et al., this volume). However, this procedure has not been widely applied because of the non-selectivity of electron-capture detection and the need for formation of suitable volatile halogenated derivatives for separation and detection. Gas chromatography with mass fragmentographic detection (GC-MS) has offered a highly sensitive and specific means of determining a wide variety of BACs [5-7], but involves costly instrumentation which is complex to operate and requires chemical transformation of BACs into suitably volatile derivatives for determination.

The recent introduction of high-performance liquid chromatography (HPLC) into the field of BAC analysis resulted in the rapid development of a number of very sensitive and specific methods for BAC measurement in biological samples. The technique of HPLC came into existence about 15 years ago, but practical application of this physico-chemical method for determination of compounds of biological

interest was delayed by the lack of suitable column and detector technology
Introduction of pellicular ion-exchange sorbents (30–40 µm particle size) and ver
sensitive low dead volume fluorescence and electrochemical detectors lead to th
rapid application of HPLC to measure BACs. The relatively low instrumental cost an
the simplicity and flexibility, particularly of reversed-phase HPLC techniques, mad
these the preferred methods for determination of polar biogenic amines such a
noradrenaline (NA), dopamine (DA), serotonin (5–HT) and a number of their respectiv
metabolites. In this review we will consider in depth the utility of HPLC, i
combination with currently available detector systems, for the assay of BACs i
biological samples. Principles of chromatographic separation relevant to th
measurement of these analytes will be considered with particular emphasis o
specific factors affecting the separation and resolution of BACs on variou
commercially available sorbents. With respect to detection systems, this revie
will focus primarily on the fluorescence and electrochemical detectors, both o
which have now been widely used for the measurement of BACs. Other optical an
vapour phase HPLC detection systems will be considered only in passing to point ou
limitations or potential future developments in detector technology. The mai
emphasis of this review will be on the practical application of HPLC to determin
BACs in a variety of biological samples. In this regard, we will also emphasiz
important issues in sample preparation and analyte isolation and purification prio
to HPLC analysis.

10.2 PRINCIPLES OF HPLC SEPARATION AND DETECTION OF BIOGENIC AMINES

   Although there has been substantial effort to simplify chemical assays of BACs
the majority of physico-chemical procedures, regardless of the mode of detection
have involved isolation from the biological matrix followed by separation of th
individual compounds prior to quantitative detection. In the main, these principle
still hold in the application of HPLC to determination of these substances. Th
introduction of high resolution bonded reversed-phase sorbents has made it possibl
to determine selected BACs directly in the supernatants of tissue homogenates or CS
without extensive sample purification prior to HPLC analysis. In spite of this, th
application of such direct procedures is in general limited to specific biologica
matrices or groups of BACs as will be considered in a later section of this review.

   The development of high sensitivity fluorescence and electrochemical detector
was a critical antecedant for the successful application of HPLC to BAC measurement
Using current detector technology the specificity of HPLC methods for BACs i
primarily dependent upon the chromatographic separation process and not th
detectors, as the latter have been employed in relatively non-selective modes o
operation. The importance of the column chromatographic separation to th
specificity and accuracy of these HPLC assays dictates a more in-depth consideratio

of the critical factors influencing the HPLC separation of BACs.

10.2.1 Chromatographic Separation

The basic mechanism of analyte separation common to all forms of liquid chromatography involves the mass transfer of analyte between the mobile and stationary phases. The mass transfer process in HPLC is dependent upon such factors as sorbent surface area and electrical charge, analyte electrical charge and polarity, and mobile phase composition and polarity. With respect to surface area, the pellicular or microparticulate sorbents of HPLC columns have considerably more uniform particle size and much greater active surface areas compared to the antecedent gravity-fed open column sorbents. As a result, zone spreading of solutes is substantially reduced due to much faster mass transfer and more uniform pathways of solute migration. This phenomenon contributes significantly to the very high column efficiencies and dramatically improved resolution of analyte peaks characteristic of HPLC. However, as alluded to above, the quality of resolution in HPLC depends not only on the column efficiency, for which the majority of commercially available HPLC columns exceed 1,000 plates per meter, but also the column selectivity and composition of the mobile phase, and these latter factors vary in the different modes of HPLC operation. Indeed, the present classification of HPLC is based on the mode of operation or the principal physico-chemical interactions underlying the mass transfer process for a particular sorbent. Thus, HPLC is generally classified into normal phase, ion-exchange and reversed-phase chromatographies, depending on whether the principal physical processes underlying analyte separation involve adsorption, ionic affinity or phase partition.

Ion-exchange HPLC was the first mode of HPLC to be successfully applied to measurement of catecholamines (CAs) and 5-HT in tissues and urine. This mode of HPLC has now been almost totally supplanted by the use of reversed-phase HPLC (RPLC) for the assay of BACs. These latter two modes of HPLC will be considered in greater detail. Normal phase chromatography, on the other hand, has not proved of significant utility for BAC measurement for a number of reasons. Firstly, normal phase sorbents are quite susceptible to deterioration when used in the analysis of biological samples. Interaction between impurities in the biological matrix and active sorbent groups results in rapid sorbent deterioration and subsequent loss of resolving capacity. The process(es) contributing to this deterioration likely involve acid or base hydrolysis, or covalent bonding of polar sorbent functional groups with biological impurities [8]. Secondly, the long column equilibration times necessary to achieve suitable operating conditions substantially limits the utility of this mode of chromatographic separation. Finally, the hydrophobic mobile phases required in normal phase liquid chromatography have poor electrical conductivity properties and in general are unsuitable for use with the electrochemical detector [9]. Despite these significant drawbacks, normal phase

systems may be suitable for very selective applications such as the separation o
some of the highly polar metabolites of biogenic amines: for example, th
separation of DA sulphate conjugates in human urine [10]. Furthermore, chemicall
bonded normal phase sorbents have been recently developed in which the supportin
matrix is derivatized with functional groups of selective polarity, such as highl
polar amino groups or the less polar cyano and diol moieties [11]. Such bonde
normal phase sorbents allow improved column selectivity and stability and in th
future may further extend the utility of normal phase HPLC in BAC analysis.

(i) Ion-exchange HPLC. As in conventional ion-exchange chromatography, solute
which readily ionize in buffered systems may be separated by their relative ioni
affinity to a sorbent which has functional groups of opposite charge under the sam
conditions. The BACs may be categorized into basic (amines), acidic (acidi
metabolites) and neutral (glycol and alcohol metabolites) groups. Some of the basi
and acidic compounds may be resolved by cation- and anion-exchange high-performanc
techniques, respectively. However, ion-exchange HPLC is not suitable fo
determination of unconjugated neutral metabolites such as 3-methoxy-4-hydroxy
phenylethyleneglycol (MOPEG), 3,4-dihydroxyphenylethyleneglycol (DOPEG) o
5-hydroxytryptophol (5-HTOL), which do not readily ionize unless extreme alkaline p
is used.

(a) Parameters affecting separation and resolution. In the application o
ion-exchange HPLC to BAC measurement, column selectivity is more important than th
mobile phase composition and conditions in the control of solute retention an
separation. Column selectivity depends on the structural and physico-chemica
characteristics of the particular sorbent. For the most part, determination o
basic and acidic BACs by ion-exchange HPLC has been performed using one of fou
bonded phase, strong ion-exchange sorbents (Table 10.1). Structurally, thes
sorbents have functional groups derivatized on silica-based particles throug
siloxane bonding. These sorbents consist of three types: totally porou
microparticulate particles (Nucleosil), superficially porous spherical particle
(Vydac and Zipax), and non-porous pellicular particles (Corasil). As shown in Tabl
10.1, the particle size of these sorbents varies from 10 μm (Nucleosil) to 50 μ
(Corasil).

The major chemical interaction between these sorbents and the basic and acidi
BACs involves ion exchange by virtue of the complementary functional side groups
However, other poorly understood physico-chemical interactions may also participat
in the separation (mass transfer) process [12]. It has been suggested that th
separation of organic compounds by ion-exchange HPLC involves both electrostatic an
hydrophobic interactions, leading some workers to suggest that this mode of HPLC i
a subcategory of RPLC [13]. This latter notion may account in part for the simila
elution sequence of biogenic amines from cation-exchange and reversed-phase column
as described later.

TABLE 10.1

Common bonded-phase ion-exchange sorbents used for determination of
basic and acidic BACs

Name	Supplier	Particle Size (μm)	Particle Type[a]	Ion-Exchange Capacity[b]
Cation-exchange resins:				
Corasil CX	Waters	37-50	P	0.03-0.04
Nucleosil SA	Machery & Nagel	10(Average)	MP	1
Vydac SCX	Separation Groups	30-44	SP	0.1
Zipax SCX	DuPont	25-37	SP	5
Anion-exchange resins:				
Vydac SAX	Separation Groups	30-44	SP	Not available
Zipax SAX	DuPont	25-37	SP	12

[a] P = Pellicular; MP = Microparticulate totally porous;
   SP = superficially porous;

[b] Meq/g dry weight.

In ion-exchange HPLC, the ionic strength and pH of the buffer are the principal
mobile phase parameters affecting solute retention and resolution. Increase in the
buffer ionic strength decreases solute retention (or capacity factor, k') due to the
decreased ability of analyte ions to compete with the mobile phase ionic species for
active sites on the sorbent. Modification of the mobile phase pH affects retention
and separation of adjacent eluate peaks. However, as explained by Snyder and
Kirland [12], changes in selectivity as a function of pH are difficult to predict
and usually have to be determined on a trial-and-error basis. The specific
contribution of each of these factors to the separation of BACs has not been studied
extensively. The low exchange capacity of bonded phase columns (see Table 10.1)
constrains the upper limit of mobile phase ionic strength, thus affecting the use of
this parameter to manipulate the retention of BACs. Furthermore, since the
mechanisms governing retention and resolution of BACs in the ion-exchange separation
mode are incompletely understood, the effect of altering the mobile phase pH to
modify the resolution of analytes may also be unpredictable. Consequently, the
types and characteristics of the mobile phases used and the choice of sorbent
employed for BAC assay have been primarily empirically derived.

In general, citric acid/sodium acetate mobile phase buffers (pH 5.2) have been
employed in cation-exchange HPLC assays of basic BACs, whereas sodium acetate buffer

(pH 4.4) has been used for the assay of acidic BACs on anion-exchange HPLC columns.
At these pH values the amines and their acidic metabolites are predominantly i
their respective ionized forms, favourable for ion-exchange interactions. Using th
same citrate/acetate mobile phase (pH 5.2), baseline separation of brain and plasm
NA, adrenaline (A) and DA has been demonstrated on Corasil CX [14,15] and Vydac SC
columns [16,17]. However, the commonly used internal standard 3,4-dihydroxy
benzylamine (DHBA) co-chromatographs with A on these columns. In comparison, NA
DHBA, A and DA are well resolved from each other on Nucleosil 10-A columns (Fig
10.1A) [18,19]. These latter authors did not provide an explanation for th
successful separation of this group of CAs on Nucleosil 10-A columns, but their dat
indicate that the totally porous microparticulate sorbent (average particle size 1
μm) offers superior resolution and selectivity compared to the pellicula
non-porous and superficially porous spherical sorbents. Speculatively, th
mechanisms contributing to this superior resolution may involve physica
interactions between the derivatized silica sorbent matrix and these CAs beside
ion exchange (e.g. hydrophobic interactions).

The simultaneous separation of 5-HT along with the NA, A and DA has not bee
achieved using any of the three ion-exchange sorbents considered above. It i
possible, however, to separate DA and 5-HT on Zipax SCX (Fig. 10.1B) [20,21], bu
on Vydac SCX the 5-HT elutes with a long retention time (> 30 min) and is poorl
resolved [22]. In comparison, Vydac SCX provides good separations of NA and D
(Fig. 10.1C), but on Zipax SCX the NA elutes very close to or merges with th
solvent peak [22].

There are only a few reports on the use of anion-exchange HPLC for th
determination of acidic metabolites of biogenic amines. Urinary homovanillic aci
(HVA) and 3,4-dihydroxyphenylacetic acid (DOPAC) have been determined using Vyda
SAX [23] and Zipax SAX [24] anion-exchange HPLC, respectively. The CSF HVA, DOPA
and 5-hydroxyindoleacetic acid (5-HIAA) have been assayed with Zipax SAX [25] an
brain 5-HIAA with Aminex 5-A anion-exchange columns [26]. Unfortunately, resolutio
of these acidic metabolites on the bonded phase Vydac and Zipax SAX columns is poor
This leads to insufficient separation of these compounds from other unknown acidi
solutes. The swelling effect of the polystyrene divinyl benzene Aminex 5-A in it
mobile phase and its low efficiency at room temperature [11] have limited th
general utility of this type of anion-exchange sorbent for the assay of acidi
metabolites.

(b) Merits and limitations of ion-exchange HPLC. Despite the significant rol
of ion-exchange HPLC in promoting and establishing the use of HPLC for determinatio
of BACs, this mode of HPLC has major limitations as may be gleaned from th
foregoing considerations. Obviously the application of a particular sorbent i
limited to separation of ionically compatible substances. Secondly, optimur
chromatographic separation of analytes on ion-exchange HPLC depends primarily o

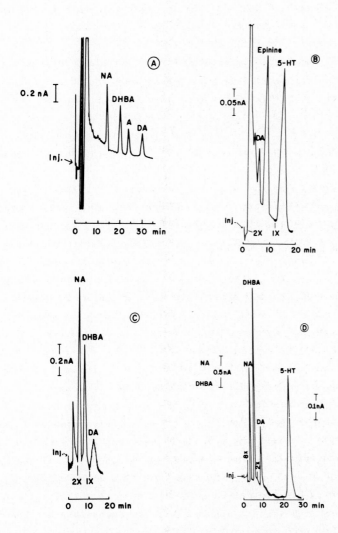

Fig. 10.1 Separation of BACs by cation-exchange HPLC. Chromatographic
systems employed: (A) Nucleosil 10-A (redrawn and reproduced
with permission from Allenmark and Hedman [19]); (B) Zipax
SCX (redrawn and reproduced with permission from Warsh et al.
[135]); (C) Vydac SCX (chromatographic conditions same as
(B)); (D) LiChrosorb C_8 (redrawn with permission from Warsh
et al. [45]).

column selectivity. This sorbent-dependent factor further limits the flexibilit
and utility of this mode of HPLC. Thirdly, there is still not a complet
understanding of all of the factors affecting the interaction between solutes an
ion-exchange sorbents. Thus the rational prediction of analyte retention o
ion-exchange sorbents is difficult. Finally, from a practical perspective, th
working lifetime of ion-exchange HPLC columns is generally short (< 400 sampl
analyses) and techniques for column regeneration have not yet been developed o
described.

The merits of ion-exchange HPLC lie principally in low cost, reliability an
accuracy for selective applications such as measurement of CAs and 5-HT [21]
However, the use of ion-exchange HPLC for the assay of BACs has been virtuall
outdated by the advances in the reversed-phase mode of HPLC. In contrast t
ion-exchange HPLC, this latter technique permits superior resolution (Fig.10.1D)
versatility and efficiency and a clearer rationale for sorbent and mobile phas
selection.

(ii) Reversed-phase HPLC (RPLC). Although introduced over thirty years ag
[27], the application of reversed-phase techniques for quantitative assay o
biological compounds only occurred after the necessary technology was harnessed t
produce chemically-bonded microparticulate reversed-phase sorbents of consisten
uniformity. Unlike normal phase HPLC this mode of chromatography involves the use o
non-polar stationary phases and polar eluents. The RPLC offers significan
advantages over both ion-exchange and normal phase modes of operation, particularl
with respect to the ability to separate compounds of varying polarity using the sam
chromatographic system and to permit predictable solute elution patterns.

(a) Types of RPLC. The active sites of all commercially availabl
reversed-phase sorbents are of the same alkylsilane type, although carbon chai
length varies. The appropriate classification of RPLC is based not on sorbent chai
length, but on the characteristics of the mobile phase, as the mobile phas
composition and conditions are major determinants of solute separation in RPLC
Six types of reversed-phase techniques have been described [11] including: 1
regular, 2) ionization control, 3) ion suppression, 4) ion-pair, 5) secondar
equilibrium and 6) non-aqueous. To date, only the techniques of ionization control
ion suppression and ion pairing have been demonstrated to be of utility in th
analysis of BACs. For this reason, the discussion of the mechanisms of retentio
and separation involved in RPLC assay of BACs will be limited to these latter three
techniques. The remaining types of RPLC have been considered in a recent review b
Majors [11].

(b) Parameters affecting analyte separation and retention. Despite som
controversy regarding the exact mechanism(s) of analyte retention in RPLC
hydrophobic interaction between sample solutes and the non-polar derivatives of th
sorbents is generally considered the major retentive mechanism [28-30]. The BACs as

a group are relatively polar organic compounds. In RPLC, modifying the polarity of these analytes relative to the mobile phase also modulates the solute mass transfer into the stationary phase and this may be successfully accomplished through several techniques. These include: 1) modification of the mobile phase pH to control or suppress analyte ionization, 2) addition of organic modifiers to the mobile phase to reduce surface tension between sorbent and mobile phase, as well as to modify the net polarity of the mobile phase, 3) formation of ion pairs to reduce analyte polarity, and 4) a combination of the above methods.

By employing ionization control techniques with neat aqueous phosphate buffer (pH 2.1), CAs and closely related compounds are eluted from octadecyl silane (ODS, C_{18}) columns in the order of decreasing polarity of these compounds, that is, NA, A, normetanephrine (NMN), DA, metanephrine (MN) and tyramine (TA) [31,32]. Using the same mobile phase conditions, acidic CA metabolites elute in the sequence of 3,4-dihydroxymandelic acid (DOMA), vanillylmandelic acid (VMA), DOPAC and HVA [32]. With citrate/acetate buffer at pH 4.1, 5-HT elutes from ODS columns following the less hydrophobic NA, A and DA [33]. In this latter system, the internal standard DHBA is usually well resolved from A and DA [34,35]. Although these neat aqueous systems permit simultaneous separation of a variety of BACs, they do have certain limitations which may be overcome by using other modes of RPLC. With neat aqueous systems the retention and separation of BACs are primarily a function of sorbent and analyte characteristics and are little influenced by the mobile phase. In addition, peak asymmetry is commonly observed with these systems [36,37]. Another consideration, which applies to all systems utilizing an aqueous mobile phase, is the requirement to buffer in the pH range of 2 to 8, the optimal range for sorbent stability. Within this pH range, buffers such as citrate, acetate, phosphate or a combination of citrate/acetate or citrate/phosphate are commonly used. However, some of these buffers should be used with care as they may affect column life or efficiency. For example, citrate buffer has been reported to cause permanent deterioration of ODS columns [38]. It has been suggested that acetate buffer gives relatively low column efficiency, possibly due to formation of nonpolar complexes with the analyte molecules [37]. As well, potassium salt is not suitable for use in mobile phases containing sodium dodecylsulphate due to insoluble complex formation with this organic detergent ion-pairing reagent [36].

The addition of water-miscible organic solvents such as methanol, acetonitrile or tetrahydrofuran to the mobile phase decreases solute retention in RPLC. Organic modifiers alter the solvation properties of the mobile phase for analytes through decreases in both the polarity [30] and surface tension [39] of the eluents. Addition of an organic modifier generally decreases column retention and the selectivity of the mobile phase for the analytes, but these effects are also dependent on the nature of the organic modifiers [29]. For example, in the analysis of plasma CAs using a methanol and phosphate buffer (48:52 v/v) mobile phase on ODS

212

columns, unknown peaks co-eluted with NA and DA [40]. Addition o
N,N-dimethylformamide (5% v/v) or use of a combination of acetonitrile (10% v/v) an
methanol (40%, v/v) in the phosphate buffer enhanced the separation of these unknow
peaks from the above amines. In another instance, a mobile phase comprised o
acetate buffer (pH 4.0)/acetonitrile (90:10 v/v) was found unsatisfactory for th
separation of CSF HVA and 5-HIAA [41,42]. However, changing the mobile phas
composition to acetate (pH 4.0)/methanol (85:15 v/v) gave baseline separation o
these acidic metabolites. Unfortunately, there is still an incomplete understandin
of the factors that affect the selectivity of these organic modifiers with respec
to the separation of BACs and what is known has in many cases been derive
empirically.

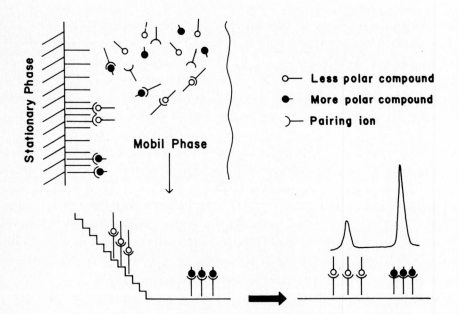

Fig. 10.2 Schematic illustration of the primary mechanism of ion-pair RPLC.
The solid and unfilled circles represent polar moieties of two
different molecules; the complementary receptacle indicates polar
moiety of the pairing ion; the arms represent hydrophobic
components of solute molecules and the reversed-phase sorbent.

The technique of ion-suppression RPLC is very useful in reducing the tailing of chromatographic peaks and in optimizing column retention and separation of structurally similar polar organic compounds containing ionizing moieties, as is the case for the biogenic amines and their acidic metabolites [31,32,37]. The k' of the acidic biogenic amine metabolites, which are weak acids in solution, varies markedly with changes in mobile phase pH below 6 [31,39]. For example, when citrate/acetate (pH 5.2) is used as the mobile phase [43] DOPAC elutes before NA from ODS columns and is poorly resolved. Using this same mobile phase buffer, but at pH 4.1, DOPAC elutes after NA, DA and 5-HT and is completely resolved from the latter amines [43]. The k' values of biogenic amines on ODS columns show significant change once a mobile phase pH of 5 is exceeded [31,44]. However, to our knowledge the application of ion-suppression RPLC to separate biogenic amines has not been described. Perhaps the reason for this lies in the facility of separating biogenic amines by ion-pair RPLC at mobile phase pH values below 5, values at which ion-suppression effects may be used to concomitantly optimize the separation of the acidic metabolites [45].

In the technique of ion-pair RPLC, the retention of biogenic amines or metabolites may be augmented through the inclusion in the mobile phase of an ion species (the counter or pairing ion) of opposite electrical charge to these analytes [46-48]. Coulombic association of the analyte ions with a compatible pairing ion results in the formation of electrically neutral ion-pairs. The enhanced hydrophobicity of these neutral complexes augments the interaction with the non-polar sorbent (Fig. 10.2) [46,49].

Two types of counter ions, inorganic and hydrophobic organic, have been employed to form ion-pairs with BACs. Separation of CAs on ODS columns has been achieved using inorganic counter ions such as nitrate, sulphate or phosphate [44]. However, with inorganic ions, peak symmetry and resolution are usually poor for CAs such as NA, α-methylnoradrenaline (α-MNA) and DHBA which posses small k' values (< 1) [44]. There may be several reasons for the above observations. Ion-pair formation between CAs and the inorganic pairing ion is poor in mobile phases with high dielectric constants, such as aqueous buffer [49]. The lipophilic nature of CAs is little enhanced by pairing with inorganic counter ions in comparison to counter ions with organic moieties. This is exemplified by the improved peak symmetry and separation of NA, α-MNA and DHBA when trichloroacetic acid is used as the counter ion in comparison to sulphate or phosphate [44].

Hydrophobic counter ions containing higher aliphatic moieties offer even better control over retention and selectivity for BACs [45,48,50]. The following general factors should be considered when using aliphatic hydrophobic counter ions such as alkylsulphonates or alkylsulphates. While retention of analytes increases with aliphatic carbon chain length of the counter ion [46,48], this parameter is also influenced by the hydrophobicity of the alkylsilane sorbent. For example, in a

mobile phase containing phosphate buffer (pH 3.0)/methanol (92:8, v/v), NMN is
better resolved from DA on octylsilane (C_8) columns when the pairing ion
octanesulphonate is substituted for heptanesulphonate. The tendency toward
lengthening the retention times of the analytes with octanesulphonate as compared to
heptanesulphonate may be offset by increasing column temperature from ambient ($22^{\circ}C$
to $35^{\circ}C$ and mobile phase flow rates from 1.8 to 3.0 ml/min [50]. In contrast, using
octylsulphate in a citrate/phosphate buffer at ambient column temperature and at a
flow rate of 1 ml/min, DA elutes in less than 12 minutes [47]. Furthermore, in
mobile phase systems employing organic modifiers, alkylsulphates with carbon numbers
longer than 22 are difficult to use because of their low solubility in the mobile
phase. For the same reason, aliphatic sulphates with carbon chain lengths longer
than 14 are difficult to use in neat aqueous mobile phase systems [46]. The values
of k' increase in direct relation to the concentration of alkylsulphonate or
alkylsulphate in the mobile phase [46]. Using heptanesulphonate, the k' values for
NA, A, DHBA and DA show no further change above a concentration of 5mM [48]. The
concentrations of alkylsulphonate employed to optimize the separation of selected
biogenic amines usually vary between 0.1 and 5 mM [45,47,48,50].

Another factor to consider in ion-pair RPLC is the effect of mobile phase pH;
this may influence the ionization of both analytes and counter ions [46]. At pH
values below 5, biogenic amines exist predominantly in ionized form conducive to
ion-pair formation with alkylsulphonates. However, the addition of strong inorganic
acids such as H_2SO_4 or H_3PO_4 to manipulate mobile phase pH can affect the
ion-pairing process, as high concentrations of the latter inorganic ions will
compete with the organic counter ions for ion-pair formation with analytes [46].
This problem should also be considered in relation to the use of high salt
concentrations for buffering in the mobile phase [46,51]. For example, the use of
$(NH_4)_2SO_4$ as a mobile phase buffer has been shown to decrease retention times of NMN
and DA in a system using dodecylsulphate as the counter ion [46]. This effect is
caused by competition between the $SO_4^=$ and organic pairing ions for ion-pair
formation with biogenic amines, a phenomenon similar to that occurring with the use
of high concentrations of strong inorganic acids in the mobile phase.

Finally, the presence of organic modifiers in the mobile phase may also affect
ion-pair RPLC. Organic modifiers such as methanol or acetonitrile decease the
retention of biogenic amine ion-pairs regardless of the counter ion [46]. This
effect is analogous to the decrease in k' for DOPAC and HVA which occurs with
increasing concentrations of methanol in the mobile phase [52]. Enhanced solvation
and decreased surface tension of the eluent likely account for the effect of mobile
phase organic modifiers on ion-pair retention.

As detailed above, the hydrophobicity of polar organic analytes is easily
enhanced through modification of mobile phase parameters, whereas the hydrophobic
nature of the non-polar stationary phase varies mainly with sorbent aliphatic chain

length, a parameter which is difficult to alter. Moreover, the performance of reversed-phase columns with sorbents of the same aliphatic chain length may also be influenced by the manner and uniformity of sorbent derivatization and the packing technique. For sorbents of the monometric bristle-type, retention, selectivity and maximum sample load increase with the length of the bristles from C_1 to C_{22} [30,53, 54]. Derivatization of silica with alkylsilane is never total and the residual underivatized silanol groups interact electrostatically with the polar moieties of the analytes. This contributes to peak asymmetry and modifies the stability of sorbents with respect to changes in mobile phase composition and conditions. The above effects are most apparent for columns with carbon chain length less than 6 [54].

It has been suggested that the greater the carbon content of the reversed-phase sorbent, the better the shielding of the unreacted silanol groups [54]. Some of the unreacted silanol groups can be removed (end-capped) by further treatment of the reversed-phase sorbents with trimethylsilanizing agents [28,54]. There is some suggestion that end-capped columns give improved performance and durability, but this remains to be demonstrated definitively. For alkylsilane derivatives, retention times of analytes increase with increased surface coverage of the silica [11]. Stability of sorbents is also enhanced since the unreacted silanol groups are better shielded.

It is generally held that column efficiency increases with smaller particle size of sorbents, but specific data on the effect of this parameter on BAC analysis is not available. It should be kept in mind, however, that the improvement in column efficiency with the smaller sorbent sizes (e.g. 5 μm or less) occurs at the expense of higher column back pressure and lower column flow rates [54]. The use of shorter columns to offset these latter effects and shorten retention times also reduces the gain in resolving power of these smaller particle sorbents. It is also argued by some [11] that sorbent particle size and shape have little effect on analyte retention in RPLC, but this view remains to be substantiated unequivocally.

In addition to the above sorbent factors, the manner in which columns are packed may also affect analyte separation. Columns packed with the same type of sorbent but obtained from different suppliers may show variations in selectivity. For example, we have found the separation between DHBA and A and between NMN and DA to be quite satisfactory on LiChrosorb (10 μm) RP-8 columns (250 x 4.6 mm i.d.) supplied by some manufacturers, but not on columns containing this same sorbent packed by other suppliers [22].

(c) <u>Merits and limitations of RPLC</u>. In comparison to the present state of the art of ion-exchange and normal phase HPLC, the advantages of RPLC for determination of BACs are substantial and the limitations few. This statement is clearly supported by the almost exclusive application of RPLC in the abundant reports of the last two years on HPLC assay of BACs. The superiority of RPLC is evident in three

principal areas: 1) resolution and separation, 2) flexibility and ease o
application and 3) stability. The three preferred modes of RPLC, ionizatio
control, ion suppression and ion-pair, used in conjunction with octadecyl o
octylsilane columns, permit separations of the majority of known BACs and in man
cases permit the simultaneous assay of 2,3 or occasionally more of these compounds
Moreover, the analytes separated may possess different ionic characteristics, i
contrast to what is observed in ion-exchange HPLC. This often permits the assay o
the parent biogenic amine along with its precursor and/or principal metabolites -
e.g. 5-hydroxytryptophan (5-HTP), 5-HT and 5-HIAA [55]; this is an importan
consideration for applications such as turnover studies. In addition, th
separation of biogenic amine analytes can be achieved with shorter elutio
times (< 20 min), thus increasing assay efficiency.

In RPLC, analyte retention and separation are easily optimized throug
modification of the mobile phase composition and conditions. Recent advances i
instrumentation, such as electronically controlled binary and ternary solvent mixing
and programming systems, allow rapid change of mobile phase composition and the us
of gradient elution systems without interrupting the chromatographic process. This
permits rapid and convenient development of methods when used in conjunction wit
suitable detectors. The general stability and rapid equilibration of reversed-phase
columns also facilitate the use of solvent programming techniques for development o
methods or to maintain consistent day-to-day separation of analytes. Finally, the
stability of reversed-phase columns permits large numbers (> 500) of samples to be
assayed over the life of a column [45]; suitable column rejuvenation techniques ca
extend column lifetimes even further [45].

The two principal limitations in use of RPLC for BAC assays are 1) the inability
to resolve and separate all of these substances from a complex biological matri
using either single or multiple modes of RPLC operation and 2) the restriction i
use of extremes of pH in aqueous mobile phases. The use of appropriate sample
pre-purification and compound-selective detection conveniently circumvents the
resolution problem. The latter limitation is only a problem in the separation of
polar analytes which have pK values > 8 and are not amenable to separation using the
ion-pair mode of RPLC. Current reversed-phase sorbents are subject to irreversible
deterioration in the presence of alkaline pH > 8.

10.2.2 HPLC Detectors for Biogenic Amine Analysis

The ultimate sensitivity and in part the specificity of HPLC procedures for BAC
determination depend upon the characteristics and mode of application of the
detection system employed. Various detector systems have been developed and applied
to analyte detection in HPLC. These include the optical systems such as ultraviolet
(UV) absorbence, fluorescence and refractive index detection, electrochemical
(redox) detectors and the vapour phase detectors, such as thermo-conductivity, flame

ionization, electron capture and the very recent mass spectrometric detector system. UV absorbence detectors have been of limited value for BAC determination, being used primarily for analyses of biological samples having high concentrations of these analytes, e.g. urine. The current modes of HPLC detection of greatest utility in BAC analysis are the fluorescence and electrochemical detectors. The use of vapour phase detector systems for BAC determination has not been described, probably because of inherent difficulties in effectively adapting these detectors to HPLC. However, a promising new vapour phase system, the liquid chromatography-mass spectrometer (LC-MS), may prove very valuable in BAC analysis. This device will be considered briefly in a following section. The current practical modes of quantitative detection of BACs following separation by HPLC are UV, fluorescence and electrochemical detection.

(i) _UV detector_. The most widely used detector system in liquid chromatography is UV detection. Most BACs absorb energy in the UV range of 200–300 nm. The absorption spectra of BACs in this UV range are broad and the absorption peaks show high extinction coefficients. However, in biological matrices other compounds containing conjugated double bonds, ketones, aldehydes, aromatic and heterocyclic rings, also have similar spectral characteristics. Unless these are removed, the contribution of absorption from such compounds severely limits the use of UV detection for the determination of BACs in biological samples. In addition to these substances certain solvents not containing the above groups (e.g.N,N-dimethyl-formamide) also absorb in the UV range and hence are unsuitable for use in the mobile phases. However, the mobile phases used in RPLC which contain aqueous buffers, aliphatic alcohols and acids are well suited for use in UV detection. At the higher end of the UV spectrum, above 340 nm, neither BACs nor most interfering compounds show any appreciable absorption of UV light. Since most BACs show absorption in the UV range of 200–300 nm, derivatization, either at pre- or post-column stages, is of little value to enhance selectivity and sensitivity.

In HPLC most analytes are eluted in very small volumes of the mobile phase. In order to avoid peak broadening or mixing of two or more closely eluted and well resolved analytes, modern detector cells have very small internal volumes (5–30 μl). The reduction in the cell volume results in decreased intensity of UV light reaching the photomultiplier tube. The cell construction is aimed at maximizing the transport of such low levels of energy to the photomultiplier tube while minimizing the loss of light due to scattering. Two types of UV detectors are commercially available. Fixed wavelength detectors use optical sources such as the zinc, low pressure- or high pressure-mercury lamps, with or without filters. These detectors are characteristically simple to operate, stable in response characteristics and offer more sensitivity than the scanning type of UV detector. However, fixed wavelength detectors lack the ability to scan the absorption spectrum for compound

identification. The multi-wavelength or the scanning instruments use xenon o
deuterium light sources which have broad emission spectra. Grating monochromator
are used in wavelength selection and scanning is performed in the stopped-flow mode
Recently, UV detectors have been introduced which perform a simultaneou
multi-wavelength scan in less than 1 sec. In these instruments xenon or deuteriu
light passes through a flow cell and its defraction spectrum is obtained with
grating monochromator. Multiple photodiodes positioned at 1-2 nm of wavelength pe
photodiode are employed to detect the total spectrum. During a HPLC run a compute
is required to collect and process the large amount of data emanating from th
detector array [56].

The manufacturers' specifications for high efficiency detectors cit
sensitivities for analyte-detection of 1-10 ng injected on column. However, lack o
selectivity of UV detection limits its use to biological samples having at least 10
ng of BACs [57]. Nevertheless, UV detectors are very useful for applications such a
standardizing analytical conditions because of their stability during solven
programming and temperature changes [58]. A limited application of the UV detecto
has been the use in HPLC assays of urinary VMA and HVA [57]. Post-column, on-lin
oxidation of VMA to vanillin and its subsequent detection at 340 nm has also bee
reported for the assay of urinary VMA [59]. For the reasons discussed above n
other biological components were detectable at this wavelength, thus enhancing th
selectivity of the method.

(ii) Fluorescence detector. Fluorescence detection employs the property o
certain analytes to emit higher wavelength (low energy) photons upon activation wit
low wavelength (high energy) photons in the UV-visible range. The ability of BAC
to fluoresce has been known for some time and, as noted (chapter 4), is the basis o
a variety of fluorescence detection assays for a number of these compounds. Thi
group of substances exhibit characteristic activation-emission spectra in th
UV-visible range (200-800 nm) which are related to molecular structure. Th
specificity and sensitivity of fluorescence detection depends not only on thes
activation-emission spectral characteristics, but are also influenced by detecto
characteristics such as the nature and intensity of the light source and th
transmission characteristics of the optical system used. The interaction of each o
these latter factors as they affect fluorescence detector sensitivity an
selectivity has been considered in depth by Bratin et al. [60]. As with U
detectors, fluorescence detectors are also available in two types, the fixe
wavelength and scanning varieties. In general, the filter fluorescence detector
have been more widely applied because of their low cost, operating stability
simplicity in use and high sensitivity. Scanning fluorescence detectors are able t
generate excitation and emission spectra through the use of grating monochromator
or graded interference filters. This capability accounts for greater selectivity o
the latter detectors as optimal activation and emission wavelengths can be employe

for compound detection. However, in order to acquire excitation and emission spectra in conventional scanning detectors, the scanning must be performed in a stopped-flow mode, a technique which is disruptive to routine analysis. Several new fluorescence detector designs aimed at permitting on-line acquisition of fluorescence spectra are under intensive investigation. With their fast scanning capabilities, the diode array detection system [61] or video-fluorometer [62] may fulfill these needs, but these new approaches will be much more costly as they require computers to process the complex signal outputs for quantitation.

Indoles and some CA compounds have good native fluorescence characteristics which permit highly sensitive HPLC determinations of these substances [63-66]. For example, detection limits of 5 pg for indoleacetic acid (IAA) have been reported by Anderson and Purdy [67]. The detection limits for other BACs such as CAs [66] and octopamine [68] were reported in the order of 5 ng using native fluorescence. For other compounds with either poor or no native fluorescence characteristics, pre- or post-column derivatization with fluorogenic compounds is required for sensitive detection. Several fluorogenic reagents have been utilized in this respect. These include o-phthalaldehyde (OPT) with 2-mercaptoethanol, fluorescamine and dansyl chloride for amines and amino acids, OPT with strong acid for indoles and post-column oxidation to trihydroxyindoles for NA and A.

Pre-column derivatization tends to reduce the chemical differences between analytes and may significantly affect the HPLC separation of these compounds. However, pre-column derivatization permits optimal yields of fluorescent products and separation of these reaction products from excess fluorogenic reagents. Post-column derivatization requires fast reaction rates compatible with the rapid analyte elution characteristics of the HPLC. In addition, the fluorogenic reagents must not themselves be fluorescent since this would result in a high signal background and low sensitivity. Finally, post-column derivatization requires intricate mixing, heating and pumping devices to control the delivery of reagents and product formation. Despite these problems, this technique can offer significant convenience in the automated analysis of column eluates.

Derivatization with OPT and 2-mercaptoethanol converts primary amines to fluorescent isoindole compounds [69]. This reaction is rapid but the fluorescent product is unstable in aqueous media [70]. The indole derivatives of CAs, indoleamines, histamine, putrescence and cadaverine are soluble in ethyl acetate and are stable at $4^{o}C$ in this solvent for several weeks [70,71], thus allowing these derivatives to be stored for some time prior to analysis. This reagent is suitable either for pre-column [70,71] or post-column [72,73] derivatization of the above amines.

Fluorescamine can also be used for pre- and post-column derivatization of primary amines and amino acids because of its fast reaction rates and lack of fluorescence. Despite these advantageous properties, this reagent has not been widely used for BAC analysis. In a few examples of the use of this reagent in HPLC, urinary NA, A [74] and TA [75] as well as brain CAs [76] have been analyzed by pre-colum derivatization.

Dansyl chloride also reacts with primary and secondary amines to produce highl fluorescent derivatives [77]. Being fluorescent itself, dansyl chloride can only b used in pre-column derivatization, and the fluorescent products of the analytes mus be separated from excess reagent prior to HPLC analysis. This reagent also react with other molecular groups such as imidazoles and phenols, but the reaction rate with these moieties are very slow [77]. Moreover, these sluggish reactions resul in formation of multiple derivatives and by-products which reduce sensitivity an possibly selectivity. For the above reasons dansylation procedures have bee employed infrequently for the assay of BACs by HPLC. One example of the use o dansylation for the analysis of plasma CAs is described by Frei et al. [78], wh identified some difficulties involved in the dansylation reaction.

It is well known that NA and A are oxidized and cyclized to tri-hydroxyindole using potassium ferricyanide and strong alkali [79]. As trihydroxyindoles ar strongly fluorescent compounds, this transformation has been used to assay these tw CAs by an automated post-column reaction following HPLC separation [80]. However this technique is not applicable to the determination of DA as this CA does no form a fluorescent trihydroxyindole derivative. Using trihydroxyindole derivatives sensitivity limits in the order of 10-20 pg/sample have been obtained in the assa of plasma NA and A [81].

Post-column reactions of NA and A with alkaline (pH 9.7) borate buffer [82] o with ethylene diamine [83] also result in formation of fluorescent products but th structure of these has not been identified. The products obtained were no particulary sensitive to HPLC detection; thus the assay procedure was restricted t the analysis of NA and A in urine only.

In summary, fluorescence detection is a very sensitive technique and permits th detection of BACs in low picomole or in some cases femtomole ranges. However fluorescence procedures are molecular species dependent, and the selectivity o detection depends upon the fluorescence characteristics of the analytes or thei derivatives, the manner in which the detector is used, as well as the separatin characteristics of the HPLC system employed. In the majority of cases the HPL separation is the principal process contributing to the assay specificity as th detectors have been used in relatively non-selective, high sensitivity modes o operation. For those procedures which employ formation of fluorescent derivative of BACs, the HPLC separation is the only basis for assay specificity. Finally ther are a variety of CA metabolites, both acidic and neutral, which cannot b quantitated by these methods because they possess poor or no native fluorescence o they do not form fluorescent derivatives with known reagents. For the latte compounds alternative methods such as electrochemical detection must be used.

Chemiluminescence, a phenomenon related to fluorescence, is a new and relatively unexplored mode of luminescence detection which appears to have potential as a very sensitive method of BAC detection in HPLC. In this technique, chemical reaction between two or more substances provides the necessary energy to excite a fluorophore which returns to the ground state by releasing this energy, thus producing a characteristic emission spectrum. For example, solutions of dansyl derivatives of amino acids or BACs react with 1,2-dioxetane-dione to produce molecular chemiluminescence with maximum emission at 495 nm [84]. The advantage of chemiluminescence over photoluminescence (e.g. fluorescence) is that the former procedure does not require an external energy source. Baseline noise related to stray light from the activation source is absent and hence chemiluminescence may permit higher sensitivities (in order of one magnitude) than current fluorescence techniques. However, the emitted light induced in this manner has a very short half-life, thus requiring fast rates of reaction with the chemiluminescent reagents. For this reason chemiluminescent reagents must be applied in the post-column mode and the delay between the chemical induction of chemiluminescence and detection maintained within the time constants of the decay phenomenon. Dansyl derivatives of urinary CAs have been separated on HPLC columns and detected using this process [85]. The detection limits in this latter procedure were found to be 5–10 pg, comparable to the sensitivities of the trihydroxyindole method of CA assay. The method, however, would likely suffer from the same deficiencies as those mentioned for fluorimetric detection of dansyl derivatives. Other chemical or biochemical luminescence reagents such as luminol or firefly luciferase have potential as very sensitive reacting media in chemiluminescence detection, but these remain to be explored for BAC analysis.

(iii) Electrochemical detector. Compounds containing catecholic, phenolic, indolic and aldehyde groups possess low oxidation potentials (< 1.5 V) and are amenable to electrochemical detection in the oxidation mode in totally or partly aqueous media. As well, compounds containing nitro or keto groups possess low reduction potentials and are also amenable to electrochemical detection but in the reductive mode. However, this latter mode of electrochemical detection has no application in the quantitative assay of BAC which are not reducible in their native states. For this reason, the present discussion is restricted to the oxidative detection of BACs only.

When a solution of an oxidizable compound is placed in an electrical field generated by the application of a suitable voltage between two electrodes, oxidation occurs demonstrated schematically as follows:

$$\text{(catechol with HO, HO, R)} \xrightarrow{+0.6\ V} \text{(quinone with O, O, R)} + 2H^+ + 2\bar{e}$$

The abstraction of two electrons from the catechol compound produces a current
the flow of which is facilitated in solvents having high dielectric constant.
Similarly, solutions of phenols may be oxidized to benzoquinones, thereby generating
an electron current in an appropriate electric field (0.6-0.8 V). The current may
be amplified for detection by potentiometric measurements.

Although the concept of detection of electroactive substances in
chromatographic column-effluent was described as early as 1952 [86], it was not
until 1973 that Kissinger and his co-workers [87] successfully applied this
technique to the assay of CAs in biological samples. In the system used by these
latter investigators the HPLC effluent flowed through the detector in a thin layer
over an anode or working electrode comprised of graphite and mineral oil paste. A
auxiliary (cathode) and a reference electrode were located further downstream (Fig.
10.3). In such a detector the oxidation which occurs at the working electrode
surface results in the generation of a current, the intensity of which depends on
the potential applied to this electrode. The relation between the current and the
applied voltage of the working electrode is shown by a hydrodynamic voltammogram in
Fig. 10.4. Here E_{min} and E_{max} are voltages required for minimum and maximum
oxidation respectively and the half wave potential (E/2) for a substance is the
voltage at which half maximal oxidation occurs. Easily oxidizable substances are
characterized by low half-wave potentials. However, the oxidation reaction also
depends on several other factors such as pH, ionic strength and flow rate of the
mobile phase. High media-pH values facilitate the dissociation and oxidation of
catechols and phenols, resulting in lowering of half-wave potentials for these
compounds. In non-polar and acidic media the opposite phenomenon is observed, that
is, dissociation and oxidation are impaired and there is an increase in the
half-wave potentials. Higher ionic strength of the medium reduces the current
generated due to decreased diffusion of the oxidizing material towards the
working electrode. Similarly, higher flow rates of the column effluent

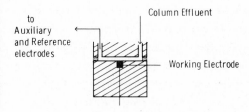

Fig. 10.3   Schematic illustration of an
            electrochemical cell.

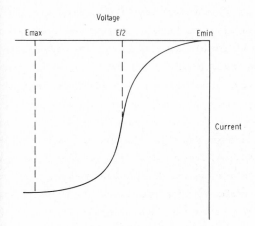

Fig. 10.4   Hydrodynamic voltammogram
            of an oxidizable compound.

reduce the diffusion and oxidation of analytes at the working electrode. Moyer and Jiang [48] have outlined the optimization of the above factors for liquid chromatography-electrochemical detection (LC-EC) assays of CAs. For other mobile phase systems and other BACs, similar consideration should be given to optimizing the above factors during development of the analytical method.

The intensity of the background current produced in this amperometric mode depends on the conductivity and hence the composition of the mobile phase as well as the potential applied to the working electrode. A change in the mobile phase composition or the working electrode potential results in a new level of background current (or baseline) and noise. In practice the baseline requires 5-10 min to settle if the mobile phase is changed and up to 30-60 min if the working electrode potential is altered. These changes, as they occur during solvent programming or voltammogram scanning, will interfere with the detection and quantitation of analytes.

As amperometry is a surface-dependent process, factors such as the nature, surface area and configuration of the working electrode have received considerable attention, especially as they relate to detector sensitivity and stability. Inertness of the working electrode toward oxidation and/or formation of salt films are important considerations in the choice of electrode material, as these processes shorten the electrode life. Various materials such as dropping mercury, gold, platinum and carbon have been used but only the last of these has been applied widely because of its resistance to these processes. Carbon has been employed as a powdered graphite paste, fused graphite discs and in mixtures with polymers such as polyethylene or teflon. Powdered graphite paste requires the use of a binder such as mineral oil, silicone-grease or wax. The need for the above binder to hold the graphite powder allows this working electrode material to be used in mobile phases

containing no more than 10% organic solvent. At high solvent concentrations th
binder is substantially leached, leading to cracking and rapid deterioration c
detector performance. However, the carbon paste electrode is easy to pack ar
polish, gives a low and relatively stable background current, permits goo
sensitivity and has a relatively long life in aqueous mobile phases. Electrode
made of solid graphite such as glassy carbon discs, isotropic carbon, basal plane c
pressure-annealed pyrolytic graphite as well as the polymer-mixed graphite are al
resistant to mobile phases containing organic solvents. The most frequently use
material has been glassy carbon, which is inexpensive and easy to maintain.

In commercial electrochemical detectors the surface area of the working electroc
is generally 2-4 sq mm and its oxidation capability is 1-10% under optima
conditions. Increasing the surface area by using a laminar, tubular or multipl
fibre working electrode increases analyte oxidation and hence the detectio
response. However, such electrodes also give increased background current and nois
and thus have little net effect on the ultimate sensitivity of the detector.

Various configurations of the positioning of the working electrodes in detecto
cells have been evaluated in attempts to increase detector sensitivity and electrod
life. Positioning the auxiliary electrode in close proximity to the workin
electrode would reduce the ohmic resistance across the two electrodes and thu
prevent a possible fall in the anodic voltage caused by the electron flux resultin
from high concentrations of oxidizable analytes. The linear dynamic range for BA
analysis was shown to be increased using such a configuration [88]. In anothe
configuration, the wall jet design, the column effluent enters the cel
perpendicular to the working electrode before turning $90°$ to flow parallel to th
latter [89]. This design increases the diffusion of the analytes by turbulence an
yields a slightly better sensitivity than the widely used laminar flow detecto
described by Kissinger et al. [87]. Finally, the rotating disc electrode concep
was recently introduced in electrochemical detection [90]. Here, the colum
effluent flow enters the cell in a manner similar to that in the wall jet design
but the working electrode is mechanically rotated to obtain further increase i
diffusion. This design also gives some slight improvement in sensitivity.

In spite of the various cell designs and working electrode materials, the norma
operating detection limit for most commercial detectors has been 1-10 pg injected c
column. One of the main reasons for this is that the detector functions as a
integral part of the LC-EC system and as such the sensitivity is also dependent c
the analyte elution volumes and peak widths, factors which are in turn dependen
on the HPLC conditions. Electrochemical detectors are more selective at lowe
operating potentials of the working electrode. At the potential of 0.5-0.7
catechols such as NA, A, DA, DOPA and DOPAC and hydroxyindoles such as 5-HT, 5-HIA
and 5-HTP are easily detectable [45], whereas at 0.8 V or higher the phenolic (e.g
MOPEG, TYR, TA, p-hydroxyphenylacetic acid) and indole (e.g. tryptophan) compound

are also detectable [91,92]. Hence at higher voltages most BACs are detectable and the detector is least selective. However, when used in conjunction with ion-exchange and reversed-phase chromatography, which effectively separate and resolve BACs, the electrochemical detector provides an excellent and versatile tool for quantitating phenolic and indole BACs. The use of electrochemical detection for BAC analysis has already been detailed in recent reviews [60,93] and only selected examples will be considered in a later section to demonstrate its capabilities and limitations in the analysis of biological samples.

Current commercial electrochemical detectors have significant advantages compared to sensitive fluorescence detectors in terms of simplicity of operation, lower cost of instrumentation and a wider applicability to BAC analysis without the need for derivatization. Several new approaches to enhance the application of electrochemical detectors include the use of serial working electrodes held at different potentials, pulsed voltammetry and the formation of electroactive derivatives of non- or poorly-oxidizable compounds. Selective monitoring of two or more closely eluting analytes with different oxidation potentials may be performed by using two working electrodes or two detectors in tandem [94]. Each working electrode is held at a different potential. The differential chromatogram obtained from the responses of the two electrodes signifies the elution characteristics of the substance with higher oxidation potential while the chromatogram obtained from the working electrode with lower potential signifies the elution characteristics of the analyte with lower oxidation potential. Similar results are obtainable by applying two or more potentials to the same working electrode in a pulsed form [95] and recording the fast output from the electrode. However, such data handling requires the use of a computer.

Some electroactive derivatives under investigation include the isoindole derivatives of primary amines formed by reaction with OPT and mercaptoethanol [96] and those of organic acids formed by reaction with N-hydroxysuccinimide [97]. At present there is insufficient data available to indicate the specific advantages and limitations of these derivatives.

(iv) Vapour phase HPLC detector. Flame ionization, electron capture, thermo-conductivity and mass spectrometric detectors require the sample analyte to be delivered to the detector in a gaseous (vapour) phase. A basic difficulty in interfacing HPLC to these detectors is the need for removal of the relatively large amounts of non-volatile mobile phase prior to introduction of the sample into the detector. A variety of LC-MS interfaces are available in commercial instruments [98]. For example, the moving belt design is at present the most appropriate interface for introducing the polar, aqueous and buffered eluants used in BAC analysis. Although the mass spectrometric detector offers high specificity and sensitivity, the detection limits of LC-MS at the present time are only in order of 10 ng because of inefficient transfer of analyte from the moving belt into the ion source. Furthermore, there are no reported data on the determination of BACs by this technique to assess. Finally, the cost of LC-MS is about ten times that of HPLC with any of the conventional detectors; this would likely be a major factor deterring its routine use.

10.3 DETERMINATION OF BIOGENIC AMINES IN BIOLOGICAL SAMPLES
10.3.1 Choice of Sample Preparation

There are now numerous demonstrations of the utility of HPLC in combination with fluorescence or electrochemical detection for determination of BACs in mammalian tissues, body fluids and tissue perfusates. In applying HPLC to measure these compounds certain common trends in methodology are evident. Firstly, reversed-phase separating systems have become the chosen chromatographic technique in the field, primarily because of the superior resolution and separation, as discussed above. Secondly, for readily oxidizable BACs electrochemical detection has become the preferred mode of detection because of its high sensitivity as well as simplicity and reliability. Thirdly, for certain biological matrices assays of selective BACs may be performed with minimal sample pre-purification.

The high degree of resolution and separation of BACs by HPLC have been the major factors leading a number of investigators to use HPLC without substantial pre-purification of these analytes from biological matrices. This approach has been applied successfully to assays of a variety of BACs in crude supernatants derived from brain homogenates [67,99,100,101]. The advantages of reducing sample manipulation lie primarily in minimizing analyte loss and assay variability that would otherwise be incurred by more extensive processing. This can lead to improved assay sensitivity and reproducibility [41] and reduces the importance of internal standardization in direct-injection assays. In reports providing data with respect to coefficients of variation (C.V.) and recoveries for direct-injection assays of BACs, these have been less than 3% and greater than 98%, respectively [67,102]. However, one must bear in mind that such C.V. values vary with analyte concentration and are not substantially better than those obtained in some procedures using more extensive sample processing along with suitable internal standards [21].

Factors influencing the successful application of direct-injection assays include analyte concentration, the nature of the biological matrix, the mode of analyte detection and the ability of a particular chromatographic system to adequately separate the analytes of interest. For example, while CAs can be readily assayed in brain supernatants by direct injection [99], the determination of low CA concentrations in human plasma (<400 pg/ml) requires prior isolation, purification and concentration procedures [103,104]. Simply to increase the sample volume is not a solution to determining low analyte concentrations by direct-injection techniques, as this will only be accompanied by increased background noise

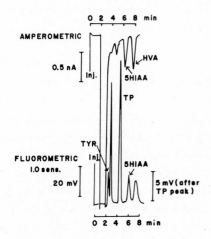

Fig. 10.5 Simultaneous determination of selected BACs by the in-line fluorescence and electrochemical detectors (Redrawn and reproduced with permission from Anderson et al. [41]).

without significant improvement in the signal to noise ratio for the analyte. In another example, the complex chemical composition of urine creates substantial difficulties in direct-injection assay of CA compounds [41]. However, this is not always the case. Using a reversed-phase system with tributylphosphate as the stationary liquid phase, Wahlund and Edlen [105] were able to assay 5-HIAA and IAA in urine by direct injection.

Several groups have demonstrated clearly the utility of LC-EC in direct-injection assays of BACs in a number of biological matrices. Indoles such as 5-HTP, 5-HT or 5-HIAA have been determined in brain supernatants using this approach in conjunction with electrochemical detection or fluorescence detection [33,67,100,102]. Catechol compounds such as DA, DOPA and HVA in brain supernatants [106], as well as HVA, MOPEG and MOPET in CSF [41,101] have also been assayed by LC-EC using direct injection. However, the direct-injection assay of NA, and sometimes DA, simultaneously with indoles such as 5-HTP and 5-HT cannot be undertaken with current reversed-phase systems and electrochemical detection since NA and DA elute rapidly and merge into the solvent front with other unretained compounds using these separating systems. This deficiency may be overcome in some situations by the use of fluorescence detection which often does not show the same broad solvent front [41,99] (Fig. 10.5). Despite the encouraging results of the direct-injection approach to assay BACs, more than 90% of the recently published HPLC methods for these substances involve at least some degree of prior sample purification. This supports the value of more extensive sample pre-purification in HPLC assays of BACs. However, the additional sample manipulation required in isolation and pre-purification will reduce sample yields in many cases. This may be associated with a decrease in absolute sensitivity of the assays and increased inter-sample variability, as a result of uncontrolled processing losses. Therefore, the use of chemically and structurally similar internal standards becomes more important in the latter approach to alleviate methodological deficiences arising out of prior clean-up of the sample. On the other hand, sample purification prior to HPLC enhances resolution and specificity of HPLC assays of BACs by removing unknown interfering compounds. Furthermore, HPLC column life can be extended due to removal

of undesirable particulate material from the samples. In some instances, sampl
clean-up procedures may actually enhance overall assay sensitivity by concentratin
the compounds of interest before separation by HPLC and detection.

10.3.2 Approaches to Sample Pre-treatment

(i) Sample deproteinization. Brief centrifugation is required to remove debri
in the less complex biological matrices such as CSF, primarily to prolong colum
life [41,67,101]. In LC-EC assays of BACs, anti-oxidants such as ascorbic aci
should also be excluded in sample collection because of the electrochemicall
reactive nature of these compounds [42]. Deproteinization is essential for th
assay of complex biological matrices such as brain tissue. Reagents such a
$ZnSO_4/NaOH$ [102], $ZnSO_4/Ba(OH)_2$ [41] and perchloric acid [106] have been employed fo
adequate precipitation of brain proteins and are also suitable for LC-EC or liqui
chromatography-fluorescence detection (LC-F) determination of BACs in brai
supernatants by direct-injection assay.

(ii) Sample pre-purification and concentrating procedures. The three mos
commonly used sample clean-up procedures involve organic extraction, selectiv
adsorption of catechol compounds on alumina, or adsorption on conventiona
ion-exchange sorbents. Acidic (e.g. 5-HIAA, DOPAC, HVA and VMA) and neutra
(MOPEG) metabolites of biogenic amines readily partition from acidified aqueou
media (pH < 3) into organic solvents such as ethyl acetate and ether [107-111]. Th
organic phase containing these metabolites may be subsequently evaporated to drynes
under a stream of nitrogen and the sample residues reconstituted in the mobile phas
buffer for HPLC assay. Organic extraction procedures are less preferred fo
isolation and purification of CAs and 5-HT because of their low extractio
efficiency (e.g. low tissue recoveries [20]) and lability of these compounds in th
basic media required for optimal extraction. Isolation of catechol compounds suc
as NA, A, DA and DOPAC has been achieved by selective adsorption of these compound
on alumina at neutral pH, followed by their elution in acidic eluents [34,35,43,47]
This procedure is suitable for assays of BACs with an intact catechol moiety, bu
not for indole compounds and methoxy metabolites of CAs which are not adsorbed o
alumina.

In RPLC systems which allow simultaneous assay of CAs and indoleamines
purification of these compounds may be achieved by adsorption on smal
cation-exchange columns such as Amberlite CG-50 [45,112], Bio-Rex 70 [113] and Dowe
50W [114]. An advantage of using cation-exchange columns for sampl
pre-purification is that concurrent isolation of secondary biogenic amines, such a
A and methoxy-phenylethylamines such as NMN, MN, 3-methoxytyramine (3-MTA), is als
possible [50,114,115]. Because of the carboxylic moiety, acidic metabolites o
biogenic amines may be purified by adsorption on anion-exchange columns such as AG
(BioRad) [116]. Furthermore, by differential elution on small sephadex G-1

columns, NA, DA and 5-HT and some of their acidic metabolites (DOPAC, HVA and 5-HIAA) have been conveniently purified from tissue samples [117,118].

(iii) <u>On-line sample enrichment</u>. This technique is derived from the combined use of a small, gravity-fed, ligand-solid extraction column with an analytical reversed-phase column to achieve serial sample purification and HPLC assay of BACs [119]. In this approach, brain supernatants were passed through the extraction columns (Amberlite CG-50 for 5-HT; Sephadex G-10 for 5-HIAA) and the unretained compounds discarded. Through timed switches the pre-column carrying the indole compounds was connected to the analytical column and the retained compounds were eluted into the RPLC mobile phase. The 5-HT and 5-HIAA were subsequently quantitated electrochemically and, at the same time, the pre-column was ready for the next sample isolation. The utility of on-line enrichment techniques requires the precise control of timed events in order that the desorption of analyte from the pre-column coincides directly with the loading phase of the HPLC injector. Any variation in the pre-column elution profile of the analyte may result in decreased yield of solute into the HPLC. Despite these potential technical problems, on-line enrichment offers the advantages of improved assay specificity, as with techniques employing sample pre-purification, but with higher yields, sensitivity and reduced variability, characteristic of the direct-injection techniques.

10.3.3 Accuracy, Reliability and Sensitivity of HPLC Assays

The accuracy of any quantitative procedure depends upon the specificity of determination, the precision of the estimation and the ability to estimate the absolute recovery in each determination. Specificity of determination can be achieved through the isolation technique, compound specific detection or a combination of these, whereas precision of estimation (reproducibility) requires control of variability between the determinations. Finally, estimation and correction for recovery losses necessitates the use of appropriate internal standards in each determination. For the assay of BACs in biological matrices these conditions have only been achieved using GC-MS assays [21].

As discussed in section 10.2 most of the commonly used detectors employed in HPLC assays provide only partial selectivity for positive identification of BACs. For this reason, the accurate determination of these compounds in biological samples is critically dependent on adequacy of separation and resolution achieved by the HPLC system. Various methods have been applied to validate the selectivity of HPLC separation of specific compounds in biological samples. These include identification of compound peaks indirectly by: 1) comparing the retention times of the putative substances with authentic standards or relative to an internal standard, preferably using two or more HPLC separating systems, 2) demonstrating an increase in the unknown peak intensity when co-chromatographed with authentic standards, 3) obtaining the expected variation in unknown peak intensities following predictable manipulations of the in vivo sample concentration [102,106,120] and 4 using multiple modes of detection in-line, each monitoring a differen physico-chemical property [42,110].

In most cases identification of BACs has been performed expediently by compariso of the elution profile of unknown peaks with that of authentic standards in single system rather than using direct chemical verification of the compound assayed. Moreover, in perusing the recent HPLC literature on BAC assays, ful elution spectra of chemically and structually similar BACs are not always reported Thus interfering peaks originating from related BACs or unknown compounds may easil affect the specificity and quantitation of the analytes of interest. This situatio is more prone to occur in direct-injection assays in which the compounds to b assayed are not isolated in advance from possible interfering compounds. In thes assays the specificity of BAC determination is more likely to be affected whe electrochemical detection is used rather than fluorescence detection, as there is greater abundance of interfering substances which are easily oxidized in compariso to their ability to fluoresce (see Fig. 10.5) [41,106]. In contrast, sampl pre-purification improves assay specificity by removing some or most of th endogenous interfering substances in biological samples, but the use of appropriat internal standards is crucial for accurate quantitation for reasons discussed Typically a chemically and structually similar compound is used as an interna standard to control for procedural losses for the various analytes in HPLC, bu accurate correction for recovery losses depends on the chemical similarities betwee internal standard and analyte [21,45]. Stable and radioisotopic congeners whic have chemical and physical characteristics identical to the analytes can only b employed as internal standards when appropriate detectors are used capable o discriminating the tracer internal standard and unlabelled analyte. For example, deuterium labelled internal standards may be used in LC-MS.

Perusal of the published methods for determination of BACs by LC-EC or LC-F assays indicates that recoveries of authentic standards added to biological sample vary substantially, from 23 to 100%. In procedures involving liquid-liquic partitioning for purification of biogenic amines (e.g. NA, DA and 5-HT) or their respective acidic and neutral metabolites (e.g. DOPAC, HVA, 5-HIAA and MOPEG) prior to HPLC determination, recoveries varied from 23 to 50% for the former compounds [20,92,121] and from 31 to 98% for the latter compounds [110,111,122]. Similarly, in methods employing conventional adsorption (e.g. alumina) or ion-exchange (Amberlite CG-50, Sephadex G-10 or Dowex AG 1) chromatography to isolate certair BACs, recoveries ranged from 40 to 85% [15,34,35,38,40,43,45,118]. The highest recovery values (> 98%) have been obtained by direct-injection assays in whick sample manipulation was minimal [41,102]. In procedures with recoveries for BACs of less than 80%, correction for procedural losses is advisable to prevent underestimation of true analyte concentrations. In the majority of reported HPLC procedures for BACs, this has been accomplished by the use of internal standards. For HPLC assays involving simultaneous determination of multiple BACs, accuracy of estimation of each analyte may be optimised if the individual compounds are referenced against separate internal standards with chemical characteristics very similar to the individual analytes. However, this mode of internal standardization is not always possible since the detectors used in HPLC assay are only partially selective and because multiple internal standards may affect the separation of the analytes by overriding the latter peaks. Thus, in most recent LC-EC and LC-F assays, single internal standards have been employed for concurrent determination of several BACs. In this case, careful selection of an appropriate internal standard is critical to the accuracy of the assay and the validity of the data.

The best way of demonstrating the importance of an appropriate internal standard to the accuracy of HPLC procedures has been through direct comparison to GC-MS methods. The latter are almost ideal reference procedures because they employ compound specific detection and physically identical (mass labelled) internal standards for each analyte. Such direct comparison of HPLC with GC-MS has been reported for determination of DA, 5-HT, and selected CA metabolites, HVA and MOPEG [21,122]. For example, employing epinine as the internal standard in the ion-exchange LC-EC assay of rat hypothalamic DA and 5-HT, recoveries (> 95%) and C.V. values (< 2%) for these two BACs were not statistically different from those obtained by GC-MS methods [45]. Despite the slightly, but statistically significant, higher concentrations of DA determined by LC-EC compared to GC-MS, epinine appeared to be suitable as an internal standard in this ion-exchange mode of HPLC assay [21]. However, in a reversed-phase system, epinine was found to be less suitable than DHBA as an internal standard for the assay of selected BACs [45]. When epinine was used, the C.V. values for NA, DA and 5-HT were 27%, 25% and 13%, respectively. In contrast, the C.V. values obtained when DHBA was employed were 3%, 17% and 10% for NA, DA and 5-HT, respectively. Such findings clearly show the importance of the selection of an appropriate internal standard to the accuracy and reproducibility of the assay. They also indicate that internal standards having chemical characteristics closely similar to certain BACs in one HPLC system may not retain the same chemical similarities to the analytes in another HPLC system. Thus the internal standard may not prove useful in the latter instance.

Concerning the variability of HPLC procedures for BACs, reported C.V. values range from 2 to 10%, with majority of procedures yielding values falling within the 5% level [19,21,34,35,41,42,45,50,106,110,118]. This degree of variability compares favourably with C.V. values for GC-MS assay of these same substances [5-7].

In HPLC methods employing an electrochemical or an improved fluorescence detector, assay sensitivies ranging from 5 to 200 pg are common [21,45,50,67]. This sensitivity is also quite comparable to that of GC-MS assay [5-7]. Although radioenzymatic assays provide slightly better sensitivity (e.g. 1 to 50 pg CA)

[123-125] than LC-EC assay, these enzyme methods usually have lower specificity because of cross-contamination with chemically similar BACs. At present, LC-EC and LC-F methods offer assay sensitivities adequate for determination of a majority of BACs in most applications. The sensitivity of these assays is likely to be further increased in the future to match that of radioenzymatic assay for measurement of at least some BACs. This will be achieved through optimizing the detector designs and materials as well as improvements in column efficiency.

10.3.4   HPLC for Enzyme Assays and Turnover Studies

The specificity and high sensitivity of HPLC assays of BACs offer a means to study the activities and kinetics of enzymes involved in the biosynthesis of biogenic amines. For example, there have been reports on the assay of brain TYR hydroxylase [126,127], brain and serum aromatic L-amino acid decarboxylase [128-130] and liver and red blood cell catechol-0-methyl transferase [131]. Although these HPLC enzyme assays are time consuming, their specificity and sensitivity provide an excellent alternative to the radioenzymatic assays.

HPLC techniques have also provided a major contribution to studies of the turnover and biosynthesis of biogenic amines [26,132,133]. With the development of low dead-volume non-destructive detectors used in combination with micro-fraction collection of the column eluates, it is possible to undertake rapid determination of BAC specific activities without using multiple assays for the precursors and products involved. This latter approach has recently been demonstrated as a potentially valuable technique to study CA biosynthesis and turnover in small brain areas [134].

10.4   SUMMARY

High performance liquid chromatography is now clearly a major tool in the methodological armamentarium for BAC analysis. Current developments in chromatographic and detector technology, such as the introduction of RPLC and electrochemical detection, played a pivotal role in raising this methodology to its present stature. Present LC-EC and LC-F procedures for BAC determination provide sensitivity, precision and accuracy which is comparable to that demonstrable previously only with GC-MS techniques. Moreover, this has been achieved with considerably greater flexibility and simplicity and much lower cost than technologies such as GC-MS. In view of these attributes it is not unexpected that in a very short time span HPLC methods have supplanted other approaches to BAC measurement as the preferred measurement tool. Current HPLC assay sensitivities have reached levels of 1 pmol or less for some BACs. With newer developments in HPLC detector technology one can expect to see even further gains in sensitivity, possibly in the order of one magnitude, and improved selectivity of detection. HPLC is a tool which lends itself easily to automation as is already quite evident in the degree of computerization of many commercial instruments. This again will lead to further gains in flexibility and utility of HPLC procedures. However, as with any tool, HPLC should not be viewed as a panacea of measurement. There are definite limitations in certain applications of HPLC, the recognition of which will allow the investigator to realize the full power and utility of this tool for BAC measurement.

ACKNOWLEDGEMENTS

We would like to thank Mrs Susan McNally for assistance in the preparation of this manuscript.

REFERENCES

1   J.A.R. Mead and K.F. Finger, Biochem. Pharmacol., 6 (1961) 52-53.
2   G. Brownlee and T.L.B. Spriggs, J. Pharm. Pharmacol., 17 (1965) 428-433.
3   R.M. Fleming, W.G. Clark, E.D. Fenster and J.C. Towne, Anal. Chem., 6 (1965) 692-696.
4   T. Kariya and M.H. Aprison, Anal. Biochem., 31 (1969) 102-113.
5   S.H. Koslow, F. Cattabini and E. Costa, Science, 176 (1972) 177-180.
6   S.H. Koslow, G. Racagni and E. Costa, Neuropharmacol., 13 (1974) 1123-1130.
7   F. Karoum, J.C. Gillin and R.J. Wyatt, J. Neurochem., 25 (1975) 653-658.
8   L.S. Snyder and J.J. Kirland, in introduction to Modern Liquid Chromatography, Wiley-Interscience Publication, 1974, 273 pp.
9   P.T. Kissinger, G.S. Bruntlett, K. Bratin and J. Rice, National Bureau of Standards Special Publication 519, 1979, p. 705-712.
10   Y. Arakawa, K. Imai and Z. Tamura, J. Chromatogr., 162 (1979) 311-318.
11   R.E. Majors, J. Chromatogr. Sci., 18 (1980) 488-511.
12   L.R. Snyder and J.J. Kirland, Introduction to Modern Liquid Chromatography, Wiley-Interscience Publication, 1974, 286 pp.
13   P.T. Kissinger, C.S. Bruntlett and R.E. Shoup, Life Sci., 28 (1981) 455-465.
14   R. Keller, A. Oke, I. Mefford and R.N. Adams, Life Sci., 19 (1976) 995-1004.
15   P.M. Plotsky, D.M. Gibbs and J.D. Neill, Endocrinology, 102 (1978) 1887-1894.
16   H. Hallman, L.-O. Farnebo, B. Hamberger and G. Jonsson, Life Sci., 23 (1978) 1049-1052.
17   M.J. Cooper, R.F. O'Dea and B.L. Mirkin, J. Chromatogr., 162 (1979)
18   P. Hjemdahl, M. Daleskog and T. Kahan, Life Sci., 25 (1979) 131-138.
19   S. Allenmark and L. Hedman, J. Liq. Chromatogr., 2 (1979) 277-286.
20   S. Sasa and C.L. Blank, Anal. Chem., 49 (1977) 354-359.
21   J.J. Warsh, A. Chiu, P.P Li and D.D. Godse, J. Chromatogr., 183 (1980) 483-486.
22   J.J. Warsh, A. Chiu, D.D. Godse, unpublished observations.
23   L.J. Felice and P.T. Kissinger, Anal. Chem., 48 (1976) 794-796.
24   L.J. Felice, C.S. Bruntlett and P.T. Kissinger, J. Chromatogr., 143 (1977) 407-410.
25   R.M. Wightman, P.M. Plotsky, E. Strope, R. Delcore and R.N. Adams, Brain Res., 131 (1977) 345-349.
26   L.M. Neckers and J.L. Meek, Life Sci., 19 (1976) 1579-1584.
27   G.A. Howard and A.J.P. Martin, Biochem. J., 46 (1950) 532-538.
28   H. Colin and G. Guiochon, J. Chromatogr., 141 (1977) 289-312.
29   S.R. Bakalyar, R. McIlwrick, E. Roggendorf, J. Chromatogr., 142 (1977) 353-365.
30   G.E. Berendsen and L. De Galan, J. Chromatogr., 196 (1980) 21-37.
31   I. Molnar and C. Horvath, Clin. Chem., 22 (1976) 1497-1502.

234

32 I. Molnar and C. Horvath, J. Chromatogr., 145 (1978) 371-381.

33 I.N. Mefford and J.D. Barchas, J. Chromatogr., 181 (1980) 187-193.

34 Y. Maruyama, R. Nakamura and K. Kobayashi, Psychopharmacology, 67 (1980)187-193.

35 I.N. Mefford, M.M. Ward, L. Miles, B. Taylor, M.A. Chesney, D.L. Keegan and J.D. Barchas, Life Sci., 28 (1981) 477-483.

36 J.P. Crombeen, J.C. Kraak and H. Poppe, J. Chromatogr., 167 (1978) 219-230.

37 P.R. Brown and A.M. Krstulovic, Anal. Biochem., 99 (1979) 1-21.

38 C.R. Freed and P.A. Asmus, J. Neurochem., 32 (1979) 163-168.

39 C. Horvath, W. Melander and I. Molnar, Anal. Chem., 49 (1977) 142-154.

40 T.P. Davis, C.W. Gehrke, C.W. Gehrke, Jr., T.D. Cunningham, K.C. Kuo, K.O Gerhardt and H.D. Johnson, J. Chromatogr., 162 (1979) 293-310.

41 G.M. Anderson, J.G. Young and D.J. Cohen, J. Chromatogr., 164 (1979) 501-505.

42 G.M. Anderson and J.G. Young, Life Sci., 28 (1981) 507-517.

43 I. Mefford, M. Gilberg and J.D. Barchas, Anal. Biochem., 104 (1980) 469-472.

44 P.A. Asmus and C.R. Freed, J. Chromatogr., 169 (1979) 303-311.

45 J.J. Warsh, A. Chiu and D.D. Godse, J. Chromatogr., in press.

46 C. Horvath, W. Melander, I. Molnar and P. Molnar, Anal. Chem., 49 (1977) 2295-2305.

47 L.J. Felice, J.D. Felice and P.T. Kissinger, J. Neurochem., 31 (1978) 1461-1465.

48 T.P. Moyer and N.-S. Jiang, J. Chromatogr., 153 (1978) 365-372.

49 E. Tomlinson, T.M. Jeffries and C.M. Riley, J. Chromatogr., 159 (1978) 315-358.

50 A. Chiu, D.D. Godse and J.J. Warsh, Prog. Neuropsychopharmacol., in press.

51 C. Horvath, W. Melander and I. Molnar, J. Chromatogr., 125 (1976) 129-156.

52 J.L.M. Van de Venne, J.L.H.M. Hendrix and R.S. Deelder, J. Chromatogr., 167 (1978) 1-16.

53 R.P.W. Scott and P. Kucera, J. Chromatogr., 142 (1977) 213-232.

54 I. Halasz, Anal. Chem., 52 (1980) 1393a-1396a.

55 A.P. Graffeo and B.L. Karger, Clin. Chem., 22 (1976) 184-187.

56 M.J. Milano, S. Lam and E. Grushka, J. Chromatogr., 125 (1976) 315-326.

57 A. Yoshida, M. Yoshioka, T. Tanimura and Z. Tamura, J. Chromatogr., 116 (1976) 240-243.

58 A.G. Ghanekar and V.D. Gupta, J. Pharm. Sci., 67 (1978) 1247-1250.

59 J.G. Flood, M. Granger and R.B. McComb, Clin. Chem., 25 (1979) 1234-1238.

60 K. Bratin, L. Felice, P.T. Kissinger, D.J. Miner, C.P. Preppy and R.E. Shoup, in Introduction to Detectors for Liquid Chromatography, BAS Press, U.S.A., 1980, pp. 6.1-6.23.

61 J.R. Jadamec, W.A. Saner and Y. Talmi, Anal. Chem., 49 (1977) 1316-1321.

62 I.M. Warner, J.B. Callis, E.R. Davidson and G.D. Christian, Clin. Chem., 22 (1976) 1483-1492.

63 D.D. Chilcote and J.E. Mrochek, Clin. Chem., 18 (1972) 778-782.

64 G.M. Anderson and W.C. Purdy, Anal. Lett., 10 (1977) 493-499.

65 H.R. McKim and W.G. Dewhurst, Proc. West. Pharmacol. Soc. 23 (1980) 291-294.

66 G.A. Scratchy, A.N. Masoud, S.J. Stohs and D.W. Wingard, J. Chromatogr., 169 (1979) 313-319.

67 G.M. Anderson and W.C. Purdy, Anal. Chem., 51 (1979) 283-286.

68 T. Flatmark, T. Skotland, T.L. Jones and O.C. Ingebretsen, J. Chromatogr., 146 (1978) 433-438.

69 S.S. Simons and D.F. Johnson, J. Amer. Chem. Soc., 98 (1976) 7098-7099.

70 L.D. Mell, Jr., Clin. Chem., 25 (1979) 1187-1188.

71 T.P. Davis, C.W. Gehrke, C.W. Gehrke, Jr., T.D. Cunningham, K.C. Kuo, K.O. Gerhardt, H.D. Johnson and C.H. Williams, Clin. Chem., 24 (1978) 1317-1324.

72 P.M. Froehlich and T.D. Cunningham, Anal. Chem. Acta, 97 (1978) 357-363.

73 M. Roth and A. Hampai, J. Chromatogr., 83 (1973) 353-356.

74 K. Imai and Z. Tamura, Clin. Chim. Acta, 85 (1978) 1-6.

75 J. Scaro, J.L. Morrisey and Z.K. Shihabui, J. Liq. Chromatogr., 3 (1980) 537-543.

76 K. Imai, M. Tsukamoto and Z. Tamura, J. Chromatogr., 137 (1977) 357–362.
77 N. Seiler, Methods Biochem. Anal., 18 (1970) 259–337.
78 R.W. Frei, M. Thomas and I. Frei, J. Liq. Chromatogr., 1 (1978) 443–355.
79 U.S. Von Euler and I. Floding, Acta Physiol. Scand., suppl., 118 (1955) 45–56.
80 G. Schwedt and Z. Fresenius, Anal. Chem., 293 (1978) 40–44.
81 Y. Yiu, T. Fujita, T. Yamamoto, Y. Itokawa and C. Kawai, Clin. Chem., 26 (1980) 194–196.
82 N. Nimura, K. Ishida and T. Kinoshita, J. Chromatogr., 221 (1980) 249–255.
83 T. Seki, J. Chromatogr., 155 (1978) 415–420.
84 S. Kobayashi and K. Imai, Anal. Chem., 52 (1980) 424–427.
85 S. Kobayashi, J. Seiko, K. Honda and K. Imai, Anal. Biochem., 112 (1981) 199–104.
86 W. Kemula, Rosz. Chem., 26 (1952) 281.
87 P.T. Kissinger, C. Refshauge, R. Dreining and R.N. Adams, Anal. Lett., 6 (1973) 465–477.
88 K. Stulik and V. Pacakova, J. Chromatogr., 192 (1980) 135–141.
89 B. Fleet and C.I. Little, J. Chromatogr. Sci., 12 (1974) 747–752.
90 B. Oosterhuis, K. Brunt, B.H. Westerink and D.A. Doornbos, Anal. Chem., 52 (1980) 203–205.
91 D.D. Koch and P.T. Kissinger, J. Chromatogr., 164 (1979) 441–445.
92 S. Ikenoya, T. Tsuda, Y. Yamano, Y. Yamanishi, K. Yamatsu, M. Ohmae, K. Kawabe, H. Nishino and T. Kurahashi, Chem. Pharm. Bull., 26 (1978) 3530–3539.
93 I.N. Mefford, J. Neurosci. Methods, 3 (1981) 207–204.
94 C.L. Blank, J. Chromatogr., 117 (1976) 35–46.
95 S.C. Rifkin and S.H. Evans, Anal. Chem., 48 (1976) 2174–2180.
96 M.H. Joseph, Abstracts of 8th Meeting of the International Society for Neurochemistry, Nottingham, U.K., September, 1981.
97 K. Shimada, M. Tanaka and T. Nambara, Chem. Pharm. Bull., 27 (1979) 2259–2260.
98 W.H. McFadden, J. Chromatogr. Sci., 18 (1980) 97.
99 A.M. Krstulovic and A.M. Powell, J. Chromatogr., 171 (1979) 345–356.
100 J.F. Reinhard, Jr., M.A. Moskowitz, A.F. Sved and J.D. Fernstrom, Life Sci., 27 (1980) 905–911.
101 P.J. Langlais, W.J. McCentee and E.D. Bird, Clin. Chem., 26 (1980) 786–788.
102 Z. Lackovic, M. Parenti and N.H. Neff, Eur. J. Pharmacol., 69 (1981) 347–352.
103 D.S. Goldstein, G. Feuerstein, J.L. Izzo, I.J. Kopin and H.R. Keiser, Life Sci., 28 (1981) 467–475.
104 G.C. Davis, P.T. Kissinger and R.E. Shoup, Anal. Chem., 53 (1981) 156–159.
105 K.-G. Wahlund and B. Edlen, Clin. Chim. Acta, 110 (1981) 71–76.
106 O. Magnusson, L.B. Nilsson and D. Westerlund, J. Chromatogr., 221 (1980) 237–247.
107 L.J. Felice and P.T. Kissinger, Clin. Chim. Acta, 76 (1977) 317–320.
108 A.M. Krstulovic, C.T. Matzura and L. Bertani, Clin. Chim. Acta, 103 (1980) 109–116.
109 L.B. Saraswat, M.R. Holdiness and J.B. Justice, J. Chromatogr., 222 (1981) 352–362.
110 A.J. Cross and M.H. Joseph, Life Sci., 28 (1981) 499–505.
111 F. Hefti, Life Sci., 25 (1979) 775–782.
112 C. Atack, Acta Physiol. Scand. suppl., 451 (1977) 1–99.
113 P.A. Shea and R.K. Jackson, Transact. Amer. Soc. Neurochem., 10 (1979) pp. 183.
114 A.M. Diguilio, A. Groppetti, S.A. Algeri, F. Ponzio and C.L. Galli, Anal. Biochem., 92 (1979) 82–90.
115 R.E. Shoup and P.T. Kissinger, Clin. Chem., 23 (1977) 1268–1274.
116 S.J. Soldin and J.G. Gilbert, Clin. Chem., 26 (1980) 291–294.
117 C.J. Earley and B.E. Leonard, J. Pharmacol. Methods, 1 (1978) 67–79.
118 B.H.C. Westerink and T.B.A. Mulder, J. Neurochem., 36 (1981) 1449–1462.

119 D.D. Kock and P.T. Kissinger, Anal. Chem., 52 (1980) 27-29.
120 A. Oke, R. Keller and R.N. Adams, Brain Res., 148 (1978) 245-250.
121 S. Sasa and C.L. Blank, Anal. Chim. Acta, 104 (1979) 29-45.
122 A. Chiu, J.J. Warsh and D.D. Godse, in preparation.
123 T. Muller, E. Hofschuster, H.J. Kuss and D. Welter, J. Neural
     Transmission, 45 (1979) 219-225.
124 C.-H. Cheng and G.F. Wooten, J. Pharmacol. Methods, 3 (1980) 333-344.
125 L. Bauce, J.A. Thornhill, K.E. Cooper and W.L. Veale, Life Sci., 27 (1980)
     1921-1928.
126 C.L. Blank and R. Pike, Life Sci., 18 (1980) 859-866.
127 T. Nagatsu, K. Oka and T. Kato, J. Chromatogr., 163 (1979) 247-252.
128 J. Wagner, M. Palfreyman and M. Zraika, in A. Frigerio and M. McCamish
     (Eds.), Vol. 10, Recent Developments in Chromatography and Electrophoresis,
     Elsevier, Amsterdam, 1980, pp. 201-209.
129 B. Tabakoff and R.F. Blank, J. Neurochem., 34 (1980) 1707-1711.
130 M.K. Rahman, T. Nagatsu and T. Kato, Life Sci., 28 (1981) 485-492.
131 R.E. Shoup, G.C. Davis and P.T. Kissinger, Anal. Chem., 52 (1980) 483-487.
132 D.K. Sundberg, B. Bennett, O.T. Wendel and M. Morris, Res. Comm. Chem.
     Pathol. Pharmacol., 29 (1980) 599-602.
133 B.H.C. Westerink and J.C. Van Oene, Eur. J. Pharmacol., 65 (1980) 71-79.
134 B.A. Bennett and D.K. Sundberg, Life Sci., 28 (1981) 2811-2817.
135 J.J. Warsh, A. Chiu, D.D. Godse and D.V. Coscina, Brain Res. Bull., 4
     (1979) 567-570.

Chapter 11

VOLTAMMETRIC TECHNIQUES FOR THE ANALYSIS OF BIOGENIC AMINES.

R. MARK WIGHTMAN

Department of Chemistry, Indiana University, Bloomington, Indiana 47405
(United States)

and

MARK A. DAYTON

Medical Sciences Program, Myers Hall, Indiana University, Bloomington, Indiana
47405 (United States)

In the past few years the use of amperometric techniques for the analysis
of biogenic amines has increased dramatically. Liquid chromatography with electrochemical detection (LCEC) has been developed and refined to such an extent
that it has become the method of choice for many neurochemical analyses. *In
vivo* electrochemistry, the direct measurement of easily oxidized species in the intact mammalian brain, is also being investigated by a wide number of research
groups. Other amperometric methods are just beginning to appear which should
further increase the utility of electrochemical methods for the analysis of biogenic amines. This review will examine the different amperometric methods useful for biogenic amine analysis and their underlying principles, evaluate them
with respect to other methods, and attempt to indicate the areas where amperometric methods can provide new unique methods for analysis.

11.1 VOLTAMMETRY

11.1.1 Introduction to voltammetry

Voltammetry is probably familiar in one form or another to many scientists
as an analytical method. The Clark, or "polarographic", oxygen electrode oper-
ates under this principle as does conventional polarography with a dropping
mercury electrode, a very sensitive method for the trace measurement of certain
metal ions. In these techniques and in all of the electrochemical experiments
described in this chapter, a voltage is applied between a reference electrode
which maintains a constant potential and a working electrode exposed to the
solution containing the species to be measured. The current which flows
between these two electrodes is measured and can be related to events occurring
at the working electrode surface. In an amperometric experiment, the voltage
difference between the two electrodes is kept at a constant value and the current is simply measured as a function of time. The term voltammetry generally

implies that the voltage between the electrodes is altered during the measurement and the current is then recorded as a function of voltage. Various waveforms for the applied voltage in voltammetry have been investigated with the goal of obtaining the most useful current information.

All of these techniques, and many others that have not yet been used in the analysis of biogenic amines are cataloged under the general term of electrochemistry. This is a very large area of research with many excellent treatises that are a useful guide to the field. The book by Ralph Adams, Electrochemistry at Solid Electrodes (1), provides an excellent introduction to this area. The author of this book was one of the pioneers in the use of carbon electrodes and also introduced electrochemical techniques for the analysis of biogenic amines. Other books that are necessary for interpreting and doing electrochemistry are the textbook by Bard and Faulkner, which provides an excellent insight into electrochemical theory (2), and the work by Sawyer and Roberts (3) which contains all the useful experimental details that are usually gained only by a painful hands-on learning experience. In the following section, a brief description of the more common electrochemical methods is given which should provide a starting point for understanding voltammetric methods.

11.1.2 Voltammetric principles

As mentioned above, electrochemical experiments measure the current between two electrodes immersed in solution while a voltage is applied between the electrodes. The overall events that occur in the experiment are schematically illustrated in Fig. 11.1. In the ideal experiment, the measured current arises

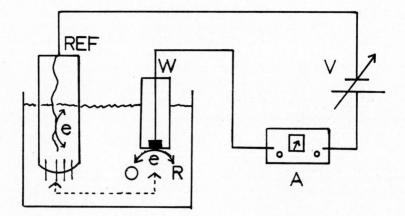

Fig. 11.1. Apparatus for simple electrochemical experiment. V = variable voltage; A - ammeter; W - working electrode; REF - reference electrode. Electron transfer (e) occurs at W and REF. Dotted line indicates ionic current in solution. O and R are an oxidant and its corresponding reductant, respectively.

from oxidation (loss of electrons) or reduction (gain of electrons) of a molecule in solution at the working electrode surface. Current must be passed
through the solution to complete the circuit so electrochemical experiments are
usually done with a large concentration (0.1 M or higher) of inert ions (supporting electrolyte) in the solution. The reference electrode should be designed so that it can pass a large current without a change in composition.
Suitable reference electrodes that are commonly used are composed of a metal
(M) and its insoluble salt (MX) such as $Ag/AgCl$ or Hg/Hg_2Cl_2 [the saturated
calomel electrode (SCE)], and have a half cell reaction

$$MX + ne \longrightarrow M + X^{n-}$$

where n is the number of electrons in the half cell reaction (usually 1). The
potential of this reaction is given by the Nernst equation as

$$E = E^o - \frac{0.059}{n} \log \frac{[M][X^-]}{[MX]}$$

where E^o is the standard potential (measured at unit activity) and E is the
half cell potential. Since M and MX are solids for the examples above, the potential is only dependent on the concentration of X^-. To maintain a constant
reference electrode potential, the reference electrode should be in a compart-
ment that prevents any change in chemical composition. Porous glass, asbestos
fibers, and even wadded-up filter paper serve this role while allowing ionic
conduction with the remainder of the electrochemical cell.

Experimentally, it is difficult to make the solution sufficiently conduc-
tive to prevent artifacts and to make the reference electrode remain at a constant value while passing current through it. Therefore, most modern electrochemistry is done with a three-electrode potentiostat, a device that maintains
a desired potential between the reference and working electrode without current
passing through the reference electrode. A schematic of a three-electrode potentiostat is given in Fig. 11.2, and it can be seen that it is the same circuit used in voltage clamp experiments. The third, or auxiliary, electrode
passes sufficient current to keep the circuit balanced, with the reference electrode simply serving as a sensing device to maintain the desired potential. If
the reference electrode is placed close to the working electrode, problems with
poor solution conductance are minimized. In electrochemistry, the sign of the
reported voltage is given with respect to the working electrode. At more positive potentials, the working electrode has more oxidizing power while at negative potential, reductions are more likely to be observed. Oxidation current
(current flowing into the working electrode) is normally given a negative sign,
although this convention is not rigorously obeyed.

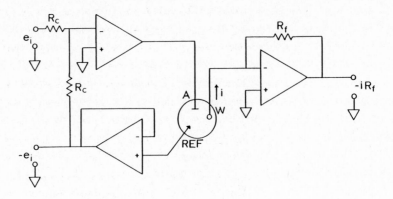

Fig. 11.2. Schematic of 3-electrode potentiostat. A - auxiliary electrode; W - working electrode; REF - reference electrode. The value of resistor R_f is selected for suitable amplification of the current (i). The other resistors (R_c) are typically 10 KΩ. The applied potential is e_i.

Assuming a stable reference electrode and a dependable potentiostat (they are commercially available from several companies), the key to a successful electrochemical experiment is a reliable working electrode. The goal of the experiment is to measure current from the electrolysis of molecules in solution without looking at artifacts. To do this, the working electrode must be inert; for example, one does not want current flowing because the electrode is dissolving at the applied potential. The electrode material that has been shown to be most useful for biogenic amine electroanalysis is graphite. One of the reasons for graphite's superiority as an electrode material is the wide potential range at which it is useful (from approximately -1.0 to +1.2 V *vs.* SCE in aqueous solution). The potential limits are defined as the potentials where a large current is observed resulting from decomposition of the solvent (or electrode). The actual potential limits depend on the solvent composition, pH, and type of graphite employed, and should be investigated for each new electrode material or solution. Noble metals such as gold and platinum tend to have a reduced range, especially in the negative direction. Mercury electrodes are oxidized at potentials slightly positive of 0.0 V *vs.* SCE and are thus only useful at negative potentials.

The actual process of electron exchange between a solution molecule and an electrode surface is an extremely complex process (4). To understand the result, one must consider that the important part of the potential difference applied between the reference and working electrode is really localized to a

small region of molecular dimensions at the working electrode surface. For
electron transfer to occur to a molecule at the surface of the working electrode, sufficient energy (or potential) must be applied to the electrode surface. To maintain the passage of current across the electrode-solution interface, the supply of electrolyzable molecules at the surface must be maintained--
in other words there must be a mechanism to bring more molecules to the electrode surface. These two principles, sufficient energy at the electrode for
electron transfer and transport of molecules to the surface, are key points in
understanding the electrochemical result.

The transport of molecules to the interface will be considered first. An obvious way to induce this process is to stir the solution or rotate the electrode.
In this case, and when the potential is fixed at a value sufficient to electro-
lyze all the molecules reaching the working electrode surface, the current (i)
is given by

$$i = \frac{nFADC}{\delta} \qquad [1]$$

where n is the number of electrons transferred per electrolyzed molecule, F is
the faraday, a scaling factor, A is the electrode area, D is the diffusion co-
efficient of the electrolyzed molecules, and C is the concentration of the electrolyzed molecule (1). The term in the denominator, δ, is referred to as the
diffusion layer thickness. This is a conceptual term which predicts the average distance from the electrode where the solution is essentially static with
respect to the electrode. Evaluation of the actual value of δ requires complex
hydrodynamic equations and a statement of the electrochemical cell geometry,
but it is always inversely proportional to the degree of stirring. Equation [1]
is the relationship of current to concentration for amperometry in flowing
streams. The important aspect of this equation is that the current is directly
proportional to the concentration as long as the solution movement is maintained
at a constant rate.

If both the electrode and the solution are kept stationary, then the only
method of transport to the electrode surface is diffusion. Under these
conditions and with the same potential as above, the current at a disk shaped
electrode is approximated by

$$i = \frac{nFADC}{\pi} \left[\left(\frac{\pi}{tD}\right)^{\frac{1}{2}} + \frac{4}{r} \right] \qquad [2]$$

where t is the time that the potential has been applied, and r is the radius
of the electrode (5). Typically, the current in the experiment is only measured
for 1 sec or less since vibration and other sources of convection tend to cause
deviations from this equation. The significance of the "steady-state" or

second term of this equation was only recently recognized, and this has been the subject of a recent review (6). Experimentally, one finds that the second term of equation [2] is significant only when the electrode radius is <100 μm. For larger electrodes, the current decays inversely with $t^{\frac{1}{2}}$ because the molecules are depleted in concentration by electrolysis at a faster rate than they can diffuse to the electrode surface. This experiment is referred to as chronoamperometry--the measurement of current as a function of time. Time-independent currents are easier to measure, so electroanalysis in flowing streams or with very small electrodes (∿10 μm diameter) gives simpler results.

Having considered how the molecules get to the electrode for electrolysis, the next problem is to consider how the electrolysis proceeds at different potentials. For the ideal case of a simple and rapid electron transfer from the molecule to the electrode (an oxidation), the current is given by

$$i = \frac{i_1}{\exp\left[\frac{nF}{RT}(E_{\frac{1}{2}} - E)\right] + 1} \qquad [3]$$

where R is the gas constant, T is the absolute temperature, and i_1 is given by equation [1] or the limiting form of equation [2] for small electrodes depending upon the mode of transport to the electrode. The curve for the oxidation of dopamine (DA) at a carbon fiber electrode is essentially described by equation [3] (Fig. 11.3), and is referred to as a voltammogram (a hydrodynamic voltammogram if the solution is stirred).

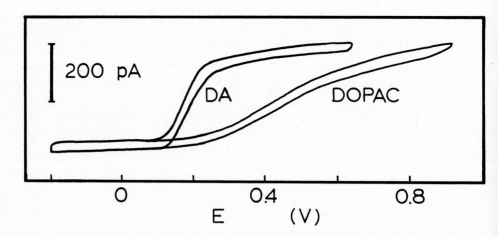

Fig. 11.3. Voltammograms at 100 mV s^{-1} sweep rate at a carbon fiber microelectrode in pH 7.4 buffer. A) Oxidation of 0.1 mM DA. B) Oxidation of 0.1 mM DOPAC.

Experimentally, these curves are obtained by simply changing the voltage in a linear fashion (a voltage ramp) and measuring the current using the circuit of Fig. 11.2. The position on the voltage axis of this curve for a particular compound is determined by the molecule's ability to donate or accept electrons from the electrode surface. This value is expressed $vs.$ a reference electrode by $E_{\frac{1}{2}}$, and is very similar in value to E^{o}. However, $E_{\frac{1}{2}}$ is affected by pH and other electrochemical parameters, so it should be determined for each compound under the solution conditions in which the compound will be analyzed. Under constant solution and electrode surface conditions, $E_{\frac{1}{2}}$ is a constant and can be used to aid in the identity of the compound.

For large (d > 100 µm) electrodes in stationary solution, the rate of diffusion of molecules to the electrode is much slower than the rate of electrolysis, so the curve predicted by equation (3) is distorted by a decay in the measured current. Equations to describe this complex behavior have been derived (7). In this type of electrochemical experiment, the products generated electrochemically tend to accumulate at the electrode surface, so a triangular voltage waveform permits observation of these products on the reverse sweep. This experiment is referred to as cyclic voltammetry (CV) and is shown in Fig. 11.4. In a voltammogram at a small electrode or under hydrodynamic conditions, the electrochemically generated products are transported from the electrode before they can be observed on the reverse sweep.

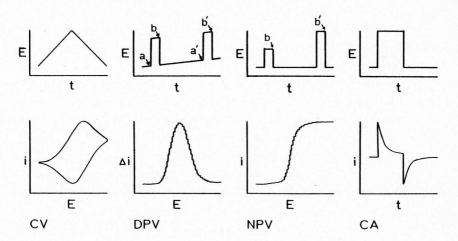

Fig. 11.4. Voltage wave form and current response for different electrochemical experiments. CV - cyclic voltammetry; DPV - differential pulse voltammetry; NPV - normal pulse voltammetry; CA - chronoamperometry. The current is sampled at times a and b in DPV and only at time b in NPV.

These two basic facets of electrochemistry--the mass transport of molecules to the surface and the current-voltage relationships about $E_{\frac{1}{2}}$--are the predominant prerequisites for understanding most electrochemical experiments. However, to improve the signal-to-noise ratio, several specialized electrochemical techniques have been developed and have seen utility in the analysis of biogenic amines (Fig. 11.4). Cyclic voltammetry is probably the most widely used method to characterize the electrochemical behavior of new compounds. Differential pulse voltammetry (DPV) employs a ramp shaped voltage with pulses superimposed (Fig. 11.4). The current is sampled before and during the pulse, and the output is the difference of the two currents. This gives a result which is close to the derivative of equation [3]. The peak shaped output of this technique makes it useful for resolving closely overlapping compounds. In normal pulse voltammetry (NPV), the applied waveform is a series of pulses of increasing voltage amplitude. Current is only sampled at a fixed time during the voltage pulse. This approach tends to minimize the time-dependent effects of equation [2] so that normal pulse voltammograms are sigmoidal in shape. We prefer this technique for *in vivo* electrochemical analyses because, for the majority of the time, the potential is maintained at a value where no electrochemistry is occurring. This prevents filming of the electrode surface by electrochemically generated products, with subsequent electrode failure. Chronoamperometry (CA) has also been used extensively in *in vivo* experiments. The method, as previously discussed (equation [2]), is useful for concentration methods, but current-voltage information is not obtained.

This brief introduction to electrochemistry has given a simplified theory behind current-voltage curves arising from faradaic events--the process of molecules exchanging electrons with the electrode surface. However, other effects are also observed in electrochemistry, and it is important that these be distinguished from the faradaic effects. Assuming that the electrochemical cell has been designed properly, as mentioned earlier, all of the non-faradaic effects arise at the working electrode. One of these effects is charging current; current that passes when the potential of the electrode is changed. In the absence of electrolyzable species, the electrode-solution interface can be thought of as a capacitor and this interface is referred to as the double layer. When the potential of the electrode is changed, this results in current to charge the capacitance. For a voltage pulse, this charging current decays exponentially with time. Measurements of faradaic electrochemistry are normally made after 1-10 msec to allow the charging current to decrease in magnitude to an insignificant value. Another source of current is the electrolysis of functional groups or molecules adsorbed on the electrode surface. These occur at all solid electrodes including gold and platinum, but seem to be least problematic at carbon electrodes. When possible, it is advisable to obtain a voltammogram in a background solution so that these effects can be easily distinguished from faradaic electrochemistry. The magnitudes of double layer charging current and current from the electrolysis of surface compounds are both proportional to electrode area. For this reason, a porous, cracked, or rough electrode surface is much more likely to exhibit troublesome effects from these sources.

In the previous discussion of current-voltage curves, it was stated that the position of the current-voltage curve on the voltage axis is determined by E^O. However, E^O is a thermodynamic quantity and the current-voltage curves are also affected by the rate of electron transfer to the electrode surface. If this rate is slow (the reaction is electrochemically "irreversible") the current-voltage curve will be shifted on the voltage axis and the voltammogram will become more drawn out. Predictions of these rate effects are difficult to make but they can drastically affect the voltammograms. For example, ascorbate is approximately 400 mV more difficult to oxidize than predicted on thermodynamic grounds at carbon electrodes. Another example that we have studied (8) is the oxidation of DA and 3,4-dihydroxyphenylacetic acid (DOPAC) at carbon fiber microelectrodes (Fig. 11.3). Although these two molecules have almost identical E^O values, their voltammograms are very different. We have attributed these effects to the difference in charge of the side chains of these two molecules which apparently alters the rate of their oxidation at carbon fiber surfaces. A promising method to control these rate effects and use them to build specificity for a particular compound is to chemically modify the electrode surface (9).

Another effect important to the shape of current-voltage curves is the state of the electrode surface. If species are adsorbed on the electrode surface, they can block sites for electron transfer and thus reduce the efficiency of the electrode. Protein adsorption on solid electrode surfaces is a severe problem (10,11). Adsorption of products generated at the electrode surface can create a film that totally stops electron transfer (12,13). Obviously, both of these problems are acute in *in vivo* electrochemistry and one has to accept the fact that the electrode is not going to work ideally. The best approach is to design electrochemical strategies to maintain the electrode surface at a constant, albeit deteriorated, condition.

As can be seen, electrochemistry is relatively simple theoretically, however many problems can arise experimentally. All of the problems described above result from something changing the properties of the electrode surface. If this problem can be avoided, electrochemistry is an extremely sensitive technique. The constant flow of fresh, clean solvent past the electrode surface in the LCEC experiment seems to clean the electrode surface, and, thus, it is probably the most reliable electrochemical technique with solid electrodes.

11.1.3   Voltammetry of biogenic amines

Voltammetry is a useful technique for the analysis of biogenic amines because many of them are easily oxidized. At a carbon paste electrode, the catecholamines, 3,4-dihydroxyphenylalanine (DOPA), and the dihydroxy acid and alcohol
metabolites are oxidized at approximately 0.2 V *vs*. SCE at physiological pH.
Electrochemically, these compounds appear very similar and cannot be resolved
(DOPAC is an exception at carbon fiber electrodes). At slightly more positive
potentials, 5-hydroxytryptamine (5-HT) and 5-hydroxyindoleacetic acid (5-HIAA)
are oxidized. At approximately 0.4 V *vs*. SCE, the methoxylated catecholamine
derivatives are oxidized [homovanillic acid (HVA), *etc*.], and at approximately
0.6 V *vs*. SCE in physiological buffer, tyrosine is oxidized. All of these potentials are pH dependent, with each of these compounds being oxidized at more
positive potentials at lower values of pH. Variations in pH cannot be used to
distinguish between a class of compounds; for example, all catechols oxidize at
approximately the same potential at a fixed pH (14).

Remarkably, very few other compounds in the mammalian brain are as easy
to oxidize electrochemically as the compounds discussed above. This is one of
the key features of voltammetric techniques that has made it especially useful
for analysis of biogenic amines. The other easily oxidized species that are
present include ascorbic acid, reduced nicotinamide adenine dinucleotide (NADH),
and uric acid. However, proteins and large peptides are faradaically inactive
at voltammetric electrodes. This is also true for most of the amino acids and
other neurotransmitters such as γ-aminobutyric acid (GABA) and acetylcholine.

The biogenic amines that are electrochemically active are oxidized in a
two-electron, two-proton process. For example, DA is oxidized as indicated in
the following scheme:

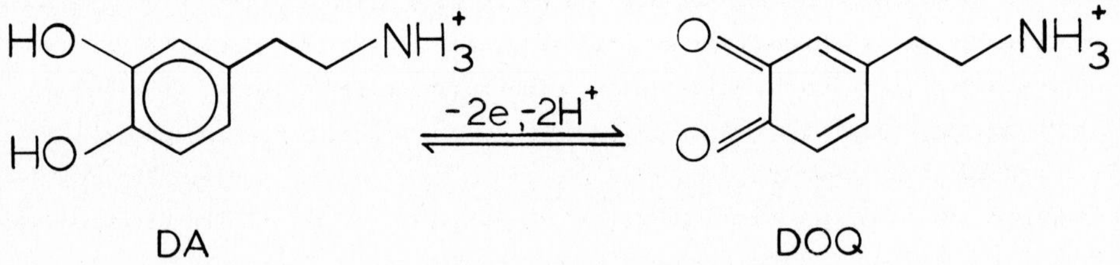

$$\text{DA} \quad \underset{}{\overset{-2e^-,-2H^+}{\rightleftharpoons}} \quad \text{DOQ}$$

The oxidation products formed in this type of reaction are, in themselves, very
reactive compounds--they can easily be reduced back to the original compound
or react with nucleophiles such as amines and sulfhydryl groups. The reactivity of the oxidized forms is generally not a problem since only a very small
percent of the total molecules in the electrochemical cell are oxidized in an
electroanalytical experiment.

11.2 LIQUID CHROMATOGRAPHY WITH ELECTROCHEMICAL DETECTION

11.2.1 Introduction to LCEC

In the previous chapter, the principles of liquid chromatography for the analysis of biogenic amines were discussed. Obviously, this technique is useful for separating mixtures of the biogenic amines. An electrochemical cell at the end of the column serves as the detector in the LCEC method. The electrochemical detector is advantageous over other detection schemes because it is extremely sensitive and it is selective; it only detects compounds that can be electrolyzed by the electrode. The degree of selectivity is adjustable by selection of the applied potential. Although the LCEC method has been around for a number of years, its current popularity was initiated by a paper by Kissinger *et al.* (15) in which it was demonstrated that the LCEC method could be used to determine DA and NA in brain tissue. Since that time, the field has grown exponentially. Two recent reviews give full details of the use and applications of LCEC for biogenic amines (16,17).

Most LCEC determinations are done at a fixed potential (amperometry). Since current-voltage curves are not obtained, identification of compounds is based on comparison of chromatographic retention times. Chromatographic separations, especially with the small particle stationary phases commercially available, give far more resolution elements than could ever be obtained in a voltammogram. The identical electrochemical response of all the catechols is not a problem since they can be made to elute at different times in the chromatogram. The concentration of a particular compound is calculated by comparing the peak height of the unknown with a calibration curve from authentic standards. Since the response of the electrode may alter from day to day, quantitative experiments usually employ an internal standard.

A typical LCEC chromatogram is shown in Fig. 11.5. The chromatogram is similar to that obtained with any other on-line detector. As can be seen, the signal-to-noise ratio is excellent. Indeed, femtomole detection limits have been reported by several investigators (18-20). The factors which determine detection limits have been considered (21,22). The most obvious is flow fluctuations at the electrode surface, which can be minimized by using a very smooth electrode surface and a constant flow rate.

The LCEC method is so successful, despite all of the shortcomings of electrochemistry pointed out in Section 11.1.2, because the design of the experiment circumvents many electrochemical problems. Transient phenomena from charging currents and residual current are eliminated by applying a constant potential and relying on the chromatography to provide resolution. Electrode filming is minimized because the chromatographic mobile phase constantly washes the electrode surface with clean solution. The low concentrations which are analyzed also minimize electrode filming by electrogenerated products. In most

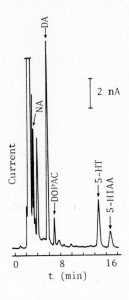

Fig. 11.5. LCEC of 100 µL of whole brain homogenate. Stationary phase: Whatman Partisil ODS (PXS5/25). Mobile phase: pH 3.0 McIlvaine buffer with 10% methanol and 25 mg L^{-1} sodium octyl sulfate, flow rate 1 mL min^{-1}. Bioanalytical TL-3/CP-O detector. (Courtesy of Bioanalytical Systems, Inc.)

electroanalytical experiments, the electrode is resurfaced daily, or, preferably, before each experiment. This is in contrast to LCEC where electrode resurfacing may not be necessary for several months. The effects of electrochemical irreversibility are masked because the potential is constant. Compounds that are very irreversible may give reduced peak heights but, in many cases, this can be circumvented by increasing the applied potential.

The LCEC method has been directly compared with a number of other methods. Electrochemical detection is more sensitive than UV (23,24) or fluorescence (24) detection following LC separation, and it has the added advantage that it is less susceptible to interferences. In a direct comparison with a gas chromatography-mass spectrometry (GC-MS) method for DA and 5-HT, the LCEC was found to have comparable detection limits (25). The GC-MS method has the advantage that identification of the nature of the chemical compound is absolute. However, the LCEC method allows a higher throughput of samples since several chromatographs can be run simultaneously with much less expense for equipment or operator training. The LCEC method for analysis of biogenic amines does not require (but does allow) derivatization, a step which may introduce error into an analysis. The LCEC method has been proven to be sufficiently specific-- in the GC-MS study and in comparison with radioenzymatic assays (26,27), good agreement of values for biogenic amine concentrations in physiological samples have been found. The radioenzymatic methods appear to be more sensitive, however they are time-consuming and require the use (and subsequent disposal) of radioactive material.

11.2.2 LCEC instrumentation

The chromatographic instrumentation necessary for the LCEC experiment has been described in the previous chapter. Actually, very inexpensive pumps can be used since LCEC assays rarely employ gradient elution because of the electrode sensitivity to changes in solution composition. Almost all investigators use stainless steel tubing, injectors, and columns. Microparticulate (5 μm) reverse phase columns provide the best resolution for biogenic amines, and have become the column of choice. Their chief drawback is that they require more care than the pellicular resins; however, with proper maintenance their resolution can be maintained for several months (28). Three-electrode potentiostats (Fig. 11.2) are almost universally employed.

Several different electrochemical detectors and working electrode materials have been proposed. Most detector designs are similar to that shown in Fig. 11.6.

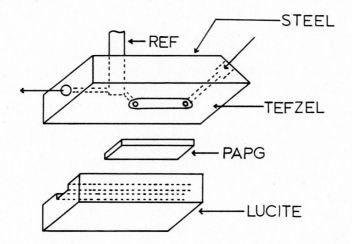

Fig. 11.6. Exploded view of an electrochemical cell for liquid chromatographic detection with a pressure-annealed pyrolytic graphite (PAPG) working electrode. Tefzel is used to form the duct and the steel top serves as an auxiliary electrode. The reference electrode (REF) is downstream.

The liquid chromatographic effluent is directed into a duct-shaped electrochemical cell. The floor of the duct contains the working electrode and the roof is the auxiliary electrode. The reference electrode is downstream. The volume of the cell is less than 5 μL so detector-induced band broadening is not a problem. Other cell designs direct the effluent perpendicularly to the working electrode, or rotate the electrode (29). Detectors with two working electrodes have been employed. The second electrode can be used to reduce background effects (30) or as a detector with a different applied potential (31). We have designed a very small volume cell for use with a capillary liquid chromatographic column, a column which provides greatly improved resolution (32).

Carbon electrodes seem to give far more reproducible results than other electrodes (33). Carbon paste (15), pressure-annealed pyrolytic graphite (34), Kel-F-graphite mixtures (35) and glassy carbon (21) have all been examined with only minor advantages shown for each material. The most widely used electrode material is glassy carbon since it is easy to work with, reliable, and has a large potential range.

11.2.3 Applications of LCEC

The LCEC method has been used to determine tyrosine and tryptophan metabolites in brain tissue, plasma, urine, and cerebral spinal fluid (16,17). Precise directions for each analysis are available in the original literature. In many cases, samples consisting of the body fluids or the fluid surrounding the homogenized tissue are directly injected onto the column. More frequently, a sample cleanup procedure is used, such as ion exchange precolumns or solvent extraction techniques. For the catecholamines, adsorption onto alumina is an ideal method to preconcentrate the sample. Extreme care must be taken in all sample handling techniques to prevent air oxidation of these compounds. This can be prevented by keeping the sample at an acidic pH or by deoxygenating with a stream of nitrogen.

Recently, the LCEC technique has also been used to measure the activity of enzymes involved in catecholamine and 5-HT regulation. In these methods, the turnover of products and reactants is measured by the LCEC method (36-39). The release of DA and 5-HT from isolated brain tissue (40,41) or perfused, intact brain tissue (42) can be monitored by the LCEC method. In fact, it has been shown that catecholamine release from perfused tissue can be directly monitored electrochemically without the use of a column (43,44). This method permits the rate of pharmacologically-induced release of neurotransmitters to be examined on a very rapid time scale.

The LCEC method in its present form is an analytical method that is simple to use, reliable, and established. Chromatographic and electrochemical conditions exist in the literature for a wide range of biogenic amines so that the novice in this field does not have to spend extensive time searching for the right conditions. However, new chromatographic methods such as microcolumns and high flow rate systems are constantly appearing and these will undoubtedly be adapted to the LCEC method. The electrochemical technology for chromatographic detection seems to have reached a plateau; detection limits have hardly altered in the last four years. The increased use of multiple electrochemical detectors may well lead to increased specificity in the LCEC method. However, the range of compounds amenable to LCEC detection is growing.

The enkephalins have been detected by this method (45), and very recently derivatization techniques that are tailor-made for electrochemical detection have
been introduced that permit brain GABA to be determined (46).

11.3 *IN VIVO* ELECTROCHEMISTRY

11.3.1 Voltammetry in the brain

Voltammetry possesses multiple attributes which are essential for *in situ*
measurements of neurotransmitters and neuromodulators in the mammalian brain.
It can provide a direct, virtually instantaneous, continuous readout of the
concentrations of endogenous electroactive compounds in the extracellular fluid
(ECF) near the electrode surface. While positive identification of the component from ECF which is responsible for the signal cannot always be made, electrochemistry does allow elimination of candidates possessing dissimilar characteristics from the measured response. In addition, while interpretation of
results requires a good symbiosis between electrochemistry and neurochemistry,
the technique itself is relatively easy to perform and inexpensive to imple-
ment. However, in contrast to LCEC, *in vivo* electrochemistry is not an established, entrenched technique. Interpretation of the voltammetric response is
often vague, and many provocative results have been reported. Since *in vivo*
electrochemistry is in a developmental stage, a more complete analysis of the
applications of the technique will be presented than was done for the review
of LCEC.

The basic experimental arrangement is the same for *in vivo* electrochemistry
as for the *in vitro* systems already discussed. A three-electrode potentiostat
(Fig. 11.2) is employed. The electrochemical cell, into which all the electrodes are placed, is the animal's head. Supporting electrolyte is provided
by the high salt content of the ECF in the brain. However, the mammalian
brain is a much more complex chemical environment than is normally studied in
a beaker. Thus, an extensive *in vitro* characterization of new working electrodes is necessary prior to their use *in vivo* to lessen the problem of complex data interpretation. To date, carbon electrodes have been most widely
employed. Detailed instruction for carbon paste (47,48), graphite-epoxy (49,
50), carbon fiber (5,8,51,52), and iodide-treated platinum (12) electrode construction can be obtained from the original literature.

An ideal sensor for biogenic amines could monitor the release of neurotransmitter at a single synapse, a concept analogous to single unit recordings
in electrophysiology. While electrodes are still huge with respect to a single
synapse, they are small enough to sample discrete regions of the mammalian brain.
An alternative method for acquiring information on biogenic amines is the push-
pull cannula technique (53). Push-pull methodology provides fluid that can be
analyzed by separation techniques for qualitative and quantitative analysis of

compounds released into the buffer. In contrast, *in vivo* electrochemistry depends on the inherent sensitivity of the electrode at characteristic potentials coupled with pharmacological manipulations to identify the species observed. On the other hand, it is not necessary with voltammetry to make assumptions concerning the effects of uptake or synthesis on the pattern of release in response to drug administration. In addition, less severe effects of tissue damage in chronic preparations are reported for *in vivo* voltammetric experiments (49,51,54,55) than for push-pull cannulation (56).

Clark first used electrochemistry in 1964 to monitor O_2 and ascorbate levels in the cortex of anesthetized cats (57). Using glassy carbon electrodes approximately 300 μm in diameter and fixed potentials between -0.6 and -1.6 V *vs.* Ag/AgCl, he was able to observe changes in the cathodic current (i_c) which correlated with the pO_2 in the atmosphere breathed by the cat (57). Oxygen tension did not affect the anodic current (i_a) when both currents were monitored by cyclic voltammetry (58). Changes in i_a could be produced by intravenous (i.v.) administration of sodium ascorbate (AA). When monitored *via* pyrolytic graphite electrodes implanted beneath the epicardium of the hearts of anesthetized cats, AA injections did not affect i_c. Since AA is a major component of the brain's redox buffer (59), it is interesting that AA and O_2 levels could not be correlated.

The idea of using *in vivo* electrochemistry to study the brain lay dormant for about 10 years until revived when Adams and co-workers began lowering carbon paste electrodes into the brains of anesthetized rats. Cyclic voltammetry in the cortex, hippocampus and caudate all demonstrated an irreversible peak at ∿0.4 V *vs.* Ag/AgCl, attributed to AA, which faded rapidly during consecutive scans unless a delay of at least 100 seconds was inserted between scans (60). Clark had seen similar trends in i_a and i_c, though i_c faded much slower than i_a (59).

Since cyclic voltammograms recorded from different regions of the non-stimulated rat brain all showed essentially the same result, McCreery *et al.* (61,62) decided to study the fate of electroactive drugs injected directly into the brains of anesthetized rats. Their injection-electrode assembly resulted in the formation of a sizeable pool (30 μL) (61) near the electrode tip, the walls of which were neural tissue. Injections of AA, NA, or DA resulted in relatively steady-state concentrations in the pool. On the other hand, 6-hydroxydopamine (6-OHDA) was rapidly oxidized to the p-quinone such that the ratio of oxidized to reduced 6-OHDA (determined by CV) was ∿40/60, apparently a function of the redox strength of the brain's buffer. The total concentration of oxidized and reduced forms disappeared and was attributed to consumption of the quinone by its reaction with the surrounding neural tissue.

These injection experiments showed that electrochemistry was usable in the environment of the mammalian brain. But, continuous monitoring of the chemical activity of neurons still had not been accomplished. So Wightman *et al.* (47,63) switched strategies, deciding to record metabolite changes in the lateral ventricles of rats and rabbits following electrical stimulations in either the substantia nigra (SN) or the Raphé nucleus. Placement of the electrodes in the ventricles provided the added advantage of being able to independently verify by LCEC the compounds detected *in vivo*. Withdrawal of CSF for LCEC analysis was simpler from rabbits since their CSF production is higher. Thus, in rabbits, the whole time course of the response could be verified, establishing the viability of *in vivo* electrochemistry. There was a substantial delay (15 to 45 minutes) following stimulation before the appearance of either HVA or 5-HIAA in the CSF. When it finally did appear, there was a sharp increase in concentration [500-700 μM HVA (47)] which eventually dispersed throughout the CSF in the ventricles and was washed away. LCEC confirmed the veracity of the time course observed by electrochemistry.

Unequivocal identification of the electrochemically detected species by an independent method is obviously much more difficult in brain tissue. However, voltammograms in the striata of rats show several waves (Fig. 11.7). Lane has used slow scan rate cyclic voltammetry of unmodified carbon paste electrodes coupled with injection of authentic compounds into the brain to aid in identifying the electrolyzed components of four similar waves (48,64,65).

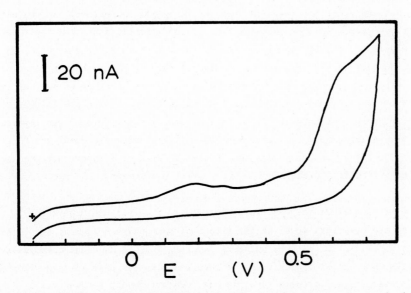

Fig. 11.7. Cyclic voltammogram obtained with a carbon paste electrode (r ≃ 300 μm) implanted in the corpus striatum of an anesthetized rat. 100 mV s^{-1} scan rate.

With slower voltage scans, the waves are more apparent than shown in Fig. 11.7. Both DA and AA caused an increase in the first peak (0.23 V *vs.* Ag/AgCl). This resulted from DA catalysis of AA oxidation (*vide infra*). The shoulder at 0.35 V was influenced by 5-HIAA (48) and 5-HT (64) injections, which implies that the first peak may also incorporate the *in vivo* voltammetric response to DOPAC. The methoxylated metabolites of DA, 3-methoxytyramine (3-MTA) and HVA, were the putative components of the ECF responsible for the generation of the third peak (0.49 V). A fourth peak (0.68 V) has not been identified. Consecutive scans 10 minutes apart for 2 to 4 hours resulted in extremely reproducible DPV peak heights (2.4%) (48), though apparently the usable lifetime of chronically implanted electrodes is shortened by 10 mV sec^{-1} scans (50).

AA interferes in the *in vivo* measurements of catecholamines, for example DA, with voltammetric electrodes. Although AA is thermodynamically easier to oxidize than DA, the peak potential for AA oxidation at carbon electrodes is displaced several hundred millivolts in a positive direction and is actually positive of the peak potential for DA (66,67). Oxidation of DA at carbon electrodes in the presence of AA results in a homogeneous catalytic oxidation of AA. The regenerated DA returns to the electrode, resulting in an enhanced current.

$$DA \xrightleftharpoons{-2e^-, -2H^+} DOQ$$

$$DOQ + AA \xrightarrow{k_c} DA + DHA$$

(DOQ and DHA represent the oxidized forms of DA and AA respectively.) The catalytic rate constant (k_c) for this reaction is sufficiently fast that the faradaic current equals the sum of the diffusion controlled currents for DA and AA (64). Thus, an increase in an electrochemical signal could be due to either DA or AA.

Three approaches have been taken to eliminating the problem of overlapping waves. The first is to use pharmacological manipulations which are thought to act on only one neurotransmitter. As will be shown, this approach has implied that *in vivo* electrochemistry can measure chemical changes from 5-HT and DA neurons. The other two approaches, which will be discussed later, are the use of chemically modified electrodes and very small voltammetric electrodes.

11.3.2 Electrochemical response to pharmacological agents and behavior

Adams and coworkers were the first to show that pharmacological agents known to affect dopaminergic and tryptaminergic systems induced an electrochemically measurable response in the brains of unanesthetized rats (49,68-71). Amphetamine was chosen for the initial studies since it has been previously documented by push-pull cannulation and by *in vitro* methods to affect DA neuronal fluxes (41, 72). They performed additional *in vitro* experiments with brain slices and synaptosomes to find the effects of amphetamine on AA. Concentrations of amphetamine ranging from 10^{-6} to 10^{-3} M demonstrated that AA was not released from striatal minces whereas DA was (49). Thus, the dose-dependent increases in chronoamperometric current measured from electrodes implanted in the caudate nucleus following amphetamine administration were attributed to the oxidation of DA released into the ECF of the brain (68). Potential steps to either +0.5 or +0.8 V resulted in equal currents measured 1 sec after the potential change, so the possibility of a contribution to the current from the methoxylated metabolites of DA (3-MTA and HVA) could be dismissed (49,68). However, DOPAC was still a viable candidate.

Similarly, p-Cl-amphetamine (p-Cl-amph) releases 5-HT and then causes a depletion of 5-HT in the brain. The electrochemical signal measured in the hippocampus reflected this pattern (49,69). Thus, 5-HT is the putative component of the signal measured after p-Cl-amph administration. The stimulated 5-HT release induced by p-Cl-amph could be prevented by prior administration of p-Cl-phenylalanine (p-Cl-phe), a 5-HT synthesis inhibitor. In addition, since the rats were unanesthetized and unrestrained, these workers could show that the time course of the behavioral manifestations of the drug and the electrochemical response were essentially parallel (69). Behavior changes were seen ∿5 minutes before any change in the electrochemistry was noted (potential steps were applied every 5 minutes) and hyperactivity lasted after the current returned to baseline. An injection of fluoxetine, a 5-HT (and p-Cl-amph) uptake inhibitor, decreased both the behavior and current induced by p-Cl-amph. Electrical stimulation of the substantia nigra (49,70) and the Raphé nucleus (69,70) both led to sharp increases in the electrochemical signal measured in the caudate nucleus and hippocampus respectively. Again, pretreatment with p-Cl-phe blocked the response from the hippocampus.

Huff and Adams (71) monitored changes in the caudate nucleus and the nucleus accumbens simultaneously to demonstrate the differential effects of DA receptor blockers. Chlorpromazine, a drug with significant extrapyramidal side effects, was shown to cause increased current in both nuclei. In contrast, clozapine only affected the nucleus accumbens, a result consistent with the claim that it causes fewer extrapyramidal side effects. Destruction of the substantia nigra catecholamine cell bodies by 6-OHDA injection eliminated the response to either receptor blocker. These lesions produced an 85% effective decrease in caudate and accumbens DA content without changing AA concentrations. While not precluding the possibility, all these data from Adams and coworkers suggest that AA is not involved in the changes in current measured by *in vivo* electrochemistry, but rather that the results reflect the fluctuations of neurotransmitter levels in the brain (73).

With the slow scanning techniques, Lane *et al*. provided further pharmacological support for their peak identifications. Haloperidol (48) and spiroperidol (65), DA receptor blockers, and DOPA (48) caused increases in both peaks 1 and 3, whereas 5-HTP resulted in more current for the second peak (64). A relatively large dose of amphetamine (10 mg kg^{-1}) produced a large change attributed to DA with no effect on the putative 5-HT response (65). Chemical lesioning via 6-OHDA injection into the SN (48), biochemical disruption of DA synthesis with α-methyl-p-tyrosine (α-MPT) (65) and electrolytic destruction of the nigrostriatal pathway (65) all resulted in a 10-30% decrease in peak 1 (consistent with a decrease in DA with no change in AA) and the complete disappearance of peak 3. Amphetamine administration subsequent to α-MPT treatment was without a noticeable effect, as was spiroperidol after electrolytic lesioning (65). Administration of pargyline decreased the peak attributed to HVA and 3-MTA to 20% of its pre-injection value (48). Production of both HVA and DOPAC are markedly reduced after MAO inhibition and peak 3 reflects the HVA decrease. However, changes in peak 1 were not reported, implying that DOPAC was not a significant contributor to the baseline electrochemical signal measured in ECF at graphoxy electrodes.

Following the premise that smaller is better, Gonon et $al.$ developed carbon fiber microvoltammetric electrodes with ∿0.5 mm of the 8 μm diameter fiber extending beyond the tip of the glass pipet (51). The 0.5 mm extension was employed to increase the current magnitude and to produce a reduction peak for DOQ on the reverse scan in CV. The relative size of the reduction to oxidation peaks and difference in potential between the peaks was taken as evidence for the irreversibility of DA electrochemistry at carbon fibers. The in $vivo$ stability of these fibers is good. Injections of amphetamine (2 mg kg^{-1}) on days 2 and 30 into rats chronically implanted with carbon fiber electrodes produced statistically equivalent rises (51). In addition, 2 amphetamine injections (5 mg kg^{-1}) 5 hours apart resulted in virtually identical responses from chloral hydrate anesthetized rats (74). Searching for evidence that the signal they monitored reflected DA release, they observed a gradual decrease of the baseline current following α-MPT administration. After this baseline depression, amphetamine no longer elicited a response (74).

Electrochemistry has also been correlated with behavior and with physical stimulation of the rat. Clark and Lyons (58) noted a sharp increase in i_c in response to a startling stimulus and Marsden et $al.$ (69) correlated 5-HT release with its associated behavior. Eating and exploring food following deprivation has been shown by Justice et $al.$ (75) to cause an increase in the electrochemical signal recorded from the caudate. Tail pinch initiates an increase in motor activity which follows a similar time course to the augmented electrochemical response (76,77). Pretreatment with α-MPT (76), 6-OHDA (77), p-Cl-amph (77), or fenfluramine (77), a 5-HT uptake inhibitor, all suppress both the voltammetric and movement increases. If p-Cl-phe is administered before p-Cl-amph or fen-fluramine, the increase in i_a and locomotion are not prevented. The authors conclude that 5-HT prevents DA release induced by tail pinch (77).

11.3.3   Interpretation of *in vivo* electrochemistry

All of the results discussed in the previous section demonstrate that chemical changes in the brain can be detected with a voltammetric electrode. The chemical changes are induced by pharmacological agents thought to affect neuronal activity and correlate with behavioral studies. As will be shown, the chemical changes also correlate with direct electrophysiological measures of neuronal firing rates. However, interpretation of the voltammetric data is very complex and many of the problems are just beginning to be realized.

As discussed in section 11.2.2, filming of the electrode surface *in vivo* tends to distort the response. Filming caused by protein adsorption or by electrochemically generated products causes waves to be less well defined and their current amplitude to diminish. Filming reduces the effective electrode area and thus concentrations measured by *in vivo* electrochemistry may be underestimated. As we have shown, maintaining the electrode for the majority of the time at a potential where electrochemistry does not occur minimizes filming (13). The type of carbon electrodes used also seems to affect the rate at which the electrode response deteriorates. Calibrating the electrode after the *in vivo* experiment seems to compensate for electrode deterioration caused by insertion in the brain.

The observed concentration of electroactive species may also differ from the true concentration because a small pool of extracellular fluid forms at the tip of an electrode implanted in brain tissue (55,78). This phenomenon is schematically represented for a 100 μm and a 10 μm radius electrode in Fig. 11.8. The response at electrodes of 50-100 μm radius is based on establishing a balance between depletion of electroactive species in the pool by electrolysis and restricted transport of new compounds into the pool from the surrounding brain tissue. Because of the restricted transport into the pool, this phenomenon distorts both the magnitude and time course of observed concentration changes. By decreasing the volume of the pool, a consequence of small electrodes, and decreasing step times, distortion caused by this pocket of extracellular fluid at the tip of the electrode can be minimized (13). The presence of a restricted diffusion pool could explain why peak shaped voltammograms are so often found in brain tissue, especially at slow scan rates. During the potential scan, the concentration of electroactive species directly adjacent to the electrode is depleted giving a rapid, but artifactual, decay in current.

To identify the actual compounds which are observed *in vivo*, modified electrodes have been used to a limited extent. The first attempt by Lane *et al.*

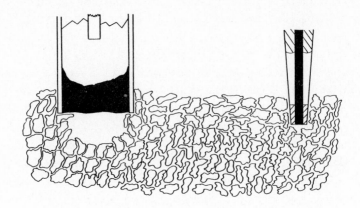

Fig. 11.8. Schematic of the pools formed at the electrode tips of a graphite-epoxy electrode (r ≃ 50 μm) and of a carbon fiber microvoltammetric electrode (r ≃ 7 μm).

was an iodide-coated Pt electrode and a modified pulse technique. This electrode separates the DA and AA voltammetric responses (12,79). The iodide layer is thought to block adsorption of physiological components. Using McCreery's injection design (61), they injected DA or AA and demonstrated the independence of the two peaks from one another. A local injection of amphetamine caused a substantial increase in the height of the peak attributed to DA with no concurrent change in the AA peak. The problem with the Pt-I electrodes is that they are not stable for long time periods, giving reproducible results for only about 1 hour in the brain. In addition, the potential range is limited to less than 0.5 V *vs.* Ag/AgCl to prevent desorption of the I⁻ (12,70).

Buda *et al.* have electrochemically modified carbon fiber microelectrodes by application of a 70 Hz triangular wave from 0.0-3.0 V to the electrode in a phosphate-buffered saline solution (PBS) (80,81). This treatment allows DPV resolution of AA (E_p ≃ -50 mV *vs.* Ag/AgCl, peak 1) from DA and DOPAC (E_p ≃ 100 mV, peak 2) by decreasing the overpotential for AA oxidation at the carbon surface. Such violent treatment also produces defects in the crystal lattice structure of the carbon fiber, resulting in a porous electrode with an increased surface area and with a high density of surface functional groups. Consequently, these modified electrodes are more sensitive probes. However, the porosity of the electrode should have a similar effect on time and concentration responses to that predicted by pool theory. Both phenomena cause marked damping of the true ECF concentration changes. They display linear calibration curves *in vitro* for DA, DOPA, DOPAC, and AA, being most sensitive to DA and DOPA, ∿10 times less responsive to DOPAC than DA and 250 times poorer for AA determinations than for DA. The discrepancy between the responses to the various compounds is not explainable by simple DPV theory (82) and thus requires the invocation of theories involving adsorption, differences in accessible area, differences in reversibility, or catalytic contributions to the catechol peak to explain the observed results.

Recording differential pulse voltammograms every 2.5 minutes from -200 to +200 mV *vs*. Ag/AgCl, they demonstrated both peaks were still observed *in vivo*, and that the two peaks were differentially sensitive to pharmacological manipulations. The peak assigned to AA, often distorted and variable (81), appeared to be partially a function of the depth of anesthesia. Peak 1 disappeared from scorbutic guinea pigs while the catechol peak decreased ∿50% (80). Lesions with 6-OHDA eliminated the catechol peak, with no effect on AA. Similarly, α-MPT, NSD1015 (a DOPA decarboxylase inhibitor) and pargyline all suppressed peak 2, with pargyline completely eliminating it (80,81). All these drugs prevent eventual DOPAC formation. Thus, peak 2 apparently reflects DOPAC dynamics *in vivo*. Making this assumption, background DOPAC levels in the rat caudate nucleus are reported to be 22 ± 4 μM (80,81).

Haloperidol and reserpine both cause large increases in the catechol peak (80,81,83). Amphetamine decreases the catechol peak in contrast to the results obtained with Lane's modified electrode (79), presumably *via* its inhibitor effects on MAO (80,81). Simultaneously, amphetamine increased the height of the peak attributed to AA. The effect of reserpine was totally abolished by pargyline pretreatment, whereas haloperidol and amphetamine under similar conditions each resulted in a small peak 2, presumably due to free DA (81). Thus, Gonon *et al*. maintain that the two major components of *in vivo* electrochemical signals from the caudate nucleus of the rat are AA and DOPAC (81). The same claim may be made for the nucleus accumbens, SN, and ventral mesencephalic tegmentum (83).

In contrast to the long, thin electrodes just described, we have designed carbon fiber electrodes (5,8) with an active surface area defined by the diameter of the fiber (8 μm). This geometry utilizes to the greatest extent the capacity of carbon fibers for spatial discrimination. The small size of the electrodes provides several other benefits in addition to spatial resolution of regions within the brain. Carbon fiber microelectrodes cause less perturbation of brain tissue during implantation and are thus not subject to the distortion of time and concentration changes resulting from pool formation at the electrode tip (13). As a consequence, we observe a baseline current corresponding to ∿500 μM AA, considerably higher than that seen within larger electrodes (49). The effect of the EC catalytic reaction between DOQ and AA to regenerate DA is minimized (8). This allows partial resolution of DA oxidation from that of AA (8,13). Filming does occur, but apparently it happens primarily during electrode implantation. In addition, it does not prevent acquisition of data from short potential pulses (13). Thus, voltammograms can be utilized for measuring the time course of changes in chemical concentrations in the brain, and also for identifying the half-wave potentials of these compounds. To demonstrate the ability of our electrodes to differentiate DA from AA and DOPAC, these compounds have been injected into the brains of anesthetized rats while recording normal pulse voltammograms. The selectivity obtained for DA over AA and DOPAC is illustrated in Fig. 11.9.

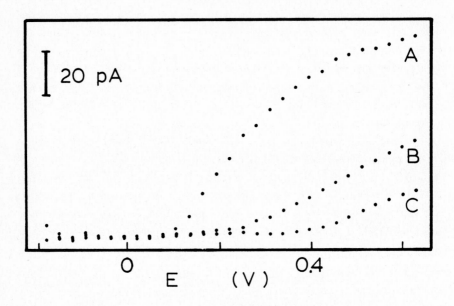

Fig. 11.9. Normal pulse voltammograms of compounds injected into the cortex of an anesthetized rat recorded as they diffused past a microelectrode. Concentrations of each were ≈30 μM, 20 mV sec^{-1} scan rate. A) α-MeDA. B) AA. C) DOPAC.

To our surprise, voltammograms recorded after amphetamine administration do not resemble those for DA. Unilateral lesions of the nigrostriatal tract with 6-OHDA altered, but did not abolish, the amphetamine-induced response in the striatum (Fig. 11.10). The chronoamperometric current reached a maximum of similar magnitude but at a later time than that for the intact side. Reinterpretation of data reported by Gonon *et al.* (74) shows a similar trend. Voltammograms recorded on the lesioned side were very similar to those obtained on the intact side and both resemble those for AA. The difference in time course observed on the lesioned and non-lesioned sides suggests that DA neurons affect the electrochemical response. However, these data also indicate that striatal

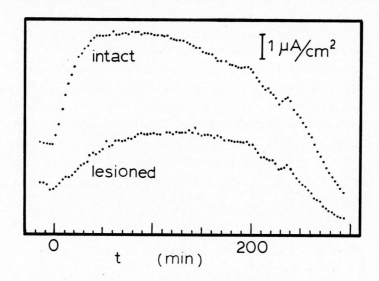

Fig. 11.10. Chronoamperometric response (0.55 V) to i.p. amphetamine administration (t = 0, 1.8 mg kg^{-1}) to an anesthetized rat previously unilaterally depleted of DA (96% lesioned) by 6-OHDA injection into the substantia nigra. The current response has been normalized to electrode areas calculated from post *in vivo* calibration with DA.

DA (or its metabolites) is not the only easily-oxidized compound which changes in concentration in the ECF of anesthetized rats following amphetamine administration. Our results indicate the need for careful interpretation of *in vivo* voltammetric results. However, even if AA is what everyone has been observing in *in vivo* electrochemistry, this compound may be playing an important, and previously unrecognized, role in neuronal activity. Recent *in vitro* data from Adams' lab (84) and data from push-pull cannula experiments (85) further support this contention.

11.3.4 Future directions

Despite the uncertainties in the interpretation of electrochemical experiments, several recent papers demonstrate new approaches and applications of *in vivo* electrochemistry. Combining electrochemistry with an expertise in iontophoretic techniques and electrophysiology, Armstrong-James *et al.* have vastly improved the reliability of iontophoretic studies (86-89). Desiring a low-noise, high impedance electrophysiology probe which could be incorporated into a multiple barrel assembly, they investigated the feasibility of employing carbon fibers (52,90-92). Iontophoresis did not affect the simultaneous acquisition of multiple unit recordings (52). In addition, by switching processing modes, fast (280 V sec^{-1}) cyclic voltammograms could be recorded through the carbon fiber after iontophoresis (87-89). Calibration curves relating the ejection current to concentration in a saline solution were non-linear for NA (87), 5-HT (88), and DA (89). Electrochemical signals *in vivo* after injection into the somatosensory cortex of cats were not predictable from the *in vitro* experiments. A larger voltammetric current *in vivo* for a given ejection current was attributed to restricted diffusion in the brain, whereas a lower value possibly reflected increased uptake of the injected substance (87-89). These data emphasize the necessity for monitoring the concentrations of compounds injected by iontophoresis if proper correlations of concentrations with effect are to be made.

Also demonstrated by these experiments is the feasibility of combining electrophysiology with electrochemistry. Armstrong-James *et al.* have elected to use the carbon fiber both for electrophysiology and electrochemistry alternately by switching processing modes (87-89). Construction of an assembly with an additional carbon fiber in a fourth barrel would allow simultaneous measurements with the two techniques.

A different approach to the concurrent application of these complementary methodologies has proven their compatibility. Schenk *et al.* (93) and Ewing *et al.* (94) have independently implemented experiments in which unit recording and electrochemical probes were implanted into the same rat. With proper care to insure electrical isolation of the two systems, it was shown that results from the combination were equivalent to those run individually, allowing for variations among rats.

Since the electrophysiology and electrochemistry were performed simultaneously, correlation of the two techniques was simplified. Thus, Schenk *et al.* (93) showed the relationship between electrophysiology and electrochemistry following 5-HT injection into the hippocampus, amphetamine administration (i.p.), and cortical depression induced by 5% KCl. Typical voltammetric results (68) were obtained following amphetamine administration, which corresponded to a decrease in the firing rate of SN DA-containing neurons. Working with George Rebec, we have measured firing rate decreases in the striatum with low doses of amphetamine that correlate with an increase in electrochemical signal (Fig. 11.11).

Refinements of the electrodes will be an integral part of the strategy to differentiate among the possible major components of the electrochemical response. Charged moieties have been shown to influence the rate of electrochemical reactions (8,95). Modification of the charge on electrode surfaces should allow a degree of specificity between cations (DA) and anions (AA, DOPAC) to be achieved. In this manner, it may be possible to establish a role played by AA in regulating neuronal activity. Indeed, the emphasis of *in vivo*

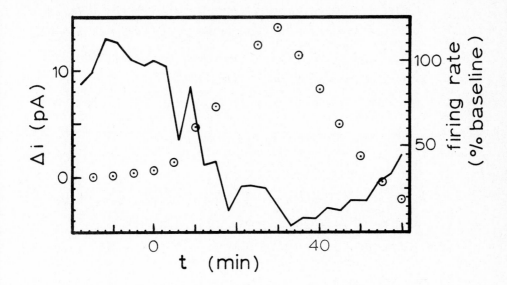

Fig. 11.11. Comparison of the electrochemical response ($\odot$, 0.55 V) to amphetamine administration (t = 0, 2.5 mg kg^{-1}) with that of the electrophysiological
single unit recording (——) from probes implanted in the corpus striatum of a
tubocurarine-immobilized rat. Voltammetric current reported as change from
baseline, unit recordings as % baseline.

electrochemistry may shift from the *in situ* analysis of catecholamines to elu-
cidating the effects of drugs on AA concentration fluctuations. Thus, electrochemistry, already a valuable component of the neurochemist's armamentarium,
has the potential for exciting new developments in the near future.

REFERENCES

1 R. N. Adams, Electrochemistry at Solid Electrodes, Marcel Dekker, New York,
   1969.
2 A. J. Bard and L. R. Faulkner, Electrochemical Methods, Fundamentals and
   Applications, John Wiley and Sons, New York, 1980.
3 D. T. Sawyer and J. J. Roberts, Experimental Electrochemistry for Chemists,
   John Wiley and Sons, New York, 1974.
4 J. O'M. Bockris and A. K. N. Reddy, Modern Electrochemistry, Vol. 2, Plenum
   Press, New York, 1970.
5 M. A. Dayton, J. C. Brown, K. J. Stutts, and R. M. Wightman, Anal. Chem.,
   52 (1980) 946.
6 R. M. Wightman, Anal. Chem., 53 (1981) 1125A.
7 R. S. Nicholson and I. Shain, Anal. Chem., 36 (1964) 706.
8 M. A. Dayton, A. G. Ewing, and R. M. Wightman, Anal. Chem., 52 (1980) 2392.
9 R. W. Murray, Acc. Chem. Res., 13 (1980) 135.
10 N. Ossendorfavá, J. Pradáĉ, J. Pradáĉova, and J. Koryta, J. Electroanal. Chem.,
   58 (1975) 255.
11 J. S. Mattson and T. T. Jones, Anal. Chem., 48 (1976) 2164.
12 R. F. Lane and A. T. Hubbard, Anal. Chem., 48 (1976) 1287.

13 A. G. Ewing, M. A. Dayton and R. M. Wightman, Anal. Chem., 53 (1981) in press.
14. A. W. Sternson, R. McCreery, B. Feinberg and R. N. Adams, J. Electroanal. Chem., 46 (1973) 313.
15 P. T. Kissinger, C. J. Refshauge, R. Dreiling and R. N. Adams, Anal. Lett., 6 (1973) 465.
16 I. N. Mefford, J. Neurosci. Meth., 3 (1981) 207.
17 P. T. Kissinger, C. S. Bruntlett and R. E. Shoup, Life Sci., 28 (1981) 455.
18 R. Keller, A. Oke, I. Mefford and R. N. Adams, Life Sci., 19 (1976) 995.
19 F. Ponzio and G. Jonsson, J. Neurochem., 32 (1979) 129.
20 I. Lacković, M. Parenti and N. H. Neff, Eur. J. Pharm., 69 (1981) 347.
21 J. Lankelma and H. Poppe, J. Chromatogr., 125 (1976) 375.
22 S. G. Weber and W. C. Purdy, Anal. Chim. Acta, 100 (1978) 531.
23 D. A. Richards, J. Chromatogr., 175 (1979) 293.
24 G. A. Scratchley, A. N. Masoud, J. J. Stohs and D. W. Wingard, J. Chromatogr., 169 (1979) 313.
25 J. J. Warsh, A. Chiu, P. P. Li and D. D. Godse, J. Chromatogr., Biomed. Appl., 183 (1980) 483.
26 P. Hjemdahl, M. Daleskog and T. Kahan, Life Sci., 25 (1979) 131.
27 D. S. Goldstein, G. Feuerstein, J. L. Izzo, I. J. Kopin and H. R. Keiser, Life Sci., 28 (1981) 467.
28 F. M. Rabel, J. Chromatogr. Sci., 18 (1980) 394.
29 K. Brunt, C. H. P. Bruins, D. A. Doornbos and B. Oosterhuis, Anal. Chim. Acta, 114 (1980) 257.
30 K. Brunt and C. H. P. Bruins, J. Chromatogr., 172 (1979) 37.
31 C. L. Blank, J. Chromatogr., 117 (1976) 35.
32 Y. Hirata, P. T. Lin, M. Novotny and R. M. Wightman, J. Chromatogr., Biomed. Appl., 181 (1980) 287.
33 P. O. Bollet and M. Caude, J. Chromatogr., 149 (1977) 625.
34 R. M. Wightman, E. C. Paik, S. Borman and M. A. Dayton, Anal. Chem., 50 (1978) 1410.
35 D. J. Chesney, J. L. Anderson, D. E. Weisshaar and D. E. Tallman, Anal. Chim. Acta, 124 (1981) 321.
36 C. L. Blank and R. Pike, Life Sci., 18 (1976) 859.
37 G. C. Davis and P. T. Kissinger, Anal. Chem., 51 (1979) 1960.
38 R. E. Shoup, G. C. Davis and P. T. Kissinger, Anal. Chem., 52 (1980) 483.
39 R. T. Borchhardt, M. F. Hegazi and R. L. Schowen, J. Chromatogr., 152 (1978) 255.
40 P. M. Plotsky, R. M. Wightman, W. Chey and R. N. Adams, Science, 197 (1977) 904.
41 M. A. Dayton, G. E. Geier and R. M. Wightman, Life Sci., 24 (1979) 917.
42 C. C. Loullis, J. N. Hingtgen, P. A. Shea and M. H. Aprison, Pharmacol. Biochem. Behavior, 12 (1980) 959.
43 R. M. Wightman, C. E. Bright and J. N. Caviness, Life Sci., 28 (1981) 1279.
44 D. J. Green and R. L. Perlman, Anal. Biochem., 110 (1981) 270.
45 J. L. Meek, H.-Y. T. Yang and E. Costa, Life Sci., 16 (1977) 151.
46 W. L. Caudill, G. P. Houck and R. M. Wightman, J. Chromatogr., Biomed. Appl., in press.
47 R. M. Wightman, E. Strope, P. Plotsky and R. N. Adams, Brain Research, 159 (1978) 55.
48 R. F. Lane, A. T. Hubbard and C. D. Blaha, Bioelectrochem. Bioenergetics, 5 (1978) 504.
49 J. C. Conti, E. Strope, R. N. Adams and C. A. Marsden, Life Sci., 23 (1978) 2705.
50 W. S. Lindsay, B. L. Kizzort, J. B. Justice, J. C. Salamone and D. B. Neill, Chem. Biomed., Environ. Instrumentation, 10 (1980) 311.
51 J.-L. Ponchon, R. Cespuglio, F. Gonon, M. Jouvet, and J.-F. Pujol, Anal. Chem., 51 (1979) 1483.
52 M. Armstrong-James and J. Millar, J. Neurosci. Meth., 1 (1979) 279.
53 G. Bartholini, H. Stadler, M. Gadea Ciria and K. G. Lloyd, Neuropharmacol., 15 (1976) 515.

54 W. S. Lindsay, B. L. Kizzort, J. B. Justice, J. D. Salamone and D. G. Neill, J. Neurosci. Meth., 2 (1980) 373.
55 W. S. Lindsay and J. B. Justice, Computers and Chem., 4 (1980) 19.
56 A. Bayon, W. J. Shoemaker, L. Lugo, R. Azad, N. Ling, R. R. Drucker-Colin and F. E. Bloom, Neurosci. Lett., 24 (1981) 65.
57 L. C. Clark, Biomed. Sci. Instrum., 2 (1964) 165.
58 L. C. Clark and C. Lyons, Ala. J. Med. Sci., 2 (1965) 353.
59 D. C. S. Tse, R. L. McCreery and R. N. Adams, J. Med. Chem., 19 (1976) 37.
60 P. T. Kissinger, J. B. Hart and R. N. Adams, Brain Research, 55 (1973) 209.
61 R. L. McCreery, R. Dreiling and R. N. Adams, Brain Research, 73 (1974) 23.
62 R. L. McCreery, R. Dreiling and R. N. Adams, Brain Research, 73 (1974) 15.
63 R. M. Wightman, E. Strope, P. Plotsky and R. N. Adams, Nature, 262 (1976) 145.
64 R. F. Lane, A. T. Hubbard and C. D. Blaha, J. Electroanal. Chem., 95 (1979) 117.
65 R. F. Lane, A. T. Hubbard and C. D. Blaha, Catecholamines: Basic and Clinical Frontiers, E. Usdin, I. J. Kopin and J. Barchas (Eds.), Pergamon Press, New York (1979) 883.
66 P. A. Loach, Handbook of Biochemistry and Molecular Biology, Vol. 1, G. D. Fasman (Ed.) CRC Press, Inc., Cleveland (1976), 122.
67 J. F. Evans, T. Kuwana, M. T. Henne and G. P. Royer, J. Electroanal. Chem., 80 (1977) 409.
68 R. Huff, R. N. Adams and C. O. Rutledge, Brain Research, 173 (1979) 369.
69 C. A. Marsden, J. Conti, E. Strope, G. Curzon and R. N. Adams, Brain Research, 171 (1979) 85.
70 R. N. Adams, J. Conti, C. A. Marsden and E. Strope, Br. J. Pharm., 64 (1978) 470P.
71 R. Huff and R. N. Adams, Neuropharm., 19 (1980) 587.
72 G. M. McKenzie and J. C. Szerb, J. Pharmacol. Exp. Ther., 162 (1968) 302.
73 R. N. Adams, Anal. Chem., 48 (1976) 1126A.
74 F. Gonon, R. Cespuglio, J.-L. Ponchon, M. Buda, M. Jouvet, R. N. Adams and J.-F. Pujol, Hebd. Seances Acad. Sci. Ser. D, 286 (1978) 1203.
75 J. B. Justice, W. S. Lindsay, B. L. Kizzort, D. B. Neill and J. Salamone, Proc. 2nd Ann. Conf. Engineering Med. Biol. Soc. of I.E.E.E., 2 (1980) 46.
76 G. Curzon, P. H. Hutson and P. J. Knott, Br. J. Pharm., 66 (1979) 127P.
77 G. Curzon, P. H. Hutson and P. J. Knott, Br. J. Pharm., 70 (1980) 132P.
78 H.-Y. Cheng, J. Schenk, R. Huff and R. N. Adams, J. Electroanal. Chem., 100 (1979) 23.
79 R. F. Lane, A. T. Hubbard, K. Fukunaga and R. J. Blanchard, Brain Research, 114 (1976) 346.
80 M. Buda, F. Gonon, R. Cespuglio, M. Jouvet and J.-F. Pujol, Hebd. Seances Acad. Sci. Ser. D, 290 (1980) 431.
81 F. Gonon, M. Buda, R. Cespuglio, M. Jouvet and J.-F. Pujol, Nature, 286 (1980) 902.
82 K. J. Stutts, M. A. Dayton and R. M. Wightman, submitted to Anal. Chem. (1981).
83 M. Buda, F. Gonon, R. Cespuglio, M. Jouvet and J.-F. Pujol, submitted to Eur. J. Pharmacol. (1981).
84 K. H. Milby, I. N. Mefford, W. Chey and R. N. Adams, submitted to Brain Research (1981).
85 J. Justice, personal communication, 1981.
86 M. Armstrong-James, Z. L. Kruk and J. Millar, J. Physiol. (Lond.), 308 (1980) 115P.
87 M. Armstrong-James, J. Millar and Z. L. Kruk, Nature, 288 (1980) 181.
88 Z. L. Kruk, M. Armstrong-James and J. Millar, Life Sci., 27 (1980) 2093.
89 J. Millar, M. Armstrong-James and Z. L. Kruk, Brain Research, 205 (1981) 419.
90 M. Armstrong-James and J. Millar, J. Physiol. (Lond.), 305 (1980) 65P.
91 M. Armstrong-James and J. Millar, J. Physiol. (Lond.), 308 (1980) 116P.
92 M. Armstrong-James, K. Fox and J. Millar, J. Neurosci. Meth., 2 (1980) 431.

93 J. O. Schenk, R. N. Adams and G. Christoph, unpublished data, 1981.
94 A. G. Ewing, M. A. Dayton, R. M. Wightman, S. Curtis, K. Alloway and G. Rebec, unpublished data, 1981.
95 R. F. Lane and A. T. Hubbard, J. Phys. Chem., 77 (1973) 1411.

Chapter 12

RADIORECEPTOR ASSAYS

JOHN W. FERKANY

Departments of Neuroscience, Pharmacology and Psychiatry, Johns Hopkins
University School of Medicine, Baltimore, Maryland 21205

and

S. J. ENNA
Departments of Pharmacology and of Neurobiology and Anatomy, University of

Texas Medical School, Houston, Texas 77025

12.1 INTRODUCTION

The contents of this volume bear witness to the importance of analytic
techniques for understanding the chemical nature of neurotransmission. During
the past century, this science has advanced from rudimentary biological assays
(1-3), to the use of highly sophisticated and expensive electronic equipment.
Sensitivity, specificity, versatility, precision, speed, simplicity and cost are
the chief considerations when contemplating the use or development of an analytical
procedure. As is evident from these chapters, no assay has yet been developed
that is superior to all others in every category.

The ability to measure trace quantities of biological chemicals and drugs
has been a driving force in the advancement of biomedical science. The development
of increasingly more sensitive assays has facilitated the study of known agents
in addition to aiding in the identification and quantification of biologically
important substances that were unknown to the last generation of scientists.
Furthermore, these advances have made it possible to detect vanishingly small
quantities of chemicals in biological fluids and in discrete regions of tissues
and organs.

One of the methods that has been particularly useful in this regard is the
radioimmunoassay (Chapter 13, this volume). The principle of this technique is
that the quantity of a radiolabeled antigen bound to a specific antibody is a
function of the amount of unlabeled antigen present in the assay medium. This
principle has now been applied for use with radioligand membrane binding assays
for neurotransmitter and drug receptors. In this latter case the method is
termed a radioreceptor assay (RRA).

For radioimmunoassay, antibodies must first be prepared against the substance
to be analysed. Because of this, these assays are best suited for quantifying

more highly antigenic molecular entities, such as proteins. However, the procedure
has also been successfully used to measure steroid hormones and cyclic nucleotides,
since these agents can be rendered more antigenic by covalently linking them to
larger structures. Nevertheless, the radioimmunoassay is of limited use for
studying most drugs, neurotransmitters and their metabolites. In the case of
drugs, covalent attachment to a larger molecule may produce an antibody which
does not recognize the parent compound.

The source of binding material for the RRA is biological tissue. Therefore,
the assay can, in theory, be employed to measure any substance which interacts
with a membrane receptor. For example, Lefkowitz et al. (6) first used this
approach to measure ACTH in tissue extracts by taking advantage of the finding
that ^{125}I-ACTH bound in a specific manner to receptors present in adrenal cortical
tissue. Subsequently, numerous receptor binding assays have been developed and
a number of these have been used as radioreceptor assays. For instance, γ-
aminobutyric acid (GABA) receptor binding to brain membranes (7) is now widely
employed as a means for measuring the GABA content of various biological tissues
and fluids, including brain, blood and CSF (8,9).

The RRA has several advantages over other analytical procedures. Because
of its simplicity, this assay can be readily mastered by inexperienced investigators
providing them with a sensitive, rapid, and accurate means of measuring a wide
variety of biological substances. Furthermore, in some instances, the RRA has
proven to be the most sensitive means of detection. While GC/MS and HPLC are,
from a qualitative standpoint, superior, quantitatively the RRA is equal or
superior to these more costly and sophisticated techniques.

This chapter discusses the fundamentals of the RRA as an analytical procedure
for biological research. While emphasis is placed on the practical aspects of
this assay, theoretical issues are considered when necessary. Both the advantages
and disadvantages of the RRA relative to other assays will be catalogued and
specific examples used for instructive purposes. Readers desiring further information on this topic are urged to consult any of a number recent reviews
(10-13).

12.2 COMPETITIVE BINDING ASSAYS

Since receptor binding assays are the basis of the RRA, knowledge of the
basic principles of ligand-receptor interactions is essential for understanding
the utility and limitations of this analytical procedure. In addition, factors
which influence receptor binding must be taken into account when developing an
RRA since subtle alterations in the binding assay can dramatically affect the
sensitivity and specificity of the method.

12.2.1 <u>Theoretical Considerations</u>

(i) <u>Saturation</u>. Receptor-ligand binding is characterized by two important
properties, both of which must be characterized to establish the validity of a
particular RRA. The first of these is saturability. That is, the number of
physiologically relevant membrane receptors is quite small for any given transmitter
or drug. In contrast, the number of nonspecific attachment sites is virtually
infinite. Accordingly, when a receptor binding assay is conducted under equilibrium
conditions, increasing the amount of radioligand in the incubation medium should
also increase the amount of receptor-specific binding up to a point at which
saturation occurs. The inflection in a binding curve represents the point at
which all of the specific binding sites are occupied with ligand (saturation).

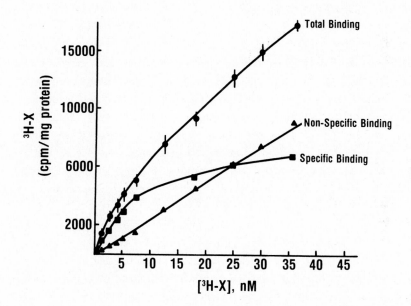

Figure 12.1 Binding curve for the attachment of ^{3}H-X to its specific receptor
site showing total, specific and non-specific binding. In an experiment of this
type, a fixed amount of tissue is incubated in the presence of increasing
concentrations of radioligand with or without an excess of unlabeled ligand.
Linear transformation of these data would reveal that only a single population
of receptors was present.

A typical binding curve for an ideal ligand is shown on Figure 12.1. In
this experiment, membranes containing receptors for the radioligand ^{3}H-X are
incubated with increasing concentrations of the isotope in the presence and
absence of a large excess of unlabeled compound X. Total binding is represented

by the amount of radioactivity bound to the membranes at each concentration of
$3H-X$ in the absence of the unlabeled ligand. Nonspecific binding (blank) is
the amount of residual radioactivity found when the incubation takes place in
the presence of excess unlabeled ligand. In contrast to receptor-specific
binding, the number of nonspecific binding sites is exceedingly large, and
therefore can accommodate both labeled and unlabeled ligand. Because of this,
nonspecific binding increases in a linear fashion with increasing concentrations
of isotope (Fig. 12.1). Specific binding therefore is the difference between
total and nonspecific binding, and it is this displaceable ligand that is presumably
attached to the biologically relevant receptor.

This type of data yields two important pieces of information. The point
at which specific binding plateaus represents the concentration of receptor
sites (B_{max}), whereas the rate at which the curve approaches this plateau reflects
the affinity (K_d) of the ligand for the receptor. More precise determinations
of B_{max} and K_d are routinely made using a linear transformation. The most popular
of these is the one devised by Scatchard (14), although other methods are equally
acceptable. Transformation of the data shown on Figure 12.1 yields a straight
line, indicating that this ligand binds to a single population of sites. In
many cases, however, binding curves have multiple components, indicating several
sites with differing affinities. Nevertheless, even in these situations, a
judicious choice of ligand concentration limits the receptor-ligand interaction
to a single portion of the binding curve. If the concentration of radioactive
tracer is substantially less than the K_d for the highest affinity sites, the
attachment of ligand can be considered selective for this population, and the
system may be treated as though it were monophasic.

(ii) Specificity. While crucial, saturability in itself is not sufficient
for conclusively demonstrating that a ligand binds to the physiologically relevant
site. For this it is imperative that the specificity of the binding site be
established. That is, proof that a substance is binding to a physiologically
relevant receptor requires that the ligand-receptor interaction be sensitive
only to those compounds known to be biologically active at this site. For example,
electrophysical studies have revealed a number of compounds that are rather selective
as GABA receptor agonists and antagonists. Agents such as muscimol, THIP and
kojic amine are all GABA receptor agonists, based on electrophysiological findings
(15, 16). In addition, these compounds inhibit $3H$-GABA binding to brain membranes
in vitro, with an order of potency that is quite similar to that found in electrophysiological tests (17). In contrast, GABA analogs such as diaminobutyric
acid and guvacine, inhibitors of GABA transport, are devoid of any appreciable
electrophysiological activity and are inactive as displacers of specifically
bound $3H$-GABA. Furthermore, (+)-bicuculline, a GABA receptor antagonist, is
active as a displacer of specifically bound $3H$-GABA, whereas the physiologically

less active isomer (-)-bicuculline is much weaker as an inhibitor of GABA binding. Likewise, (+)-butaclomol, a clinically effective neuroleptic, is more potent than (-)-butaclomol as an inhibitor of dopamine receptor binding (18).

The degree of saturability and specificity are very much dependent upon incubation time, temperature, tissue concentration, pH, and buffer. Generally, however, by the time a binding assay is used as an RRA, these conditions have been established to insure optimal binding site specificity. However, because changes in these conditions can modify receptor binding parameters, it is important that they be carefully maintained when undertaking an RRA.

12.2.2 Practical Considerations

Since receptor ligand binding is saturable and specific, it is possible to use this technique for quantifying neurotransmitter and drug concentrations. Thus, for an RRA, by incubating membranes with a radioligand that binds in a specific and saturable manner it is possible to calculate the amount of unlabeled ligand in the incubation mixture by simply determining the degree of inhibition of the specific binding, and comparing this value to a standard displacement curve for the radioligand. Since many factors influence receptor-ligand interactions, it is important to be cognizant of those which may be of particular importance with respect to radioreceptor assays.

(i) Tissue. One of the immediate challenges that must be met when conducting a binding assay is the low density of specific receptor sites relative to other cellular components. For instance, only femtomole quantities of radioligands are bound to brain receptor sites at saturating concentrations (10-12). While the absolute number of physiologically relevant receptors for a substance in tissue may be low, the number of membrane constituents that will nonspecifically bind the ligand is high. For this reason, some purification of the receptor complex is attempted. With brain, a frequent source of tissue for the RRA, methods employing differential centrifugation are used for this purpose. Even though these procedures rarely produce a homogeneous or highly purified receptor sample, by removing extraneous tissue components they serve to enhance the ratio of specific to nonspecific binding. With respect to the RRA, an increase in this ratio increases the precision of the assay. This point is illustrated on Figure 12.2. In this example, curve A represents ligand binding to an unpurified whole tissue homogenate, while curve B represents binding to a receptor-enriched tissue preparation. In both cases, specific binding is 5000 cpm. However, if the cumulative error (the sum of the experimental and counting error) of the assay is ten percent, the range of possible values estimated for a sample from curve A will be 60 percent greater than those estimated for the same sample from curve B at the midpoints of the displacement curves. The difference in accuracy occurs because nonspecific ligand binding, which contributes disproportionately to the error of the assay, has been reduced in the purified tissue preparation. In displacement curve A, nonspecific binding represents 50% of the total, whereas in curve B it is only about 30% of total binding.

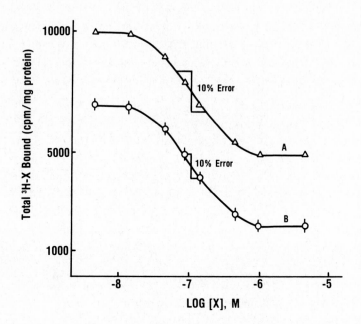

Figure 12.2. Displacement curves for a ligand binding in whole tissue homogenate (A) and a receptor-enriched tissue preparation (B). For this experiment, a fixed concentration of tissue and radioligand is incubated with increasing concentrations of unlabeled ligand. The horizontal bars represent the range of possible values obtained from each curve for a concentration of displacer which should inhibit 50 percent of the specifically bound ligand, assuming the cumulative error for each system is 10 percent.

Inspection of either curve will also reveal that the most accurate estimate of the amount of unlabeled ligand present is obtained when 50 percent of the specifically bound radioactivity is displaced. This point is better illustrated when displacement data are plotted in a linear fashion on logit paper (19) (Fig. 12.3). However, for practical purposes, radioreceptor assays are quite accurate and reproducible if the sample assayed displaces less than 80 percent, or more than 20 percent, of the specifically bound tracer.

(ii) pH. Receptor binding assays have an optimal pH at which the attachment of the ligand to the receptor complex is most avid. While for some receptors there may be a fairly wide optimal pH range, with others it may be quite limited.

Since in preparing a sample for analysis by RRA it may be necessary to extract the substance into an acidic or alkaline medium, it is important to ascertain whether the sample extract significantly alters the pH of the receptor binding assay medium. This problem can sometimes be circumvented by diluting the extract prior to assay. However, when possible, it is advisable to prepare the final sample in a neutral solution, or in the incubation buffer itself.

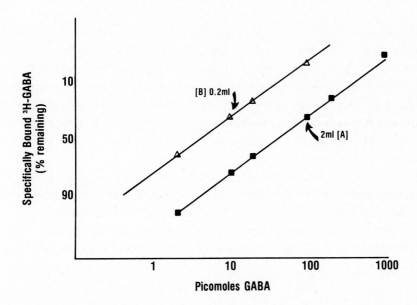

Figure 12.3. Logit plot of ^{3}H-GABA binding. Data for A obtained from Table 12.1. Line B shows the theoretical increase in assay sensitivity if the reaction is performed in a final volume of 0.2 ml.

(iii)   Contaminants. In many instances it is not necessary to purify a sample before analysing by RRA. Accordingly, the final extract contains not only the substance of interest but also a variety of other tissue constituents. Normally the presence of these contaminants has little effect on the assay, even in those situations where a major portion of the assay buffer is replaced by extract. For example, in monitoring human serum neuroleptic levels using ^{3}H-spiroperidol binding, serum is placed directly into the assay tube after the removal of an equivalent volume of buffer (20). Similarly, cerebrospinal fluid GABA is analysed without extracting the amino acid from this fluid, and a quantity of untreated cerebrospinal fluid representing up to 10% of the total volume of incubation buffer can be added to the assay medium without significant effect (21).

However, not all binding assays are resistant to the effects of extraneous substances. For instance, the serum levels of the β-adrenergic antagonist propranolol can be quantified by analysing the potency of the samples to displace specifically bound ^{3}H-dihydroalprenolol from brain membranes (22). Unlike the neuroleptic assay however, human serum must first be dialyzed before use in this RRA since minute amounts of plasma proteins significantly interfere with the binding of this tracer.

It is also noteworthy that some receptor binding assays are modified by inorganic ions. For example, when freshly prepared brain membranes are used for the assay, ^{3}H-GABA binding is dramatically altered in the presence of Na^+ (23). Indeed, when using fresh tissue and physiological concentrations of this ion, GABA is bound primarily to transport sites rather than to the postsynaptic receptor. The binding to transport sites has not only a different affinity than that found with the sodium-independent site, but also a different pharmacological specificity (23). Obviously, under these conditions, the sensitivity and selectivity of an RRA for GABA would be dramatically altered. Likewise, the binding of ^{3}H-diazepam to brain membranes is modified by Cl^- (24).

Therefore, when developing an RRA, it is crucial to ascertain the effects of possible sample contaminants on the binding assay. Should effects be noted, steps must be taken to eliminate them. Usually, a further dilution of the sample or additional extraction steps will suffice. However, if it is impossible to rid the extract of interfering substances, then a correction may be made for these nonspecific effects by including comparable amounts of vehicle in the reference (standard) curve.

12.3 ASSAY PROCEDURE

12.3.1 Equipment and Reagents

The equipment and reagents required for radioreceptor assays are minimal, and normally present in any laboratory routinely engaged in biochemical studies. In fact, the requirements are such that this method can easily be used as a teaching aid in laboratory courses.

An essential requirement is a refrigerated centrifuge. The Dupont-Sorval RC-5B, or equivalent, has proven dependable in this regard. Since the centrifuge is used for preparing tissue for the binding assay, it should be equipped with a rotor capable of accommodating multiple samples of up to 50 ml volume each. In addition, if the RRA requires termination of the assay by centrifugation, it is necessary to have a rotor designed to handle a number of samples of smaller volume.

While some RRA reactions require centrifugation, most are terminated by collecting the receptor-ligand complex on a suitable support. This is usually accomplished by passing the incubation mixture over a filter maintained under reduced pressure. Tissue with bound ligand is trapped on the filter, while the unbound ligand and buffer solution pass through.

A useful device for this purpose is a custom manufactured filter apparatus

that consists of a 45-well manifold fitted on a metal solvent tank. At the end
of the binding assay, the tissue is trapped on a 2.5 cm filter located over the
manifold wells. Similar equipment can be purchased from commercial suppliers.

In addition to the centrifuge and filter apparatus, access to a liquid
scintillation counter is essential for assaying radioactivity. Additional supplies
include items such as buffers, drugs, labeled ligands and scintillation materials,
all of which are commercially available.

12.3.2 Preparation of Tissue for the Binding Assay

This step normally entails the homogenization and centrifugation of the
tissue to remove substances which may interfere with the binding assay. By enhancing
the specific binding of the ligand, these manipulations increase the accuracy and
precision of the RRA.

Since most radioreceptor assays utilize neurotransmitter receptor binding,
brain is a popular source of receptors, and therefore many similarities exist in
the way tissue is prepared for different ligands. Nevertheless, sufficient
differences exist among the various receptor binding assays that details are
important when preparing brain membranes for an RRA.

For example, to prepare tissue for the ^{3}H-GABA binding assay, whole rat
brain is homogenized in ice-cold sucrose and the crude mitochondrial fraction
is isolated by differential centrifugation. This fraction is lysed with distilled
water and washed repeatedly by homogenization and centrifugation to remove endogenous
GABA that would interfere with the binding assay. The washed pellet is frozen for
at least 18 hours, further enhancing the ratio of specific to nonspecific binding,
and eliminating the influence of Na^+.

Prior treatment of tissue with the detergent Triton-X-100 yields an additional
increase in specific binding (23). Thus, on the day of the RRA, the frozen pellet
is homogenized in sufficient Tris-Citrate buffer (0.05 M, pH 7.1, 4°C) to yield
a protein concentration of 1 mg/ml, and Triton-X-100 (0.05% v/v) is added. The
mixture is incubated for 30 minutes at 37°C, after which the tissue is centrifuged
at 48,000 x g for 10 minutes. Following this centrifugation, the supernatant is
decanted and the pellet is resuspended in fresh buffer and centrifugation and
homogenization are repeated twice again to remove residual detergent. Finally,
the pellet is resuspended in fresh buffer for use in the binding assay.

The tissue preparation contrasts markedly with other procedures. Thus,
^{3}H-diazepam binding to brain membranes requires no prior treatment with detergent
and can be performed on either fresh or frozen tissue. In some instances, such
as for measuring serotonin$_2$ receptor binding (25), freezing the tissue may cause
a precipitous decline in specific binding, necessitating analysis with freshly
prepared brain membranes. However, for most receptor binding assays, tissue can be
frozen for long periods of time with little or no loss in receptor binding capacity.

Caution must also be exercised in certain instances when selecting particular brain regions as the receptor source. For example, the binding of ^{3}H-spiroperidol to dopamine receptors in the corpus striatum has been used as the basis of an RRA to monitor serum neuroleptic levels (20). Since this isotope is capable of binding to a number of neurotransmitter receptors (adrenergic, cholinergic and serotonergic) in other brain regions, it would be inappropriate to use a whole brain preparation for this assay as specificity would be compromised.

In summary, no single tissue or preparation procedure is suitable for all radioreceptor assays. Since each method is tailored to the individual receptor binding assay, it is imperative to follow the conditions established during the development of the binding assay.

12.3.3 Sample Preparation

The preparation of a sample for RRA analysis is generally less complex than that required for more sophisticated analytical methods. For example, to measure GABA in cerebrospinal fluid, an aliquot of untreated CSF is placed directly in the binding assay, and to quantify brain GABA content the substance is simply extracted into water prior to assay (21). Neuroleptics need not be extracted from serum prior to analysis by the dopamine receptor binding method. However, in other cases, more elaborate extraction procedures are required. Thus, GABA must be extracted from blood into perchloric acid before assay since serum can significantly alter the characteristics of ^{3}H-GABA binding. In addition, because only small quantities of GABA are found in blood (400-800 picomoles/ml), a sufficient dilution cannot be made to eliminate solvent effects, making it necessary to neutralize the acid extract prior to analysis (8). Similarly, prior to analysing serum tricyclic antidepressant concentrations by RRA, these compounds must be extracted into an organic solvent (26). The organic phase is separated from the aqueous plasma mixture, and the tricyclics are re-extracted into acid. Finally, the organic phase is removed and the acid solution is neutralized before analysis.

To extract opioid peptides from brain, the tissue is first homogenized in 0.1N HCl, and the acid supernatant lyophilized. The residue is resuspended in acid, centrifuged, the supernatant decanted, neutralized and centrifuged a final time. This final extract is again lyophilized prior to RRA (27).

In summary, as with all analytical procedures, sample preparation for RRA depends on the nature of the compound to be measured and the source of the tissue from which the substance is prepared. Once again, it is essential that preliminary experiments be performed to determine whether the extraction medium alters the binding characteristics of the assay.

12.3.4 Standard Curve and Data Analysis

The standard curve for an RRA represents the displacement of specifically

bound tracer by known concentrations of unlabeled ligand. By comparing the amount
of receptor bound radioactivity displaced by the unknown sample to the standard
curve, it is possible to determine the amount of ligand in the sample being analysed.

TABLE 12.1

Standard curve analysis for GABA inhibition of ^{3}H-GABA binding to rat brain
membranes

Condition	Total cpm Bound	Specifically Bound ^{3}H-GABA (cpm)	% of Specifically Bound
^{3}H-GABA (5 nM)	9000	8000	100
+ GABA (10^{-3}M)	1000	0	0
+ GABA (5 x 10^{-7}M)	1480	480	6
+ GABA (10^{-7}M)	2600	1600	20
+ GABA (5 x 10^{-8}M)	3640	2640	33
+ GABA (10^{-8}M)	6360	5360	67
+ GABA (5 x 10^{-9}M)	7320	6320	79
+ GABA (10^{-9}M)	8680	7680	96

Displacement curve for ^{3}H-GABA binding to rat brain membranes. Tissue (1 mg
protein) was incubated with ^{3}H-GABA (5 nM) at 4°C in 2 ml 0.05 M Tris-Citrate
buffer (pH 7.1) for 5 minutes with or without various concentrations of unlabeled
GABA. The reaction was terminated by centrifugation at 48,000 x g for 10 minutes.

The protocol and results of a standard curve analysis for ^{3}H-GABA binding
to brain membranes are shown on Table 12.1. In this example, total binding in the
presence of ^{3}H-GABA alone is 9000 cpm, and nonspecific binding (in the presence
of 1 mM GABA) is 1000 cpm. The difference between these values, 8000 cpm, represents
total specific binding. Since, under these conditions, the K_d for GABA is between
10 and 50 nM, the concentration of ^{3}H-GABA used was 5 nM, well below the apparant
affinity constant. In order to insure accuracy, concentrations of unlabeled GABA
ranging between at least 10-fold lower and 10-fold higher than the K_d are used
to generate a complete displacement curve. For analysis, these data are plotted on
logit paper to obtain a straight line (19) (Fig. 12.3). Extrapolation to the
ordinate from the appropriate percent displacement reveals the concentration of
unlabeled ligand in the sample.

As discussed above, these standard curves are most accurate at the midpoint,
and therefore attempts should be made to prepare RRA samples such that no
less than 20% nor more than 80% of the specifically bound isotope is displaced.
When feasable, it is best to analyze several dilutions such that approximately
25, 50 and 75% of the specific binding is displaced. Since the assay is linear
over this range, the mean of the three values should be more accurate than a

determination made on the basis of a single point.

12.4 SENSITIVITY

The lower limit of sensitivity for an RRA depends upon a number of factors
(28, 29). From a practical standpoint, three considerations are of greatest importance. These include, (1) the affinity of the ligand for the receptor; (2) the
final incubation volume of the assay; and (3) the specific activity of the
radioligand.

While it would appear that the higher the affinity of the radioligand for
the receptor the greater the sensitivity of the RRA, this principle applies only
when the affinity of the isotope is less than that of the compound to be detected.
For instance, while the GABA agonist ^{3}H-muscimol has a 10-fold higher affinity
for the GABA binding site than GABA itself, the use of this isotope as a ligand
for detecting GABA (30) offers little advantage in terms of sensitivity since
the affinity of GABA for the receptor, and therefore the sensitivity of the
assay, remains unchanged. In some instances, the use of high affinity ligands
may even have a negative effect on assay sensitivity. Thus, ^{3}H-spiroperidol
binds to striatal dopamine receptors with a K_d of around 0.5 nM, and this assay
has been used as an RRA to study serum levels of neuroleptics (20). While it
might be anticipated that this same system may be useful to quantify dopamine
as well, the assay is less robust in this regard since 1 μM dopamine is necessary
to displace 50% of specifically bound ^{3}H-spiroperidol. Rather, it would be better
to use ^{3}H-dopamine as the binding ligand since as little as 20 nM dopamine will
inhibit 50% of the specifically bound catecholamine (31). Accordingly, in developing
a radioreceptor assay, it is best to select a radioligand with an affinity for
the receptor similar to the substance being measured.

The sensitivity of an assay is also a function of incubation volume (Fig. 12.3).
In an incubation volume of 2 ml, with an affinity of 10 nM, the amount of
GABA required to displace 50 percent of the specifically bound isotope is 45
picomoles. A 10-fold reduction in incubation volume, tissue and radioligand
concentrations reduces the amount of unlabeled ligand required to displace the
same percentage of specifically bound ^{3}H-GABA. In this case the reduction is 10-fold,
to 4.5 picomoles.

The optimum incubation volume is dependent upon two factors. The first of
these relates to the minimum amount of tissue and radioisotope that must be used
to obtain a reliable assay. If the number of cpm bound is too low, there is a
substantial reduction in assay accuracy and precision. Secondly, as the volume of
the assay is reduced, the possibility that contaminants may interfere with the assay
is greater.

As isotopes having greater specific activities become available it becomes
more feasible to reduce the volume of incubation to increase assay sensitivity.

Thus, a two-fold increase in specific activity permits a reduction in assay volume of 50% with no loss of assay precision, since while the amount of isotope is reduced by one-half, the number of cpm added is the same because of the higher specific activity.

12.5 SPECIFICITY

The major disadvantage of the RRA relates to specificity. That is, since this is an indirect assay any substance that may interfere with ligand binding will be detected. For example, in developing a radioreceptor assay to measure blood GABA it was essential to establish that blood contains no substances other than GABA which are capable of displacing ^{3}H-GABA from its binding site. One such substance could be imidazoleacetic acid, a GABA receptor agonist and a naturally occurring compound. One way to demonstrate the specificity of the RRA for GABA in blood was to verify the values using an alternate method of analysis (Table 12.2). When the same samples were analyzed by both GC/MS and RRA, the results obtained were quite similar. While these data do not completely rule out the possibility that additional substances might be present in blood which could displace specifically bound ^{3}H-GABA, they suggest that such substances are either eliminated during the extraction procedure, or are present in such small quantities that they do not significantly interfere with the binding assay. This problem of specificity can be circumvented to some extent by expressing the amount of endogenous displacing activity as equivalents. Accordingly, when analysed by RRA, values for endogenous opioid peptides are expressed as morphine equivalents, since the assay will not differentiate between the various peptides (27).

Exogenously administered compounds must also be considered when conducting an RRA. For example, serotonin may be measured by RRA using specific binding of ^{3}H-serotonin to brain membranes as a binding procedure (32). However, this technique would be less useful following treatment of animals with ergot alkaloids, LSD or neuroleptics since all of these agents will displace ^{3}H-serotonin if present in sufficient concentrations. Accordingly, it would be necessary to demonstrate that the quantity of any injected substances is insufficient to interfere with the assay. If interference is noted, then steps must be taken to separate the offending substance from the extract. Such a precaution was taken by Childers et al. (33) during their studies aimed at examining the influence of opiate agonists and antagonists on enkephalin levels in brain. Since the drugs administered, morphine and naloxone, are both active using an opiate receptor RRA, it was necessary to separate residual drug from the brain extracts using column chromatography before measuring the opiod peptides. Even with this step however, the specificity of the RRA for met-enkephalin under these conditions was reconfirmed using a highly specific radioimmunoassay.

In some instances low specificity may be used to advantage. Thus, using highly specific methods of analysis, attempts have been made to correlate serum neuroleptic levels with clinical symptoms. However, a better correlation was found using a radioreceptor assay since the RRA measured both the parent compound as well as active metabolites (20).

TABLE 12.2

Blood GABA analysis by radioreceptor assay and gas chromotography/mass spectrometry

Sample	GABA Concentration (picomoles/ml)		
	RRA	GC/MS	RRA/GCMS
1	469	567	0.83
2	557	581	0.96
3	520	733	0.71
4	479	651	0.75
5	1022	1071	0.95
Mean $\pm$ S.E.M.	609 $\pm$ 104	720 $\pm$ 93	0.86

Rat blood was obtained by cardiac puncture from 5 animals and the GABA content of each sample was measured simultaneously by RRA and a GC/MS. Each value is the mean of two separate determinations. Adapted from Ferkany et al., (8).

12.6 APPLICATIONS

Radioreceptor assays have become popular tools for both clinical and basic research (Table 12.3). Using an RRA, cerebrospinal fluid and blood GABA levels have been measured in laboratory animals and man. Prior to the development of a suitable RRA, the only feasible way to monitor the small quantities of this amino acid in these fluids was by GC/MS or ion exchange fluorometry (34, 35). While both of these methods are extremely sensitive and precise, they involve the use of relatively sophisticated and expensive equipment, and only a few samples can be analysed at a time. With the radioreceptor assay there is approximately the same precision and sensitivity but less expensive equipment is required and it is possible to analyse dozens of samples in a day.

Radioreceptor assays have also been used to monitor serum neuroleptic concentrations. While the parent compound can be measured by radioimmunoassay, the RRA detects not only this agent but also any metabolites which possess ^{3}H-spiroperidol displacing activity. Since neuroleptics are believed to function by blocking dopamine receptors in brain, it could be argued that radioreceptor analysis yields a better indication of the total amount of active substance. In addition to neuroleptics, assays have also been developed to monitor serum levels of benzodiazepines (36) as well as antidepressants (26) and β-adrenergic antagonists (22).

The latter compounds exert powerful influences on central aminergic systems and these assays are of growing importance in correlating drug level with both clinical response and experimental observation. While to date RRA for biogenic amines themselves have not been extensively utilized, sensitive ligand binding procedures in which dopamine (38), 5-HT (25,39), norepinephrine (40) and epinephrine (40) are active are now available. Thus, these methods may represent a rapid and simple approach for the determination of biogenic amines in tissue.

TABLE 12.3

Some radioreceptor assays for neurotransmitters and drugs

Substance Measured	Binding Assay	Reference
β-Adrenergic Antagonists	^{3}H-Dihydroalprenolol	22
Anticholinergics	^{3}H-Quinuclidinyl benzilate	37
Benzodiazepines	^{3}H-Diazepam	36
GABA	^{3}H-GABA	8
Opioids	^{3}H-Naloxone	27
Neuroleptics	^{3}H-Spiroperidol	20
Tricyclic Antidepressants	^{3}H-Quinuclidinyl benzilate	26

Radioreceptor assays are also useful for discovering neurotransmitters and neuromodulators. For example, once it was discovered that ^{3}H-naloxone bound in a specific manner to brain tissue opiate receptors, it was reasoned that this organ may contain some endogenous substrate for this site. The use of the ^{3}H-naloxone radioreceptor assay facilitated the search for this substrate. Similarly, the identification of ^{3}H-diazepam binding sites in brain has lead to the suggestion that an endogenous benzodiazepine-like substance may exist. Tissue and fluid extracts are now being screened by RRA in an attempt to verify this hypothesis. If an extract does exhibit displacing activity, then the RRA can be used to examine the relative activity of the substance during purification procedures. Finally, once the agent is identified, the RRA can be used routinely to detect the compound.

12.7 SUMMARY

The development of radioactive substances having high specific activity has facilitated the development of receptor binding assays for a variety of neurotransmitters and drugs. Because the amount of receptor-bound radioligand is proportional to the amount of unlabeled ligand present, receptor binding assays can be used to measure a variety of chemical agents. These radioreceptor assays are simple to perform and compare favorably to other analytical methods with respect to sensitivity and precision. In addition, these assays are rapid, making it possible to analyse large numbers of samples at a single time. The major limitation of radioreceptor assays is their specificity. However, in some cases, the relative lack of specificity has been advantageous in that it allows for the simultaneous analysis of multiple ligands. Radioreceptor assays have been used to measure a number of neurotransmitters and drugs in a wide variety of biological tissues and fluids.

ACKNOWLEDGEMENTS

Preparation of this manuscript was made possible, in part, by USPHS grant NS-13803, a Research Career Development Award NS-00035 (S.J.E.) and an NIH-post-doctoral training fellowship MH-15330 (J.W.F.).

REFERENCES

1   O. Loewi and E. Navratil, Flugers Arch. Physiol., 214 (1926) 678-688.
2   W. Cannon and A. Rosenbleuth, Autonomic Neuroeffector Systems, Macmillan Co., New York, 1977, 229 pp.
3   U.S. von Euler, in H. Blaschko and E. Muschol (Eds.), Catecholamines, Handb. Exp. Pharmak, Springer-Verlag, Berlin, 33 (1972) pp. 186-230.
4   A. Berson, R. Yalow, A. Bauman, M. Rothschild and K. Newerly, J. Clin. Invest., 35 (1956) 170-190.
5   A.L. Steiner, C.W. Parker and D.M. Kipnis, J. Biol. Chem., 247 (1972) 1106-1113.
6   R.J. Lefkowitz, J. Roth and I. Pastan, Science, 170 (1970) 633-635.
7   S.J. Enna and S.H. Snyder, Brain Res., 100 (1975) 81-97.
8   J.W. Ferkany, L.A. Smith, W.E. Seifert, R.M. Caprioli and S.J. Enna, Life Sci., 22 (1978) 2121-2128.
9   J.W. Ferkany, I.J. Butler and S.J. Enna, J. Neurochem., 33 (1979) 29-33.
10 S.J. Enna, in H.I. Yamamura, S.J. Enna and M.J. Kuhar (Eds.), Neurotransmitter Receptor Binding, Raven Press, New York, 1978, pp. 127-139.
11 S.J. Enna, in I. Hanin and S.H. Koslow (Eds.), Physico-Chemical Methodologies in Psychiatric Research, Raven Press, New York, 1980, pp. 83-101.
12 S.J. Enna, in D.B. Bylund (Ed.), Receptor Binding Techniques, Society for Neuroscience, Washington, D.C., 1980, pp. 257-271.
13 S.J. Enna, in L.L. Iversen, S.D. Iversen and S.H. Snyder (Eds.), Handbook of Psychopharmacology, Plenum Press, New York, in press.
14 G. Scatchard, Ann. N.Y. Acad. Sci., 51 (1949) 660-672.
15 P. Krogsgaard-Larsen, T. Honore and K. Thyssen, in P. Krogsgaard-Larsen, J. Scheel-Kruger and H. Kofod (Eds.), GABA-Neurotransmitters, Academic Press, New York, 1979, pp. 201-206.
16 G.G. Yarbrough, M. Williams and D.R. Haubrich, Arch. Int. Pharmacodyn. Therap., 241 (1979) 266-279.
17 S.J. Enna, J.W. Ferkany and P. Krogsgaard-Larsen, in P. Krogsgaard-Larsen, J. Scheel-Kruger and H. Kofod (Eds.), GABA Neurotransmitters, Academic Press, New York, pp. 191-200.
18 I. Creese, D.R. Burt and S.H. Snyder, Science, 192 (1976) 481-483.
19 R.J. Tallarida and R.B. Murray, Manual of Pharmacological Calculations, Springer-Verlag, New York, 1981, pp. 150-151.
20 L.E. Tune, I. Creese, J.R. Depaulo, P.R. Slavney, J.T. Coyle and S.H. Snyder, Am. J. Psychiat., 37 (1980) 187-190.

21 S.J. Enna, J.H. Wood and S.H. Snyder, J. Neurochem., 28 (1977) 1121-1124.
22 R.B. Innis, D.B. Bylund and S.H. Snyder, Life Sci., 23 (1978) 2031-2038.
23 S.J. Enna and S.H. Snyder, Mol. Pharmacol., 13 (1977) 442-453.
24 T. Costa, D. Rodbard and C.B. Pert, Nature, 277 (1979) 315-317.
25 S.J. Peroutka and S.H. Snyder, Mol. Pharmacol., 16 (1979) 687-699.
26 R.B. Innis, L.E. Tune, R. Rock, J.R. Depaulo, D.C. U'Prichard and S.H. Snyder, Eur. J. Pharmacol., 58 (1979) 473-477.
27 R. Simantov, S.R. Childers and S.H. Snyder, Brain Res., 135 (1977) 358-367.
28 H.I. Yamamura, S.J. Enna and M.J. Kuhar, Neurotransmitter Receptor Binding, Raven Press, New York, 1978, 195 pp.
29 P. Cuatrecasas and M.D. Hollenberg, Adv. Prot. Chem., 30 (1976) 251-451.
30 R. Bernasconi, N. Bittiger, J. Heid and P. Martin, J. Neurochem., 34 (1980) 614-618.
31 D.R. Burt, S.J. Enna, I. Creese and S.H. Snyder, Proc. Nat. Acad. Sci. (USA), 72 (1975) 4655-4659.
32 J.P. Bennett and S.H. Snyder, Mol. Pharmacol., 12 (1976) 373-389.
33 S.B. Childers, R. Simantov and S.H. Snyder, Eur. J. Pharmacol., 46 (1977) 289-293.
34 S.J. Enna, L. Stern, G. Wastek and H.I. Yamamura, Arch. Neurol., 34 (1977) 683-685.
35 J.H. Wood, B.S. Glaeser, S.J. Enna and T.A. Hare, J. Neurochem., 30 (1978) 291-293.
36 R.L. Chang and S.H. Snyder, Eur. J. Pharmacol., 48 (1978) 213-218.
37 L.E. Tune and J.T. Coyle, J. Psychopharmacol. (in press).
38 D.R. Burt, S.J. Enna, I. Creese and S.H. Snyder, Proc. Nat. Acad. Sci. (USA), 72 (1975) 4655-4659.
39 J.P. Bennett and S.H. Snyder, Mol. Pharmacol., 12 (1976) 373-389.
40 D.C. U-Prichard and S.H. Snyder, J. Biol. Chem., 252 (1977) 6450-6463.

Chapter 13

RADIOIMMUNOASSAYS FOR PHENALKYLAMINES

KAMAL K. MIDHA

College of Pharmacy, University of Saskatchewan, Saskatoon, Saskatchewan,
S7N OWO (Canada)

and

JOHN W. HUBBARD

Faculty of Pharmacy, University of Manitoba, Winnipeg, Manitoba,
R3T 2N2 (Canada)

13.1   INTRODUCTION

13.1.1   A brief outline of the principles of radioimmunoassay

Radioimmunoassay (RIA) is a highly sensitive assay technique which was developed
in the 1950's by Solomon Berson and Rosalyn Yalow (1). Following the injection of
porcine insulin into humans or guinea-pigs, they observed the production of a
globulin (antibody) with an extraordinary specific affinity for insulin. Their
RIA depends upon competition between natural insulin and ^{131}I-labelled insulin
for binding sites on a limited amount of the antibody. After separation of the
antibody-bound insulins from the free, unbound insulins, the amount of radioactivity
in each fraction is determined. The ratio of radioactivity in the free fraction to
radioactivity in the bound fraction is related to the concentration of natural
('cold') insulin in the sample. RIA procedures have been developed subsequently
for other macromolecules such as growth hormone, parathyroid hormone and adreno-
corticotrophic hormone and also for a wide variety of low molecular weight compounds.

Small drug molecules, such as the phenalkylamines, are not antigenic unless
they are covalently attached to a high molecular weight 'carrier antigen'. Usually
the drug is modified chemically to provide a chain of 2-5 atoms through which
covalent linkage with the macromolecule can be established. The modified drug is
called the 'hapten', and the hapten-macromolecule complex is often referred to as
the 'conjugate'. The choice of the 'haptenic modification' is very important because it can have a profound influence on the specificity of antibodies subsequently
raised to the conjugate. For example, the antibody may be unable to distinguish
between the drug and analogs of the drug which differ in structure at the point
of derivatization of the hapten. There are a number of good examples of this
phenomenon in the review of RIA procedures for phenalkylamines which follows.

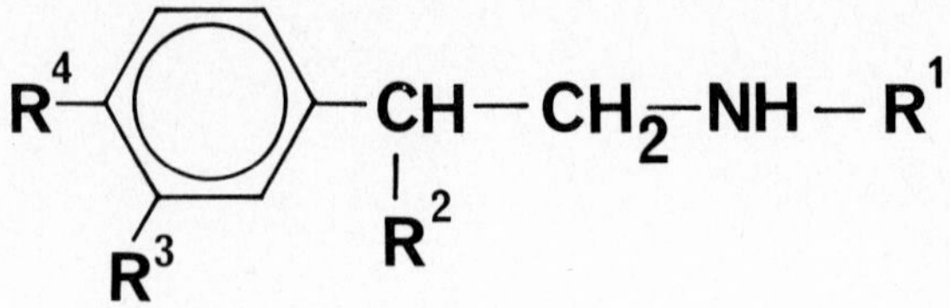

$$R^4 \!\!-\!\! \bigcirc \!\!-\!\! CH \!-\! CH_2 \!-\! NH \!-\! R^1$$
$$R^3 \qquad R^2$$

	R^4	R^3	R^2	R^1
Adrenaline	HO	HO	HO	Me
Noradrenaline	HO	HO	HO	H
Dopamine	HO	HO	H	H
Synephrine	HO	H	HO	Me
Octopamine	HO	H	HO	H
Tyramine	HO	H	H	H
Metanephrine	HO	MeO	HO	Me
Normetanephrine	HO	MeO	HO	H
3-Methoxytyramine	HO	MeO	H	H
DMPEA	MeO	MeO	H	H

Fig. 13.1. Phenylethylamines which have been studied in the development of radioimmunoassays.

A wide variety of macromolecular carrier antigens have been employed success-fully in the development of RIAs for small molecules. The choice of carrier anti-gen and the methods of coupling the hapten to the macromolecule have been descri-bed elsewhere (2-4).

Satisfactory antisera to hapten-macromolecule conjugates have been raised in many animal species, although guinea-pigs or rabbits are most often used. The immunization procedure (5) is largely empirical and the responsiveness of antisera from different animals varies a great deal (6). For these reasons, it takes a considerable amount of patience and skill to recognize when an antiserum of ade-quate sensitivity and specificity has been obtained. It is important to realize that, although the antiserum is raised to a hapten-macromolecule conjugate, the antiserum can recognize and interact strongly with the original small molecule.

The radiolabelled tracer (7) is a critical component of the RIA because it is the component which is actually analyzed by counting. The most popular radio-labelled tracer is the original small molecule in which tritium (or less frequent-ly, carbon-14) has been incorporated. These radioisotopes have long half-lives, but they have much lower specific activities than isotopes such as iodine-125 or iodine-131. When the isotopes of iodine are employed in the RIA of small mole-cules, the advantage of their high specific activities is offset by their short half-lives and by the fact that the chemical modification brought about by the incorporation of iodine into the small molecule often changes its affinity for the antibody.

For an extensive review on the development of RIA (591 references), see Skelley et al. (8).

13.1.2   Comparison of radioimmunoassay with other analytical methods

Without question, RIA is one of the most highly sensitive analytical techniques. Once the antiserum has been produced and characterized, the RIA is a very simple, rapid and reproducible procedure which requires very little manipulation of the sample. RIA can also be highly specific, although the specificities of different antisera vary enormously. Therefore the cross-reactivity profile of each anti-serum must be assessed carefully. In addition, the authors believe that a new RIA procedure should be 'validated' by direct comparison with a more rigorous chemical method of analysis such as combined gas chromatography-mass spectrometry or high performance liquid chromatography. Any discrepancy between the 'chemical method' and the RIA then invites a more searching appraisal of the cross-reactivity profile of the antiserum.

There are two common methods by which cross-reactivities are expressed: a) The amount of interfering substances which causes 50% inhibition of the binding of the radiolabelled tracer is quoted; b) The amount of drug which causes a 50% inhibition of the binding of the radiolabelled tracer is divided by the amount of interfering substance which causes a 50% inhibition of the binding of the radio-labelled tracer and multiplied by 100. This figure is often referred to as the 'percentage cross-reactivity'.

13.2   RADIOIMMUNOASSAY OF PHENALKYLAMINES

13.2.1   Phenethylamines

One of the most interesting developments in assay methodology for phenolic amines is the highly specific RIA procedure for p-tyramine (p-TA) published by Faraj et al. (9). In the preparation of the antigen, p-aminohippuric acid was coupled to methylated bovine serum albumin (BSA) by the carbodiimide method. The amino groups of the p-aminohippuric acid residues were then diazotized and coupled to p-TA. The authors assumed that the coupling took place ortho to the phenolic

group at the 3-position of the benzene ring of p-TA (Fig. 13.2). It is especially interesting to note therefore that the antiserum subsequently raised to this
antigen (rabbits) did not cross-react significantly with dopamine (DA) (2.4%),
3-methoxytyramine (3-MT) (2.4%), noradrenaline (NA) (< 1%), adrenaline (A)
(< 1%), or 3,4-dimethoxyphenethylamine (DMPEA) (< 1%), all of which have a substituent at the 3-position of the benzene ring. It frequently happens that the
antiserum is unable to distinguish between its true ligand and interfering

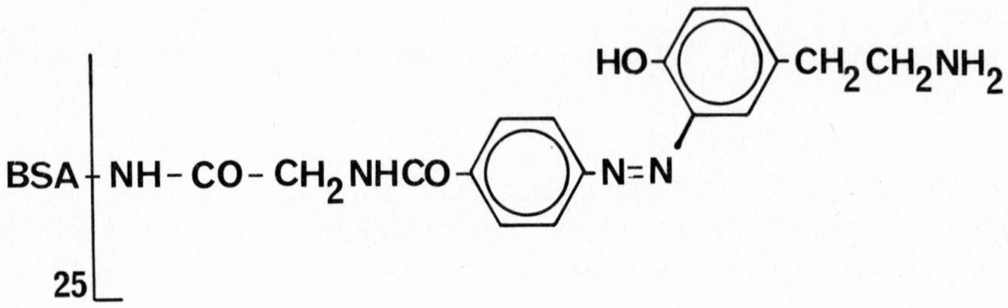

Fig. 13.2. p-Tyramine antigen, prepared using p-aminohippuric acid.

substances which differ at the point of haptenic modification. In this case however, the antiserum showed remarkable specificity when challenged with a representative range of ring-modified analogs.

As would be anticipated, the antiserum also gave very low cross-reactivities
with a variety of side-chain-modified derivatives, including octopamine (OA), p-
hydroxyphenylethanol (p-HPE), p-hydroxyphenylacetic acid (p-HPAA), p-hydroxymandeli<
acid (p-HMA), L-tyrosine and L-dopa. In fact, this antiserum showed no significant
cross-reactivity with any agent likely to interfere with the analysis of p-TA in
biological fluids. The RIA procedure, which was reported to detect 1 ng of p-TA
(0.1 ml sample), was validated by comparison with an established radiochromato-
graphic method for p-TA.

Taking a different approach, Grota and Brown (10) conjugated p-TA to BSA by
means of the Mannich formaldehyde reaction (11) which had been used previously to
couple 5-hydroxytryptamine (5-HT) (12) to BSA. The authors speculated that, accord
ing to precedent (11,12), a hapten containing a phenolic hydroxyl group could be
attached to an albumin carrier by means of a methylene bridge ortho to the phenolic
substituent. The suggestion of a haptenic modification at the 3-position of the p-
TA residues of the conjugate is supported by the observation that the antiserum sub
sequently raised (rabbits) had a stronger affinity for 3-MT (radiotracer ^{3}H-p-TA
cross-reactivity 213%) than for p-TA itself (100%). Certainly, the antiserum

showed very low cross-reactivities for the N-methyl compounds synephrine (SYN) (0.04%), A (0.04%) and metanephrine (MN) (0.01%). Furthermore, β-hydroxylation of the side-chain reduced the cross-reactivities of normetanephrine (NMN) (40%), OA (9%) and NA (1.6%), compared with their respective unhydroxylated counterparts, 3-MT (213%), p-TA (100%) and DA (22%). These cross-reactivity data all suggest that the coupling of p-TA to BSA is much more likely to have occurred on the 3-position of the benzene ring than on the amino group of the side-chain (Fig. 13.3).

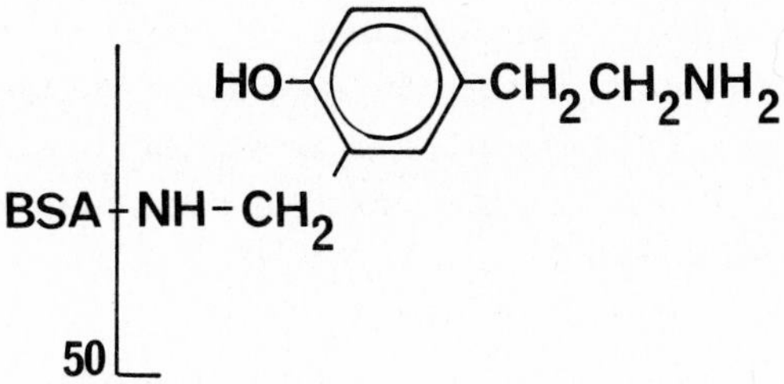

Fig. 13.3. p-Tyramine antigen prepared using the Mannich formaldehyde reaction.

Grota and Brown (13) obtained similar results when they used the Mannich formaldehyde reaction to couple SYN to BSA. In this case, the antiserum raised (rabbits) to the SYN-BSA conjugate had a stronger affinity for the 3-methoxy analog MN (radiotracer ^{3}H-MN, cross-reactivity of MN 800%), than for SYN (100%). Again the cross-reactivities of the side-chain modified derivatives tested were very low. The only compound reported to have a significant cross-reactivity was A (60%). It should be noted, however, that if the cross-reactivities are recalculated with MN as 100%, SYN becomes 12.5% and A becomes 7.5%. When viewed in this manner, the antiserum could form the basis of a reasonable RIA for MN. It is interesting to note that Grota and Brown chose to raise their antisera with "soluble conjugates with an apparent molar ratio of greater than 50". This is in contrast to the recommendations of Landsteiner (2) who concluded that with serum albumin as the carrier antigen, 10 haptenic molecules per molecule of albumin was optimal, and that too little or too much hapten led to a poor response. The MN antiserum, however, displayed remarkable selectivity amongst the compounds reported; this was combined with a sensitivity range of 25-100 pg.

Lam et al. (14) used the method of Grota and Brown (13) to develop a highly sensitive RIA for MN. The smallest quantity of unlabelled MN which produced a significant mean displacement of ^{3}H-MN relative to control was 40 pg, which corresponds to a sensitivity of 400 pg/ml.

Metanephrine and SYN displaced ^{3}H-MN equally from the antibody. The cross-reactivities of NA, NMN, 3-MT, p-TA, DA and vanillylmandelic acid (VMA) were reported to be less than 1%. The cross-reactivity of A (15%) was reduced to 1% in the RIA procedure by the addition of 2 ng of A per assay tube. The authors applied the RIA to the analysis of MN in urine, but they did not report any attempt to compare the RIA with a chemical method of analysis.

More recently, Raum et al. (15) have developed a highly sensitive RIA for MN, based on antisera produced in rabbits by the SYN-albumin method of Grota and Brown (10). The cross-reactivity profile of the antibody (SYN 20%, A 13%) was similar to that of Grota and Brown (10). None of the other 16 compounds tested reacted significantly with the antibody. These compounds included NMN, NA, DA, 3-MT, p-TA, OA, and a selection of side-chain modified metabolites of the biogenic amines. The authors included in their RIA procedure an extraction step, which removed SYN and A from the samples, and reduced their effective cross-reactivities to 5% and 10^{-5}% respectively. The RIA of MN in urine was validated by comparison with an established colourimetric method. Good correlation ($r^2 = 0.993$) was demonstrated between the two assays, although the authors point out that their RIA is faster, more precise and 1000 fold more sensitive than the colourimetric method. Part of the reason for the high sensitivity of this RIA stems from the use of [3-^{125}I]3-iodosynephrine as the radiotracer (specific activity 1000-2000 KCi/mol). The ^{125}I-ligand, which was easily prepared by established methods, attained a higher maximum binding than ^{3}H-MN, and maintained a higher binding with increasing dilutions of the antiserum.

Peskar et al. (16) produced antibodies to NMN by immunizing rabbits with conjugates which were synthesized by linking the side-chain amino group of the hapten to the carrier antigen with glutaraldehyde. As would be anticipated, the antiserum cross-reacted (106%) with MN, which differs in structure from NMN only at the point of haptenic modification. Other side-chain modified cross-reactants were 3-MT (21%) and 3-methoxy-4-hydroxyphenylglycol (MHPG) (4.5%), although the antiserum did not recognize VMA or homovanillic acid (HVA). The antiserum did not cross-react significantly with a long list of ring modified analogs, which included the catecholamines.

The RIA for NMN/MN (16) employed ^{125}I-labelled albumin-NMN conjugate as the radiotracer. Despite the use of a ^{125}I-labelled tracer however, the detection limit was disappointing (100 ng/ml). Peskar et al. suggested that if the sensitivity were improved the RIA could be applied to the measurement of NMN in cerebrospinal fluid, or in brain, where interference from MN is negligible. The present reviewers agree with this approach, but point out that the credibility of data so obtained is greatly enhanced if the method has been validated by comparison with an appropriate chemical method.

Wisser et al. (17) developed three different types of antibodies to 3,4-dimethoxyphenethylamine (DMPEA), the controversial 'pink spot' material found in the

urine of schizophrenics (18) and normal individuals (19). Their first antigen
(Fig. 13.4, A) was prepared by the reaction of the side-chain amino group of DMPEA
with succinic anhydride, followed by the carbodiimide-catalyzed coupling of the
hapten to bovine γ-globulin.

The second hapten (Fig. 13.4, B) was prepared by the nitration of the 6-
position of the benzene ring of DMPEA. The free amino group of the side-chain was
protected with a τ-butyloxycarbonyl group during the subsequent reduction of the
nitro group by hydrogen/Raney-Nickel. The new amino group was then succinoylated.
The hapten so formed was coupled to human serum albumin (HSA) or to bovine γ-
globulin. Finally the protecting group was removed by hydrolysis with dilute hydrochloric acid. The haptens and their synthetic intermediates were characterized by
infrared and proton magnetic resonance spectroscopy and by elemental analysis. The
reviewers believe that the chemical characterization and estimation of the purity
of the haptens is important because these measures greatly assist in the subsequent
evaluation of the antisera.

Antisera to the three antigens were raised in rabbits. The antiserum to antigen A had the highest titer (1:16000) against [5-^{3}H] DMPEA while the titers of the
antisera to HSA antigen B and bovine γ-globulin antigen B were 1:1000 and 1:4000.
The cross-reactivities of all three antisera to a variety of catecholamines, their
3-methoxy metabolites and side-chain degradation products and of some mono-hydric
phenolic amines were all less than 0.1%. Taken together with the observation that
all three antisera cross-reacted with 3,5-DMPEA less than 0.1%, these data signify
that the antibodies were all highly specific for substituents on the benzene ring.

The antiserum to bovine γ-globulin antigen B also showed extremely high specificity for the side-chain of DMPEA, showing cross-reactivities of 0.1% or less with
all of the side-chain modified compounds with which the antiserum was challenged.
In fact, of the 17 compounds tested, the highest cross-reactivity reported was an
insignificant 0.8% cross-reactivity of mescaline (MES). This antiserum serves as
a model of the specificity which can be achieved by a thoughtful approach to the
development of RIA and by patient and painstaking attention to detail in every step
of the production of the antiserum.

The antibody to HSA antigen B was less specific for the side-chain of DMPEA, and
cross-reacted with 3,4-dimethoxybenzyl alcohol (13.1%), 3,4-dimethoxycinnamic acid
(9.6%), 3,4-dimethoxyphenylacetic acid (11.4%) and 3,4-dimethoxybenzoic acid (1.2%).
The antiserum to antigen A, in which the hapten was also coupled through the side-
chain, cross-reacted to a much greater degree with the same four compounds (47%,
52%, 15% and 7.5% respectively).

A similar result was obtained by Riceberg and Van Vunakis (20), who used a side-
chain succinoylated DMPEA coupled to poly-lysine (Fig. 13.4 C) as their antigen.
Their antibody (raised in rabbits) showed much greater specificity for the substituted ring system of DMPEA than for the side-chain. Thus the antiserum cross-

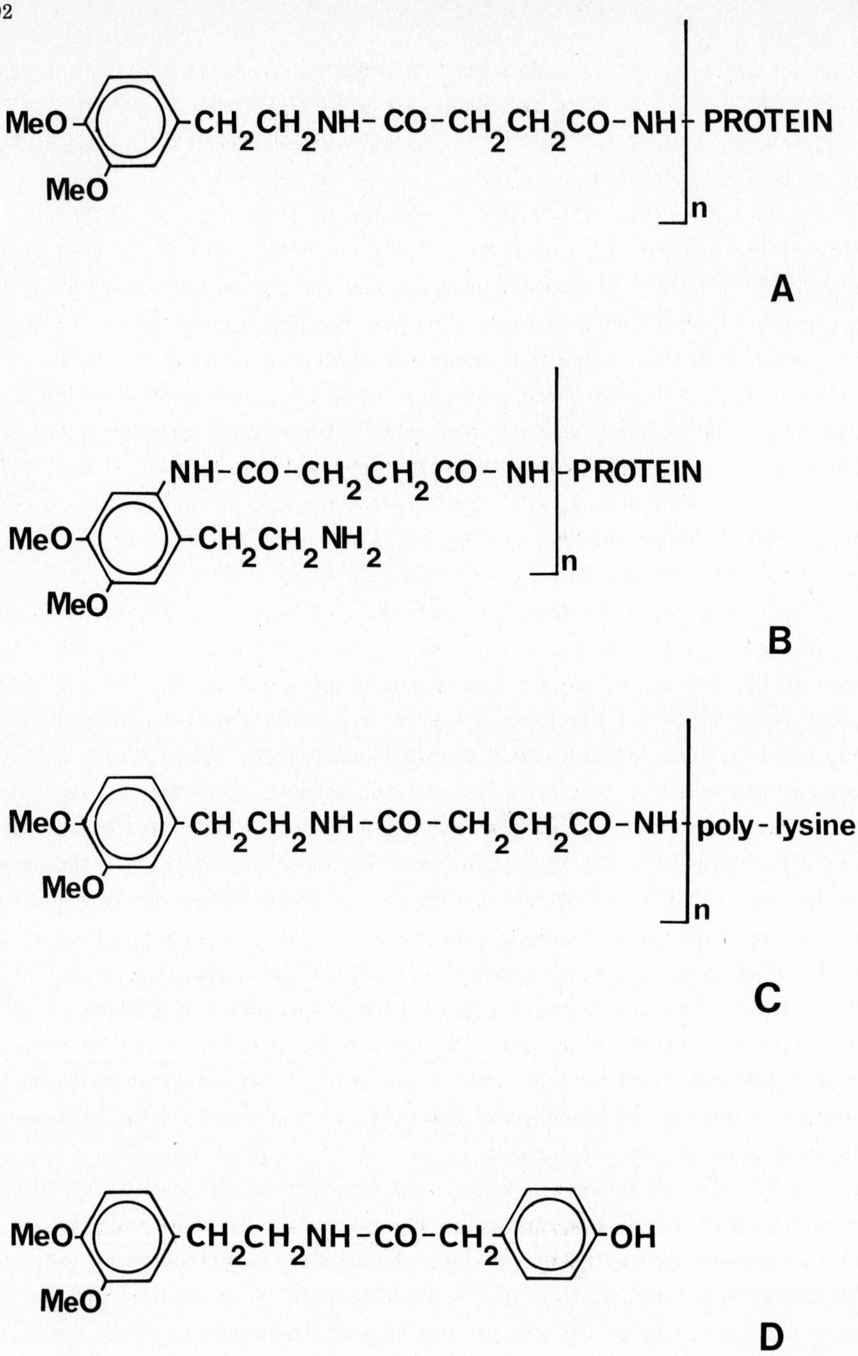

Fig. 13.4. 3,4-Dimethoxyphenylethylamine (DMPEA) antigens (A-C) and a radiotracer (D) for the DMPEA assay.

reacted with a variety of side-chain types in which the 3,4-dimethoxyphenyl group was present, but did not recognize compounds in which the ring substitution pattern was changed. This is well illustrated by the very high cross-reactivity (450%) of N,N-dimethyl-3,4-dimethoxyphenethylamine (N,N-dimethyl-DMPEA), whereas the cross-reactivity of N,N-dimethyl-3,4,5-trimethoxyphenethylamine (N,N-dimethyl-MES) was an insignificant 0.5%.

The high percentage cross-reactivity (225%) of the active metabolite N-acetyl-DMPEA (21) is perhaps understandable since the side-chain nitrogen function in both N-acetyl-DMPEA and the hapten are present as carboxamide functions. It is surprising, however, in view of the lack of specificity of this antiserum for side-chain deviants, that the DMPEA metabolite 3,4-dimethoxyphenylacetic acid had a low cross-reactivity (0.9%). Nevertheless, it is important to recognize that this metabolite may be present in the urine in much higher concentration than DMPEA (21), especially in experiments in which DMPEA is administered to man or animals.

In order to overcome the analytical problems posed by the lack of specificity of the antiserum, Riceberg and Van Vunakis (20) used high pressure liquid chromatography in conjunction with RIA to identify and measure immunologically active material in urine. The radiotracer employed in the RIA was N-2-(3,4-dimethoxyphenyl)ethyl-2-(4-hydroxyphenyl)acetamide (Fig. 13.4 D) after iodination by reaction with Na^{125}I. The RIA procedure was sensitive enough to detect 100 pg of DMPEA in a 0.1 ml sample.

In an earlier study, Van Vunakis et al. (22) produced an antigen for DMPEA by coupling the side-chain amino function directly to the γ-carboxyl groups of poly-L-glutamic acid. Unfortunately, the antiserum which was subsequently raised in rabbits was not characterized properly because an appropriate selection of potential cross-reactants was unavailable. A similar antigen for MES gave rise to an antiserum which was specific for the ring system of MES, but which showed a high degree of cross-reactivity with the two side-chain altered derivatives of MES which were examined (N-methyl-MES, 76% and N,N-dimethyl-MES, 60%).

The same group subsequently improved upon their antiserum for MES (23). They reported the preparation of two new antigens for MES, the first of which (Fig. 13.5 A) was linked through the side-chain amino function directly to carboxylic acid functions of BSA. The second antigen (Fig. 13.5 B) was prepared by coupling succinoylated MES to BSA. Antisera to both types of antigen were raised in rabbits. Both antisera were specific for the ring substitution pattern of MES, but were less specific for the side-chain. The antiserum to the directly coupled antigen (Fig. 13.5 A) cross-reacted strongly with N-methyl-MES (92%) and with N,N-dimethyl-MES (75%), which are similar to the cross-reactivities of these compounds with the antiserum to MES-poly-L-glutamic acid (22). The reason for the very high cross-reactivities of these antisera with terminal side-chain structural deviants probably stems from the lack of a 'haptenic bridge' to hold the hapten away from

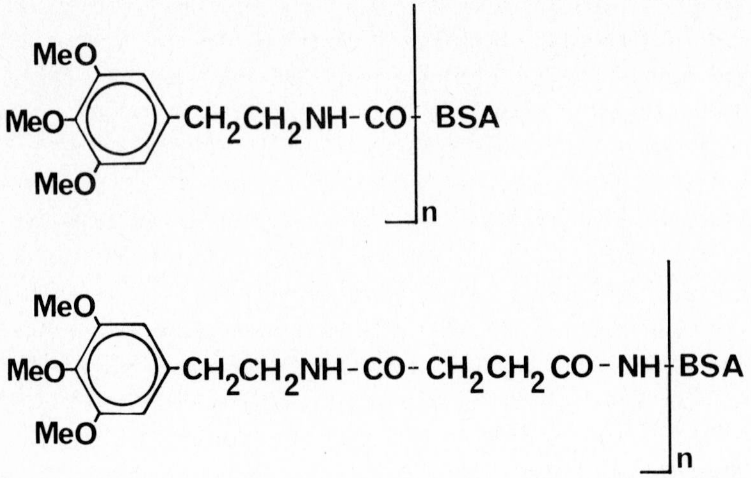

Fig. 13.5. Mescaline antigens.

the surface of the carrier antigen. Certainly, the antigen in which MES was linked to BSA through a succinoyl bridge (Fig. 13.5 B) gave rise to an antiserum which had much lower cross-reactivities to N-methyl-MES (24%) and N,N-dimethyl-MES (20%). The N-acetyl derivative of MES was not tested, but on the basis of the data presented, it must be presumed likely to cross significantly with these antisera. Neither antiserum cross-reacted significantly with 3,4,5-trimethoxyphenyl-acetic acid which is a metabolite of MES (24).

In the development of a RIA for MES, Riceberg et al. (23) employed ^{125}I-labelled N-[2-(3,4,5-trimethoxyphenyl)ethyl]-2-(4-hydroxyphenyl)acetamide, which is similar to the radiotracer for DMPEA (Fig. 13.4 D), as a radiotracer. The detection limit of the RIA was 100 pg in a 0.1 ml sample, which corresponds to 1.0 ng/ml of MES. The authors used their RIA to follow the rapid disappearance from the plasma of intravenously administered MES.

13.2.2   Catecholamines

The development of RIAs for catecholamines (25-27) has proved to be a difficult challenge because of the inherent instability or reactivity of the catecholamines during derivatization, coupling and immunization procedures. Wisser et al. (17) prepared conjugates of DA similar to their antigens for DMPEA (Fig. 13.4), but were unable to raise antisera to DA despite the fact that they protected the catechol group, by conversion into the isopropylidene derivative, during the derivatization and coupling procedures. Similarly, Grota and Brown (10) concluded that the catecholamines were too unstable to undergo hapten-protein coupling by the Mannich formaldehyde condensation.

Miwa et al. (28) recognized that the side-chain of a catecholamine can undergo cyclization (29) under the conditions of the Mannich formaldehyde reaction unless the amino function is first substituted with a suitable protecting group.

Accordingly, the N-maleoyl derivatives of L-A, L-NA, DA and L-dopa were coupled to BSA by the Mannich formaldehyde reaction. The conjugates were then reacted with 0.01 N HCl at 60° for 3 hours to remove the maleic acid residues (30). In order to investigate the position of derivatization on the catecholamine ring system, Miwa et al. (31) carried out the Mannich formaldehyde condensation with ethylamine and a model compound, 4-methylcatechol. The result, confirmed by infra-red spectroscopy and by proton and ^{13}C nuclear magnetic resonance spectroscopy, indicated that the product of the reaction was 5-ethylaminomethyl-4-methylcatechol. This suggests that the catecholamine undergoes the Mannich formaldehyde condensation para to the 3-position hydroxy function. This is in contrast to monohydric phenols such as p-TA which appear to undergo the reaction ortho to the phenolic group (Fig. 13.3).

The probable structures of the catecholamine-BSA conjugates of Miwa et al. (30, 31) are shown in Fig. 13.6. The RIA procedure for A (32) was extremely sensitive

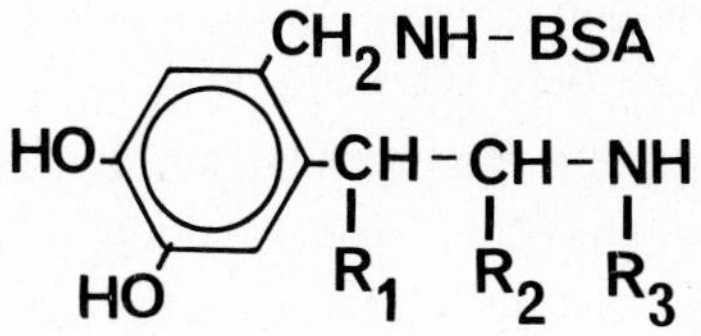

	R_1	R_2	R_3
L-Adrenaline	OH	H	CH_3
L-Noradrenaline	OH	H	H
dopamine	H	H	H
L-DOPA	H	COOH	H

Fig. 13.6. Catecholamine antigens.

(detection limit, about 20 pg) and highly specific (^{3}H-A tracer) for both catechol and side-chain functions. Cross-reactivities were extremely low to NA (0.017%), DA (0.017%), L-dopa (< 0.01%), NMN (< 0.01%), 3,4-dihydroxyphenylacetic acid (DOPAC) (< 0.001%), 3,4-dihydroxymandelic acid (DHMA) (< 0.001%), HVA (0.001%), VMA (0.001%) and isoprenaline (0.05%). The highest cross-reactivities of the compounds tested were to MN and SYN (both 1.25%), which have the same side-chain as A. Curiously, the antiserum recognized both enantiomers of A equally, although the antibody was apparently raised to a conjugate of L-A-BSA (30).

In a recently published investigation of the potential for the development of RIAs for catecholamines and their metabolites, Diener et al. (33) prepared conjugates for p-TA, OA, SYN, NMN and DA. The antigens were all prepared by methods discussed in this review. The cross-reactivity profiles of the antisera were essentially consistent with earlier experience.

Faraj et al. (34,35) developed an enzyme-RIA procedure for the measurement of L-dopa, DA and 3-MT in urine. This method was based on the incubation of urine in the presence of catechol-O-methyltransferase, aromatic L-amino-acid decarboxylase and S-adenosylmethionine. The concentration of the DA metabolite 3-MT in the urine sample was then measured by RIA (36).

13.2.3   5-Hydroxytryptamine (Serotonin; 5-HT)

In one of the earliest approaches to the development of RIA for serotonin, Ranadive and Sehon (12) pioneered the adaptation of the Mannich formaldehyde reaction to the synthesis of protein conjugates of serotonin and 5-hydroxyindoleacetic acid (5-HIAA). Antibodies subsequently raised to the serotonin conjugates were capable of inhibiting, in mice, the cutaneous reactions evoked by intradermal injection of serotonin.

Grota and Brown (13) adapted the method of Ranadive and Sehon to synthesize protein conjugates for serotonin and N-acetylserotonin, but were unable to prepare suitable conjugates for 5-methoxy-N-acetylserotonin (melatonin). The authors speculated that the failure of melatonin in the Mannich formaldehyde conjugation reaction was due to the lack of a free phenolic group in the molecule. It has been demonstrated that aminomethylation occurs ortho (or para) to a phenolic group under these conditions (11). Grota and Brown (13) therefore suggested that the coupling reaction had occurred at a position ortho to the 5-hydroxy group of serotonin or of N-acetylserotonin. This proposal is supported by the cross-reactivity profiles of the antisera raised to the conjugates. The antiserum for serotonin cross-reacted (166%) with 5-methoxytryptamine (5-MT) but did not cross-react with a variety of side-chain variants of serotonin. The inability of the antiserum to recognize chemical modifications to the phenolic group is consistent with the conjugation of the protein at one or both positions ortho to the phenolic group.

A different approach to the development of antibodies for serotonin (37) involved the immunization of rabbits with conjugates prepared by the reaction of serotonin creatinine sulfate with diazotized p-aminoacetanilide. After cleavage of the acetyl protecting group, the hapten was diazotized and coupled to the carrier antigen. The antibodies had low titres and were not used for RIA, although they were shown to be capable of inhibiting the in vivo reaction of serotonin in rats.

Peskar and Spector (38) also used a diazotization reaction to couple serotonin creatinine sulfate to BSA which had been previously enriched with p-aminophenylalanine residues. The antisera raised to this conjugate in rabbits cross-reacted (93%) with 5-MT, but did not recognize melatonin, N-acetylserotonin, 5-hydroxytryptophan (5-HTP), 5-HIAA or any of the other side-chain modified derivatives of serotonin tested. These data support the author's suggestion that diazotization had occurred ortho to the 5-hydroxy group of serotonin, leaving the aminoethyl side-chain unmodified. The RIA is capable of detecting 1 ng of serotonin and has been compared favourably with a standard spectrofluorometric method for serotonin (39, 40). Spector and his co-workers have subsequently applied the RIA to the determination of serotonin in platelets (40) and to the measurement of serum serotonin concentration in normal male and female rats after intestinal ischemia shock (41).

Jaffe and co-workers have developed a RIA for serotonin (42,43) based on the method of Peskar and Spector (38). The antiserum cross-reacted to a lesser extent (27%) with melatonin than did that of Peskar and Spector but was similar in that it did not recognize side-chain modified variants of serotonin. The RIA, which can detect 100 pg serotonin, has been validated by comparison with a spectrofluorometric method and applied to the measurement of serotonin concentrations in protein-free supernatants of human whole blood and in platelet-rich plasma and platelet-poor plasma (42,43). The authors point out that this RIA is sensitive and specific enough to facilitate studies into the physiological role of serotonin and that there are a number of potential clinical applications for the procedure.

13.2.4 Phenisopropylamines: stereospecific (RIA)

A brief section on the RIA of phenisopropylamines has been included because recent developments in this field have shown that RIA for phenalkylamines can be highly stereospecific. There have been a number of reports on the development of non-stereospecific RIAs for the potent hallucinogen 1-(3,5-dimethoxy-4-methylphenyl)-2-propylamine (DOM; STP) (22,23,44), but the first attempt to develop a stereo-specific RIA for a phenisopropylamine was reported by Faraj et al. (45). This procedure was based on a non-stereospecific RIA for amphetamine (AM)/methamphetamine (Me-AM) (46). The antigen was prepared by the reaction of (+)-Me-AM with N-(4-bromobutyl)phthalimide and the subsequent coupling of N-(4-aminobutyl)Me-AM to BSA. The antiserum displayed an encouraging preference for (+)-AM (100%) and (+)-Me-AM (118%), but did cross-react with the (-)-AM to the extent of 36%.

In order to explore the potential limits of stereoselectivity for the RIA of phenalkylamines, the present authors prepared separate antigens for (+)- and (-)-ephedrine (EPH). These drugs were chosen as models because they have two chiral centers, which afford four optical isomers to challenge the stereoselectivity of the antisera. Haptens were prepared by the reaction of the secondary amino group of (+)- or (-)-EPH with methyl acrylate (47,48), followed by mild alkaline hydrolysis of the methyl ester. This method was chosen because the reaction conditions are mild, there is little or no contamination from side products and no racemization, and because the nitrogen atom of the hapten remains as an aliphatic amino function. Thus, the electronic environment of the hapten at the position of derivatization is similar to that in the original drug. The haptens were coupled to BSA by the carbodiimide method, and the antigens so produced (Fig. 13.7) were used to immunize rabbits.

The antisera to (+)- and (-)-EPH were highly stereospecific. Some cross-reactivities of the antisera are shown in Fig. 13.7. Neither antibody cross-reacted with the optical antipode of its substrates nor with a racemic mixture of the diastereoisomers (49).

RIA procedures for (+)- and (-)-EPH based on these antisera were validated by comparison with a gas-liquid chromatography (GLC) method with electron-capture detection. The GLC method was not stereospecific and the sum of the concentrations of the enantiomers of EPH, as determined by RIA, was compared with the concentration of total EPH, as determined by GLC. Treated thus, the RIA- and GLC-generated data gave virtually identical plasma concentration/time curves after the administration of racemic EPH to two healthy volunteers. These experiments confirmed the stereospecificity of the RIAs and that they were free from interference from metabolites of EPH or from endogenous substances.

Findlay et al. (50) developed a highly stereospecific RIA for (+)-pseudo-ephedrine (PEPH). The antigen was prepared by the methyl acrylate method described above and was used to immunize rabbits. The antiserum was highly stereospecific and did not cross-react significantly with (-)-PEPH, (+)-EPH, (-)-EPH or (+)-nor-PEPH or with a variety of other related compounds. The RIA based on this antiserum was compared with a GLC procedure in the determination of PEPH concentrations in human plasma. The results were in excellent agreement, and the RIA was subsequently applied to assess the bioequivalence of immediate and sustained-release PEPH formulations in normal volunteers.

13.3   CONCLUSION

The foregoing review has illustrated that RIA can be extremely sensitive, highly specific and even highly stereospecific. It is also apparent that RIA is not a panacea. Each antiserum and each RIA procedure have their own individual characteristics and must be evaluated in the context in which they are to be applied. Criticism of RIA almost invariably centers on the cross-reactivity profile of the antiserum, and on the concern that there may be unidentified cross-reactants which might interfere with the assay. All too often this kind of criticism is invited by authors who describe their RIAs as being 'specific' when clearly there are worrying cross-reactivities associated with the procedures. These concerns can be addressed directly, if the RIA procedure, in a particular application, is compared

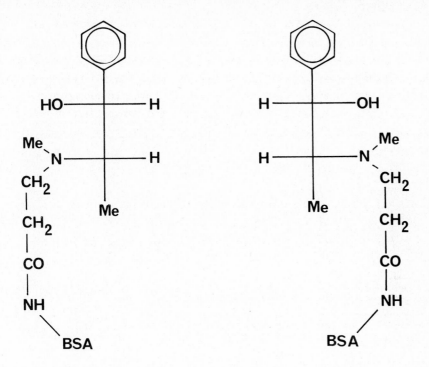

CROSS REACTIONS OF EPHEDRINE ANTISERUM

	Cross-Reaction %	
Compound Tested	(+)-Ephedrine Antiserum	(-)-Ephedrine Antiserum
(-)-Ephedrine	<2	100
(+)-Ephedrine	100	<2
(+)-Pseudoephedrine	<1	<1
(±)-Norephedrine	0	<1
(±)-Parahydroxyephedrine	0	0

Fig. 13.7. Ephedrine antigens and some cross-reactions.

with a more rigorous chemical method. Once this kind of comparison has been made, the RIA can be used to analyze thousands of samples with a great deal more credibility.

Considerable progress has been made in the development of RIAs for the analysis of low concentrations of phenolic phenalkylamines in biological fluids and tissue preparations. With a greater understanding of the immune response and of the means by which the most highly sensitive and specific antisera can be reliably reproduced, RIA seems destined to play an important part in the analysis of the biogenic amines.

REFERENCES

1 R.S. Yalow and S.A. Berson, Nature, 184 (1959) 1648-1649.
2 K. Landsteiner, The Specificity of Serological Reactions, Harvard Univ. Press, Cambridge, Mass., 1945, 310 pp.
3 D.F. Erlanger, Pharmacol. Rev., 25 (1973) 271-280.
4 N. Weliky and H.H. Weetall, Immunochemistry, 2 (1965) 293-322.
5 L.M. Freeman and M.D. Blaufox (Eds.), Radioimmunoassay, Grune and Stratton, New York, 1975, 162 pp.
6 S.A. Berson and R.S. Yalow, in E.B. Astwood and C.E. Cassidy (Eds.), Clinical Endocrinology, Grune and Stratton, New York, 1968, 699 pp.
7 Y. Kobayashi and D.V. Maudsley, Biological Applications of Liquid Scintillation Counting, Academic Press, New York, 1974, 139 pp.
8 D.S. Skelley, L.P. Brown and P.K. Besch, Clin. Chem., 19 (1973) 146-186.
9 B.A. Faraj, J.-Y. Mu, M.S. Lewis, J.P. Wilson, Z.H. Israili and P.G. Dayton, Proc. Soc. Exp. Biol. Med., 149 (1975) 664-669.
10 L.J. Grota and G.M. Brown, Endocrinology, 98 (1976) 615-622.
11 B.B. Thompson, J. Pharm. Sci., 57 (1968) 715-733.
12 N.S. Ranadive and A.H. Sehon, Can. J. Biochem., 45 (1967) 1701-1710.
13 L.J. Grota and G.M. Brown, Can. J. Biochem., 52 (1974) 196-202.
14 R.W. Lam, R. Artal and D.A. Fisher, Clin. Chem., 23 (1977) 1264-1267.
15 W.J. Raum and R.S. Swerdloff, Clin. Chem., 27 (1981) 43-47.
16 B.A. Peskar, B.M. Peskar and L. Levine, Eur. J. Biochem., 26 (1972) 191-195.
17 H. Wisser, R. Herrmann and E. Knoll, Clin. Chim. Acta, 86 (1978) 179-185.
18 A.J. Friedhoff and E. Van Winkle, Nature, 194 (1962) 897-898.
19 M. Siegel and H. Teft, J. Nerv. Ment. Dis., 152 (1971) 412-426.
20 L.J. Riceberg and H. Van Vunakis, Biochem. Pharmacol., 24 (1975) 259-265.
21 J.W. Schweitzer and A.J. Friedhoff, Amer. J. Psychiat., 124 (1968) 1249-1253.
22 H. Van Vunakis, H. Bradvica, P. Benda and L. Levine, Biochem. Pharmacol., 18 (1969) 393-404.
23 L.J. Riceberg, H. Van Vunakis and L. Levine, Anal. Biochem., 60 (1974) 551-559.
24 K.D. Charalampous, K.E. Walker and J. Kinross-Wright, Psychopharmacologia, 9 (1966) 43-63.
25 M. Yoshioka, A. Miwa and Z. Tamura, in E. Usdin, I.J. Kopin and J. Barchas (Eds.), Catecholamines: Basic and Clinical Frontiers, Proceedings of the 4th International Catecholamine Symposium, Pergamon Press, Elmsford, N.Y., 1979, 868 pp.
26 A. Miwa, Medical Pharmacy, 13 (1979) 221.
27 S. Spector, C. Dalton and A.M. Felix, Biochem. Pharmacol. Supplement, 263 (1974).
28 A. Miwa, M. Yoshioka, A. Shirahata, Y. Nakagawa and Z. Tamura, Chem. Pharm. Bull., 24 (1976) 1422-1424.
29 G. Cohen and M. Collins, Science, 167 (1970) 1749-1751.
30 A. Miwa, M. Yoshioka, A. Shirahata and Z. Tamura, Chem. Pharm. Bull., 25 (1977) 1904-1910.
31 A. Miwa, M. Yoshioka and Z. Tamura, Chem. Pharm. Bull., 26 (1978) 2903-2905.
32 A. Miwa, M. Yoshioka and Z. Tamura, Chem. Pharm. Bull., 26 (1978) 3347-3352.
33 U. Diener, E. Knoll and H. Wisser, Clin. Chim. Acta, 109 (1981) 1-11.

34 B.A. Faraj, W.R. Walker, V.M. Camp, F.M. Ali and W.B. Cobbs, Jr., J. Nucl. Med., 19 (1978) 1217-1224.
35 B.A. Faraj, D.H. Lawson, D.W. Nixon, D.R. Murray, V.M. Camp, F.M. Ali, M. Black, W. Stacciarini and Y. Tarcan, Clin. Chem., 27 (1981) 108-112.
36 B.A. Faraj, V.M. Camp, A.W. Pruitt, J.W. Isaacs and F.M. Ali, J. Nucl. Med., 18 (1977) 1025-1033.
37 G. Fillipp and H. Schneider, Acta Allergol., 19 (1964) 216-228.
38 B. Peskar and S. Spector, Science, 179 (1973) 1340-1341.
39 S. Spector, Methods in Immunology and Immunochemistry, 5 (1976) 139-147.
40 R.A. O'Brien and S. Spector, Anal. Biochem., 67 (1975) 336-338.
41 S. Halevy, S. Spector and B.M. Altura, Biochem. Med., 23 (1980) 236-238.
42 J.M. Kellum and B.M. Jaffe, Gastroenterology, 70 (1976) 516-522.
43 B.M. Jaffe, in B.M. Jaffe and H.R. Behrman (Eds.), Methods of Hormone Radio-immunoassay, Academic Press, New York, N.Y., 1979, pp. 527-539.
44 Y. Kido, K. Nagumatsu and C. Ishizeki, Yakugaku Zasshi, 94 (1974) 1290.
45 B.A. Faraj, Z.H. Israili, N.E. Kight, E.E. Smissman and T.J. Pazdernik, J. Med. Chem., 19 (1976) 20-25.
46 L.T. Cheng, S.Y. Kim, A. Chung and A. Castro, FEBS Letters, 36 (1973) 339-342.
47 J.W. Hubbard, K.K. Midha, I.J. McGilveray and J.K. Cooper, J. Pharm. Sci., 67 (1978) 1563-1571.
48 J.W. Hubbard, K.K. Midha, J.K. Cooper and C. Charette, J. Pharm. Sci., 67 (1978) 1571-1578.
49 K.K. Midha, J.W. Hubbard, J.K. Cooper and C. Macbonba, submitted.
50 J.W.A. Findlay, J.T. Warren, J.A. Hill and R.M. Welch, submitted.

SUBJECT INDEX

303

304

Date Due

			UML 735